# SEROTONIN RECEPTORS in NEUROBIOLOGY

# FRONTIERS IN NEUROSCIENCE

*Series Editors*
Sidney A. Simon, Ph.D.
Miguel A.L. Nicolelis, M.D., Ph.D.

## Published Titles

**Apoptosis in Neurobiology**
*Yusuf A. Hannun*, M.D., Professor of Biomedical Research and Chairman/Department of Biochemistry and Molecular Biology, Medical University of South Carolina, Charleston, South Carolina
*Rose-Mary Boustany*, M.D., tenured Associate Professor of Pediatrics and Neurobiology, Duke University Medical Center, Durham, North Carolina

**Methods for Neural Ensemble Recordings**
*Miguel A.L. Nicolelis*, M.D., Ph.D., Professor of Neurobiology and Biomedical Engineering, Duke University Medical Center, Durham, North Carolina

**Methods of Behavioral Analysis in Neuroscience**
*Jerry J. Buccafusco*, Ph.D., Alzheimer's Research Center, Professor of Pharmacology and Toxicology, Professor of Psychiatry and Health Behavior, Medical College of Georgia, Augusta, Georgia

**Neural Prostheses for Restoration of Sensory and Motor Function**
*John K. Chapin*, Ph.D., Professor of Physiology and Pharmacology, State University of New York Health Science Center, Brooklyn, New York
*Karen A. Moxon*, Ph.D., Assistant Professor/School of Biomedical Engineering, Science, and Health Systems, Drexel University, Philadelphia, Pennsylvania

**Computational Neuroscience: Realistic Modeling for Experimentalists**
*Eric DeSchutter*, M.D., Ph.D., Professor/Department of Medicine, University of Antwerp, Antwerp, Belgium

**Methods in Pain Research**
*Lawrence Kruger*, Ph.D., Professor of Neurobiology (Emeritus), UCLA School of Medicine and Brain Research Institute, Los Angeles, California

**Motor Neurobiology of the Spinal Cord**
*Timothy C. Cope*, Ph.D., Professor of Physiology, Wright State University, Dayton, Ohio

**Nicotinic Receptors in the Nervous System**
*Edward D. Levin*, Ph.D., Associate Professor/Department of Psychiatry and Pharmacology and Molecular Cancer Biology and Department of Psychiatry and Behavioral Sciences, Duke University School of Medicine, Durham, North Carolina

**Methods in Genomic Neuroscience**
*Helmin R. Chin*, Ph.D., Genetics Research Branch, NIMH, NIH, Bethesda, Maryland
*Steven O. Moldin*, Ph.D, University of Southern California, Washington, D.C.

**Methods in Chemosensory Research**
*Sidney A. Simon*, Ph.D., Professor of Neurobiology, Biomedical Engineering, and Anesthesiology, Duke University, Durham, North Carolina
*Miguel A.L. Nicolelis*, M.D., Ph.D., Professor of Neurobiology and Biomedical Engineering, Duke University, Durham, North Carolina

**The Somatosensory System: Deciphering the Brain's Own Body Image**
*Randall J. Nelson*, Ph.D., Professor of Anatomy and Neurobiology, University of Tennessee Health Sciences Center, Memphis, Tennessee

**The Superior Colliculus: New Approaches for Studying Sensorimotor Integration**
*William C. Hall*, Ph.D., Department of Neuroscience, Duke University, Durham, North Carolina
*Adonis Moschovakis*, Ph.D., Department of Basic Sciences, University of Crete, Heraklion, Greece

**New Concepts in Cerebral Ischemia**
*Rick C. S. Lin*, Ph.D., Professor of Anatomy, University of Mississippi Medical Center, Jackson, Mississippi

**DNA Arrays: Technologies and Experimental Strategies**
*Elena Grigorenko*, Ph.D., Technology Development Group, Millennium Pharmaceuticals, Cambridge, Massachusetts

**Methods for Alcohol-Related Neuroscience Research**
*Yuan Liu*, Ph.D., National Institute of Neurological Disorders and Stroke, National Institutes of Health, Bethesda, Maryland
*David M. Lovinger*, Ph.D., Laboratory of Integrative Neuroscience, NIAAA, Nashville, Tennessee

***In Vivo* Optical Imaging of Brain Function**
*Ron Frostig*, Ph.D., Associate Professor/Department of Psychobiology, University of California, Irvine, California

**Primate Audition: Behavior and Neurobiology**
*Asif A. Ghazanfar*, Ph.D., Princeton University, Princeton, New Jersey

**Methods in Drug Abuse Research: Cellular and Circuit Level Analyses**
*Dr. Barry D. Waterhouse*, Ph.D., MCP-Hahnemann University, Philadelphia, Pennsylvania

**Functional and Neural Mechanisms of Interval Timing**
*Warren H. Meck*, Ph.D., Professor of Psychology, Duke University, Durham, North Carolina

**Biomedical Imaging in Experimental Neuroscience**
*Nick Van Bruggen*, Ph.D., Department of Neuroscience Genentech, Inc.
*Timothy P.L. Roberts*, Ph.D., Associate Professor, University of Toronto, Canada

**The Primate Visual System**
*John H. Kaas*, Department of Psychology, Vanderbilt University
*Christine Collins*, Department of Psychology, Vanderbilt University, Nashville, Tennessee

**Neurosteroid Effects in the Central Nervous System**
*Sheryl S. Smith*, Ph.D., Department of Physiology, SUNY Health Science Center, Brooklyn, New York

**Modern Neurosurgery: Clinical Translation of Neuroscience Advances**
*Dennis A. Turner*, Department of Surgery, Division of Neurosurgery, Duke University Medical Center, Durham, North Carolina

**Sleep: Circuits and Functions**
*Pierre-Hervé Luoou*, Université Claude Bernard Lyon, France

**Methods in Insect Sensory Neuroscience**
*Thomas A. Christensen*, Arizona Research Laboratories, Division of Neurobiology, University of Arizona, Tuscon, Arizona

**Motor Cortex in Voluntary Movements**
*Alexa Riehle*, INCM-CNRS, Marseille, France
*Eilon Vaadia*, The Hebrew University, Jerusalem, Israel

**Neural Plasticity in Adult Somatic Sensory-Motor Systems**
*Ford F. Ebner*, Vanderbilt University, Nashville, Tennessee

**Advances in Vagal Afferent Neurobiology**
*Bradley J. Undem*, Johns Hopkins Asthma Center, Baltimore, Maryland
*Daniel Weinreich*, University of Maryland, Baltimore, Maryland

**The Dynamic Synapse: Molecular Methods in Ionotropic Receptor Biology**
*Josef T. Kittler*, University College, London, England
*Stephen J. Moss*, University College, London, England

**Animal Models of Cognitive Impairment**
*Edward D. Levin*, Duke University Medical Center, Durham, North Carolina
*Jerry J. Buccafusco*, Medical College of Georgia, Augusta, Georgia

**The Role of the Nucleus of the Solitary Tract in Gustatory Processing**
*Robert M. Bradley*, University of Michigan, Ann Arbor, Michigan

**Brain Aging: Models, Methods, and Mechanisms**
*David R. Riddle*, Wake Forest University, Winston Salem, North Carolina

**Neural Plasticity and Memory: From Genes to Brain Imaging**
*Frederico Bermudez-Rattoni*, National University of Mexico, Mexico City, Mexico

**Serotonin Receptors in Neurobiology**
*Amitabha Chattopadhyay*, Center for Cellular and Molecular Biology, Hyderabad, India

# SEROTONIN RECEPTORS in NEUROBIOLOGY

*Edited by*

Amitabha Chattopadhyay

Center for Cellular and Molecular Biology
Hyderabad, India

CRC Press
Taylor & Francis Group
Boca Raton London New York

CRC Press is an imprint of the
Taylor & Francis Group, an informa business

CRC Press
Taylor & Francis Group
6000 Broken Sound Parkway NW, Suite 300
Boca Raton, FL 33487-2742


Printed in the United States of America on acid-free paper
10 9 8 7 6 5 4 3 2 1

International Standard Book Number-10: 0-8493-3977-4 (Hardcover)
International Standard Book Number-13: 978-0-8493-3977-6 (Hardcover)



**Library of Congress Cataloging-in-Publication Data**

Serotonin receptors in neurobiology / [edited by] Amitabha Chattopadhyay.
p. ; cm. -- (Frontiers in neuroscience)
Includes bibliographical references and index.
ISBN-13: 978-0-8493-3977-6 (alk. paper)
ISBN-10: 0-8493-3977-4 (alk. paper)
1. Serotonin--Receptors. 2. Serotoninergic mechanisms. I. Chattopadhyay, Amitabha, 1956- II. Series: Frontiers in neuroscience (Boca Raton, Fla.)
[DNLM: 1. Receptors, Serotonin. 2. Signal Transduction. WL 102.8 S4865 2007]

QP801.S4S487 2007
612.8'042--dc22 2006038801

**Visit the Taylor & Francis Web site at**
**http://www.taylorandfrancis.com**

**and the CRC Press Web site at**
**http://www.crcpress.com**

# Contents

**Series Preface** .......... ix

**Preface** .......... xi

**Editor** .......... xiii

**Contributors** .......... xv

**Chapter 1** Quantitative Imaging of Serotonin Autofluorescence with Multiphoton Microscopy .......... 1

*S. K. Kaushalya and Sudipta Maiti*

**Chapter 2** Monitoring Receptor-Mediated Changes of Intracellular cAMP Level by Using Ion Channels and Fluorescent Proteins as Biosensors .......... 19

*Evgeni G. Ponimaskin, Martin Heine, Andre Zeug, Tatyana Voyno-Yasenetskaya, and Petrus S. Salonikidis*

**Chapter 3** Membrane Organization and Dynamics of the $Serotonin_{1A}$ Receptor Monitored Using Fluorescence Microscopic Approaches .......... 41

*Shanti Kalipatnapu, Thomas J. Pucadyil, and Amitabha Chattopadhyay*

**Chapter 4** Calmodulin Is a 5-HT Receptor-Interacting and Regulatory Protein .......... 61

*Justin N. Turner, Sonya D. Coaxum, Andrew K. Gelasco, Maria N. Garnovskaya, and John R. Raymond*

**Chapter 5** Identification of Novel Transcriptional Regulators in the Nervous System .......... 81

*Federico Remes-Lenicov, Kirsten X. Jacobsen, Anastasia Rogaeva, Margaret Czesak, Mahmoud Hadjighasem, Mireille Daigle, and Paul R. Albert*

**Chapter 6** Serotonin 2A (5-$HT_{2A}$) Receptor Function: Ligand-Dependent Mechanisms and Pathways .......... 105

*Ishier Raote, Aditi Bhattacharya, and Mitradas M. Panicker*

**Chapter 7** The 5-$HT_{1A}$ Receptor: A Signaling Hub Linked to Emotional Balance ........ 133

*Probal Banerjee, Mukti Mehta, and Baishali Kanjilal*

**Chapter 8** Do Limits of Neuronal Plasticity Represent an Opportunity for Mental Diseases, Such as Addiction to Food and Illegal Drugs? Use and Utilities of Serotonin Receptor Knock-Out Mice ........ 157

*Valerie Compan*

**Chapter 9** Use of Mice with Targeted Genetic Inactivation in the Serotonergic System for the Study of Anxiety ........ 181

*Miklos Toth*

**Index** ........ 197

# Series Preface

Our goal in creating the Frontiers in Neuroscience Series is to present the insights of experts on emerging fields and theoretical concepts that are, or will be, at the vanguard of neuroscience. Books in the series cover topics ranging from genetics, ion channels, apoptosis, electrodes, neural ensemble recordings in behaving animals, and even robotics. The series also covers new and exciting multidisciplinary areas of brain research, such as computational neuroscience and neuroengineering, and describes breakthroughs in classical fields like behavioral neuroscience. We want these books to be the books every neuroscientist will use in order to get acquainted with new ideas and frontiers in brain research. These books can be given to graduate students and postdoctoral fellows when they are looking for guidance to start a new line of research.

Each book is edited by an expert and consists of chapters written by the leaders in a particular field. Books are richly illustrated and contain comprehensive bibliographies. Chapters provide substantial background material relevant to the particular subject. We hope that as the volumes become available the effort put in by us, the publisher, the book editors, and individual authors will contribute to the further development of brain research. The extent that we achieve this goal will be determined by the utility of these books.

# Preface

A number of developments spanning a multitude of techniques make this an exciting time for research on serotonin receptors. Serotonin receptors display diverse functions and play a crucial role in the generation and modulation of cognitive and behavioral functions. It is perhaps for the first time that we are beginning to address and correlate issues related to serotonergic signaling at the molecular level to that in animal models. It is against this backdrop that this book on serotonin receptors in neurobiology has been organized.

The chapters in this book focus on various issues of serotonin receptors in neurobiology and bring together aspects of serotonin receptors from a broad-based multidisciplinary approach. The approaches described in this book vary from molecular biological techniques to fluorescence microscopy and imaging, and eventually to genetic manipulation in animal models, thereby providing a broad spectrum of approaches to study serotonergic phenomena. Although each of these approaches has its own advantages and limitations, it is expected that the synthesis of information and knowledge achieved from studies using multiple approaches will result in a comprehensive understanding of the underlying complex phenomena involved in serotonergic signaling and its implications in health and disease. I believe that this book will help the prospective reader gain an overall understanding about these receptors based on currently used methodologies and approaches. The descriptions of the specific experimental approaches should be of use to researchers interested in addressing similar problems involving other G-protein coupled receptor signaling systems.

I would like to use this opportunity to thank all the contributors who are leaders in their respective areas of research. Special thanks are due to Sidney Simon, Barbara Norwitz, and Jill Jurgensen for their cooperation and support in organizing this volume.

**Amitabha Chattopadhyay**

# Editor

**Amitabha Chattopadhyay** is a group leader at the Centre for Cellular and Molecular Biology in Hyderabad, India. He obtained his B.Sc. with honors in chemistry from St. Xavier's College, Calcutta. He then went to the Indian Institute of Technology in Kanpur from where he received his M.Sc. in chemistry. He received his Ph.D. in physical biochemistry from the State University of New York at Stony Brook. He subsequently was a postdoctoral fellow at the University of California, Davis, where he worked on the nicotinic acetylcholine receptor. He joined the Centre for Cellular and Molecular Biology in Hyderabad in 1989 and subsequently became a group leader. He is also an honorary professor at the Jawaharlal Nehru Centre for Advanced Scientific Research in Bangalore, India.

Dr. Chattopadhyay's major research interest is in the area of membrane and receptor biology. His work has addressed issues related to the organization, dynamics, and function of membrane receptors, using a variety of approaches with special emphasis on fluorescence-based techniques. Dr. Chattopadhyay received the prestigious Shanti Swarup Bhatnagar Prize from the prime minister of India in 2001 for his research accomplishments. Among his other achievements are the Sreenivasaya Memorial Award by the Society of Biological Chemists (India), Raman Research Fellowship by the Council of Scientific and Industrial Research (CSIR, India), and Wood-Whelan Fellowship from the International Union of Biochemistry and Molecular Biology. He has recently been awarded the Dozor Visiting Fellowship by the Ben Gurion University in Israel.

Dr. Chattopadhyay is a fellow of the Indian National Science Academy. He serves as a member of the editorial boards of a number of international journals. He is a member of the International Society for Neurochemistry and the Biophysical Society (U.S.) besides various scientific societies in India. He has contributed more than 100 research papers and has guest-edited special issues for a number of journals.

# Contributors

**Paul R. Albert**
Ottawa Health Research Institute (Neuroscience)
and
Department of Cellular and Molecular Medicine
University of Ottawa
Ottawa, Ontario, Canada

**Probal Banerjee**
Department of Chemistry and the CSI/IBR Center for Developmental Neuroscience
College of Staten Island–City University of New York
Staten Island, New York, United States

**Aditi Bhattacharya**
National Centre for Biological Sciences
Tata Institute for Fundamental Research
Bangalore, India

**Amitabha Chattopadhyay**
Centre for Cellular and Molecular Biology
Hyderabad, India

**Sonya D. Coaxum**
Department of Medicine, Medical University of South Carolina
and
Ralph H. Johnson Department of Veterans Affairs Medical Center
Charleston, South Carolina, United States

**Valerie Compan**
Départment de Neurobiologie
Institut de Génomique Fonctionnelle
Montpellier, France

**Margaret Czesak**
Ottawa Health Research Institute (Neuroscience)
and
Department of Cellular and Molecular Medicine
University of Ottawa
Ottawa, Ontario, Canada

**Mireille Daigle**
Ottawa Health Research Institute (Neuroscience)
and
Department of Cellular and Molecular Medicine
University of Ottawa
Ottawa, Ontario, Canada

**Maria N. Garnovskaya**
Department of Medicine, Medical University of South Carolina
and
Ralph H. Johnson Department of Veterans Affairs Medical Center
Charleston, South Carolina, United States

**Andrew K. Gelasco**
Department of Medicine, Medical University of South Carolina
and
Ralph H. Johnson Department of Veterans Affairs Medical Center
Charleston, South Carolina, United States

**Mahmoud Hadjighasem**
Ottawa Health Research Institute (Neuroscience)
and
Department of Cellular and Molecular Medicine
University of Ottawa
Ottawa, Ontario, Canada

**Martin Heine**
Université Bordeaux 2
Institut François Magendie
Bourdeaux, France

**Kirsten X. Jacobsen**
Ottawa Health Research Institute (Neuroscience)
and
Department of Cellular and Molecular Medicine
University of Ottawa
Ottawa, Ontario, Canada

**Shanti Kalipatnapu**
Centre for Cellular and Molecular Biology
Hyderabad, India

**Baishali Kanjilal**
Department of Chemistry and the CSI/IBR Center for Developmental Neuroscience
College of Staten Island–City University of New York
Staten Island, New York, United States

**S. K. Kaushalya**
Tata Institute of Fundamental Research
Mumbai, India

**Sudipta Maiti**
Tata Institute of Fundamental Research
Mumbai, India

**Mukti Mehta**
Department of Chemistry and the CSI/IBR Center for Developmental Neuroscience
College of Staten Island–City University of New York
Staten Island, New York, United States

**Mitradas M. Panicker**
National Centre for Biological Sciences
Tata Institute for Fundamental Research
Bangalore, India

**Evgeni G. Ponimaskin**
Department of Neuro and Sensory Physiology
University of Göttingen
Göttingen, Germany

**Thomas J. Pucadyil**
Centre for Cellular and Molecular Biology
Hyderabad, India

**Ishier Raote**
National Centre for Biological Sciences
Tata Institute for Fundamental Research
Bangalore, India

**John R. Raymond**
Department of Medicine, Medical University of South Carolina
and
Ralph H. Johnson Department of Veterans Affairs Medical Center
Charleston, South Carolina, United States

**Federico Remes-Lenicov**
Ottawa Health Research Institute (Neuroscience)
and
Department of Cellular and Molecular Medicine
University of Ottawa
Ottawa, Ontario, Canada

**Anastasia Rogaeva**
Ottawa Health Research Institute (Neuroscience)
and
Department of Cellular and Molecular Medicine
University of Ottawa
Ottawa, Ontario, Canada

**Petrus S. Salonikidis**
Department of Neuro and Sensory Physiology
University of Göttingen
Göttingen, Germany

**Miklos Toth**
Department of Pharmacology
Cornell University
New York, New York, United States

**Justin N. Turner**
Department of Medicine, Medical University of South Carolina
and
Ralph H. Johnson Department of Veterans Affairs Medical Center
Charleston, South Carolina, United States

**Tatyana Voyno-Yasenetskaya**
Department of Pharmacology
University of Illinois
Chicago, Illinois, United States

**Andre Zeug**
Department of Neurophysiology and Cellular Biophysics
University of Göttingen
Göttingen, Germany

# 1 Quantitative Imaging of Serotonin Autofluorescence with Multiphoton Microscopy

*S. K. Kaushalya and Sudipta Maiti*

## CONTENTS

Introduction .......... 2
Abbreviations .......... 2
Multiphoton Microscopy .......... 2
  Multiphoton Excitation: Basic Concepts .......... 2
  The Requirement for a Pulsed Laser .......... 3
  The Requirement for Tight Focusing .......... 4
  Choosing the Right Excitation Wavelength .......... 5
  Advantages of Three-Photon Microscopy .......... 7
Optical Setup .......... 8
  Point Scanning .......... 8
  Excitation .......... 8
    Coupling the Scanning Optics with the Microscope .......... 8
  Detection .......... 9
    Detecting the Fluorescence .......... 9
    Epi Collection .......... 9
    Non-Epi Collection (without a Lens) .......... 10
  Specific Optical Setup for Serotonin Imaging .......... 11
  Checklist before Imaging .......... 12
Imaging Serotonin and Other Monoamines in Live Neurons .......... 12
  Imaging Protocol .......... 12
  Establishing the Source of the Signal .......... 13
    Test for the Order of Excitation .......... 13
    Checking Co-Localization of Bright Structures with NADH .......... 13
  Applications of Serotonin Imaging .......... 14
    Exocytosis with K+ Depolarization .......... 15
    Monitoring the Differentiation of Serotonergic Cells .......... 15
    Amphetamine-Induced Serotonin Release .......... 16

Catecholamine Detection.......16
Acknowledgments.......17
References.......17

## INTRODUCTION

In this article we describe a technique to directly image vesicular serotonin in live neurons. In the context of serotonergic signaling in the brain, it is important to map out serotonergic storage and release sites, together with the distribution of serotonin receptors. Receptor mapping is possible by using endogenously labeled receptors, but serotonin vesicle mapping is more difficult. Following the early success in mapping serotonin in mast cells with three-photon excitation microscopy,[1] we have recently demonstrated that such mapping is possible also in live neurons.[2] We describe here the technique of serotonin imaging in detail, together with a few application examples. This three-photon excitation based imaging technique should also be useful in general for live cell microscopy of UV chromophores.

What we present here is intended to be a practical guide and not a review of three-photon microscopy. For the technical details described here, we have strictly relied on our own hands-on experience. The description is divided into three main sections. "Multiphoton Microscopy" presents a brief summary of the concepts of multiphoton excitation. This is followed by a description of the optical technique of serotonin imaging in "Optical Setup." Finally, "Imaging Serotonin and Other Monoamines in Live Neurons" describes a few examples of serotonin imaging in neuronal cells and also describes how this technology can possibly be extended to image other monoamines.

## ABBREVIATIONS

**MPE**, multiphoton excitation
**MPM**, multiphoton microscopy
**UV**, ultraviolet
**fs**, femto second

## MULTIPHOTON MICROSCOPY

### Multiphoton Excitation: Basic Concepts

Multiphoton excitation is a novel method for exciting monoamines and UV chromophores in general, with relatively benign infrared light. Multiphoton excitation depends on a nonlinear quantum effect first predicted by Maria Goppert–Meyer in 1931,[3] observed in practice by Singh and Bradley in 1964,[4] and put to use in biological microscopy by Webb and coworkers in 1990.[5] It allows a molecule to be photo-excited by light that has much lower energy (and hence a longer wavelength) compared to what would be predicted from its conventional (one-photon) excitation spectrum. Simply put, when the intensity of the excitation light is very high, a fluorophore molecule can absorb multiple low-energy photons simultaneously, whose total energy equals or exceeds its one-photon excitation energy.

*n*-Photon excitation depends on the *n*th-power of the intensity of the excitation light. This intensity dependence automatically confers three-dimensional (3-D) localization of the excitation (as described later), which is exploited to obtain high resolution 3-D. Also, this process requires a longer excitation wavelength ($\sim n\lambda$) compared to the conventional one-photon excitation wavelength of $\lambda$. This factor helps us in lowering both photodamage to the sample and photobleaching of the chromophore. This is also the factor that allows us to perform live cell microscopy of UV molecules using infrared light.

There are excellent reviews that describe the principle and practice of two-photon microscopy in general.[6–8] The conceptual framework of two-photon microscopy can be simply extended to describe three-photon microscopy, so we will discuss the principles only briefly. For practice, it is useful to understand how the signal and the resolution vary with the nature of the excitation light, and these aspects would be highlighted here. The practical excitation source in multiphoton microscopy is a pulsed laser, usually tunable in the near-infrared region. The repetition rate of this laser, the pulse width, and the pulse shape in time, and the size of the collimated laser beam at the back aperture of the microscope objective lens are the key parameters that determine the efficiency of excitation.

## The Requirement for a Pulsed Laser

The rate of excitation of a given molecular species depends on the repetition rate $R$, the pulse width $\tau$, the average power $P_{av}$, and the beam waist at the focus $\omega_0$ of the excitation beam. The molecular property contributes to the *n*-photon excitation process through the *n*-photon absorption cross section $\sigma_n$. The average rate of fluorescence emission (per molecule) is proportional to the average rate of absorption, which in turn is proportional to

$$\frac{\sigma_n}{R^{(n-1)}\tau^{(n-1)}}\left(\frac{\lambda_n\, P_{av}}{\pi\,\omega_0^2\, hc}\right)^n \tag{1.1}$$

For CW (continuous wave, steady) excitation, $R\tau = 1$. As the order of excitation increases excitation cross section $\sigma_n$ goes down, so for the same pulse width, the required $P_{av}$ requirement goes up. For a typical value of $\sigma_3 = 3.2 \times 10^{-96}$ m$^6$ s$^2$ photon$^{-2}$ (which is the value for serotonin at 740 nm[1]) and that of the beam waist of 300 nm, the $P_{av}$ required to get a near-saturation (i.e., near maximal) signal from a CW laser is about 400 W. This is a tremendous amount of power, and therefore only short pulsed lasers with high peak powers (but much lower average powers) are practical for this kind of microscopy.

This relationship also shows that for a three-photon process, the signal will increase by a factor of eight if the power is increased by a factor of two, and it will decrease by a factor of four if the pulse width is increased by a factor of two (the pulse shape factor has been neglected in this description; the reader is referred to

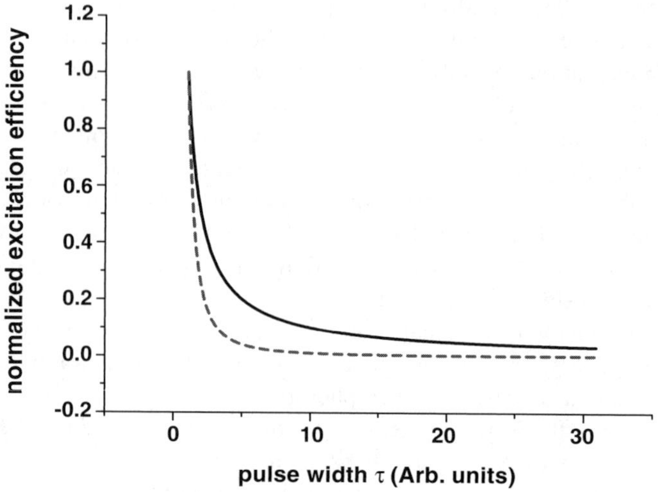


**FIGURE 1.1 The dependence of excitation efficiency on the pulse width for two- (solid) and three-photon (dash) excitation.** For $\tau = 1$, excitation efficiency for both is normalized to unity. For three-photon excitation, the efficiency drops faster than two-photon excitation as the pulse width increases.

Xu et al.[9] for details). In practice, the pulse width used is frequently about ~100 fs, as shorter pulses get severely broadened by the microscope optics through group velocity dispersion and nonlinear effects.[10]

We note here that the log–log plot of fluorescence vs $P_{av}$ would give a straight line with a slope of $n$, which is the order of the excitation. This relation holds only in the absence of saturation and photo bleaching of the fluorophore. Near saturation there is no more fluorophore that can be excited, and fluorescence counts approach a plateau.

Figure 1.1 shows the change in excitation efficiency with $\tau$ for two- and three-photon excitation. We assume the excitation rates to be the same for some arbitrary pulse width. Both decrease as pulse width increases, but the decrease for three-photon excitation is much sharper.

## The Requirement for Tight Focusing

Now we focus on the spatial properties of the laser beam. For a focused beam that has a Gaussian spatial profile, the spatial dependence of the intensity $I$ can be expressed as a function of $z$ (the distance along the beam propagation direction, measured from the focus) and $r$ (the distance perpendicular to the beam propagation direction, measured from the beam axis).

$$I(r,z) = I_0 \left( \frac{\omega_0}{\omega(z)} \right)^2 \exp\left( \frac{-2r^2}{\omega^2(z)} \right) \tag{1.2}$$

Where the dependence on $z$ is implicit through the function $\omega(z)$, $\omega(z)$ is the radius of the beam (defined as the distance where the field intensity drops to $1/e^2$ of the intensity at the beam axis) at a distance $z$ from the focus, and $\omega_0$ is the minimum radius (i.e., radius at the focus, $z = 0$). For a Gaussian beam of wavelength $\lambda$, the variation of the beam diameter with $z$ can be expressed as[11]

$$\omega(z) = \omega_0 \left[ 1 + \left( \frac{\lambda z}{\pi \omega_0^{\ 2}} \right)^2 \right]^{1/2} \tag{1.3}$$

Figure 1.2A shows the excitation intensity $I(r,z)$ around the focus spot (the beam is assumed to remain Gaussian throughout). Figure 1.2B and Figure 1.2C show $I^2(r,z)$ and $I^3(r,z)$ distribution around the focus, which are relevant for two- and three-photon excitation, respectively. From the figures it is clear that in MPE, as the order of excitation increases, excitation also becomes more and more localized. This is the origin of the 3-D resolution inherent in MPE.

We note that a higher order MPE also requires longer wavelengths to excite the same chromophore. This would widen the excitation profile somewhat, compared to that shown in these figures. An estimate can be made by considering Equation 1.1 and Equation 1.3. It is suggested that for a two-photon process, the resolution would be worse by a factor of $\sqrt{2}$, and for a three-photon process, it would be worse by a factor of $\sqrt{3}$ (compared to a single-photon confocal detection). However, frequently the excitation peak for a multiphoton process occurs at a wavelength shorter than the expected value of $n\lambda$, and this then improves the resolution. Ultimately, the resolution of a single photon confocal microscope is typically comparable to that of a multiphoton microscope.

Equation 1.3 shows that for a particular $z$, as we increase the beam diameter, $\omega_0$ decreases. This implies that as one increases the beam diameter at the back aperture of the objective lens, MPE becomes more efficient with the higher intensity produced. For typical high resolution microscopy, the beam size is increased to a size slightly bigger than the back aperture of the objective lens to get the smallest possible beam waist. In this case, the beam no longer remains Gaussian at the focus. Fraunhoffer diffraction theory predicts what the light field will be in such cases,[12–14] but the rules of thumb derived from the Gaussian description above remain valid. So for obtaining a higher signal in three-photon microscopy, and to obtain the highest possible resolution, the excitation beam should fill the back aperture of the objective lens.

In a mode locked fs-pulsed laser, the power is very high for the duration of the pulse. As it is focused with a high NA objective lens, the intensity is sufficient for MPE to occur at the focus spot. Hence, all the fluorescence originates from the focus spot only, which typically has a volume of $\sim 10^{-13}$ cc (unless the intensity is increased beyond a limit, and the saturation effects start dominating). This focal volume can be characterized, if needed, using a technique called a fluorescence correlation spectroscopy.[15–19]

## Choosing the Right Excitation Wavelength

In MPE, instead of one photon of wavelength $\lambda$, a molecule absorbs $n$ photons[9] of wavelength $\sim n\lambda$. However, the emission spectrum remains the same for almost all

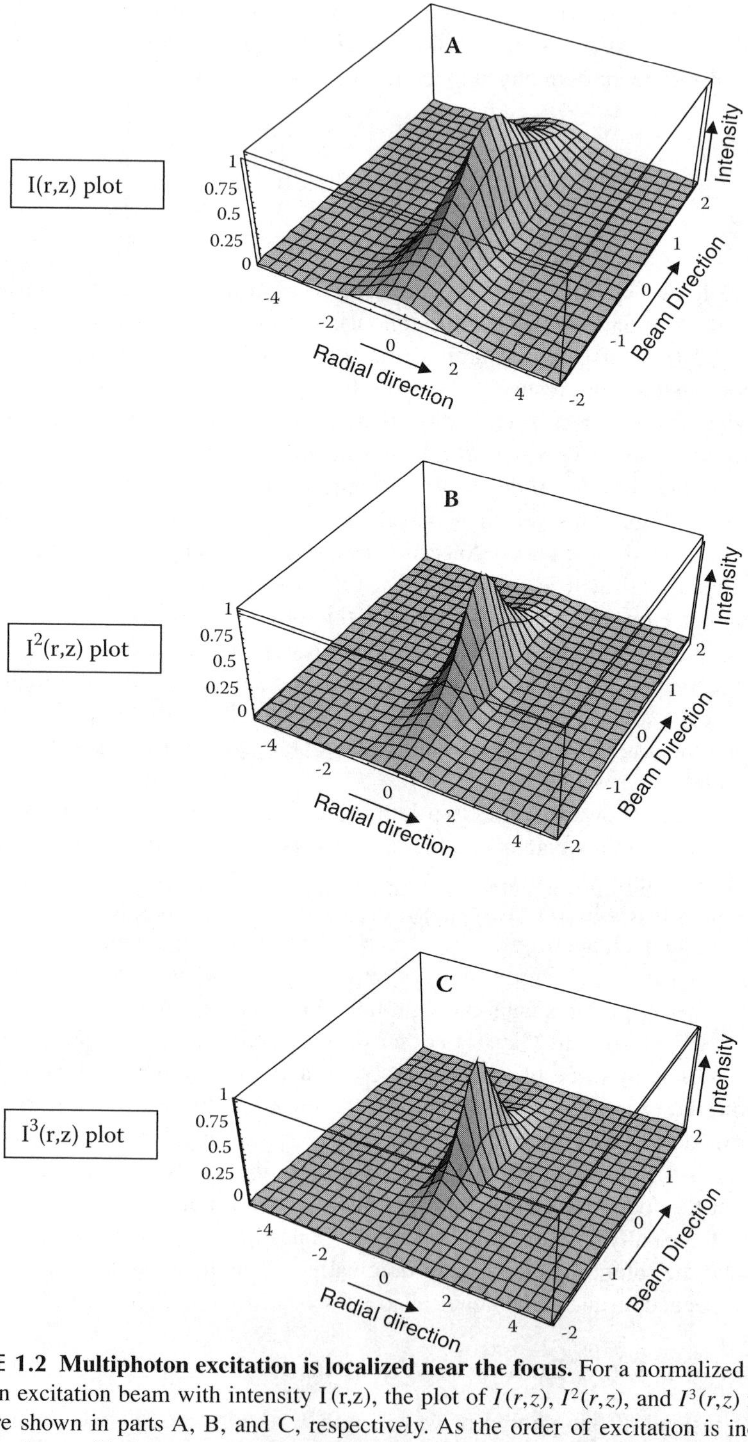


**FIGURE 1.2 Multiphoton excitation is localized near the focus.** For a normalized focused Gaussian excitation beam with intensity I (r,z), the plot of $I(r,z)$, $I^2(r,z)$, and $I^3(r,z)$ near the focus are shown in parts A, B, and C, respectively. As the order of excitation is increased, the excitation profile becomes more and more localized.

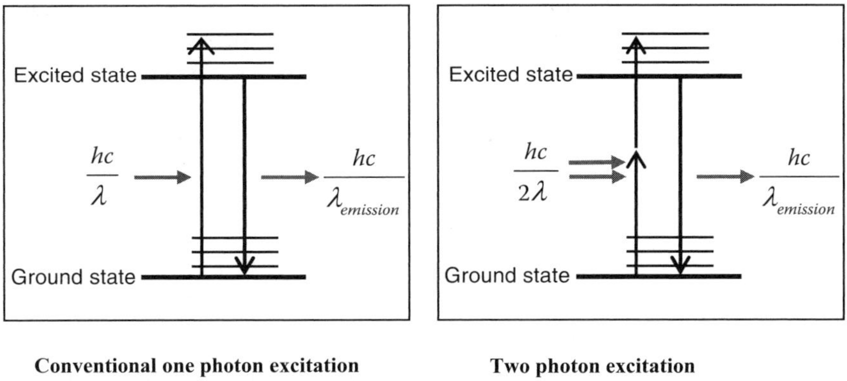


**FIGURE 1.3 The energy scheme for multiphoton excitation.** In conventional one-photon excitation, a molecule absorbs a single photon of energy $E = hc/\lambda$, which matches the energy difference between the excited and the ground state. In two-photon excitation, the same molecule can be excited by simultaneously absorbing two photons of the energy $E/2 = hc/2\lambda$. However, the fluorescence spectra remain the same.

the molecules[6] (Figure 1.3). A fluorophore whose excitation peak is at wavelength λ often has a multiphoton excitation peak at a wave length smaller than $n\lambda$ (n = 2, 3, ...) as mentioned previously. For a chromophore with an unknown excitation spectrum, one has to scan the excitation wavelength around $n\lambda$ and measure the fluorescence obtained at each $\lambda_{ex}$ to find the peak of the excitation spectrum.

## Advantages of Three-Photon Microscopy

1. **Less damaging for cells:** With three-photon excitation, serotonin can be excited using an infrared light source (e.g., a mode-locked fs laser) instead of using ultraviolet light (~270 nm). For imaging monoamines in live biological samples, this is a big advantage over one-photon confocal imaging, as infrared is far less damaging than the UV.
2. **Quantitative imaging:** As the image is a representation of the auto fluorescence intensity distribution of the molecule itself in the sample, MPM is a very quantitative method. Concentration of the fluorophore *in situ* can be estimated by calibrating with a known concentration of the same fluorophore (after taking into account all photo-physical and photo-chemical effects).
3. **Enables imaging deeper in the thicker sample:** With MPE, imaging deeper inside the biological samples (tissue) is possible with little distortion in the resolution compared to one-photon confocal imaging. In multiphoton excitation longer wavelengths are used for excitation so scattering of the excitation is much less

$$(\text{scattering of light} \propto \frac{1}{\lambda^4}).$$

The tissue absorption is also much less in the near infrared. This results in good focusing and resolution to a considerable depth (~70 $\mu$m).

4. **Exciting more than one fluorophore simultaneously with the same excitation:** MPE spectra are typically broad, which ensures that there is a considerable possibility of finding a good wavelength for simultaneously exciting multiple types of fluorophores. As only a single beam is used, no colocalization errors can creep into these measurements.

## OPTICAL SETUP

### Point Scanning

So far we have discussed the fluorescence obtainable from the MPE of a small stationary volume. To know the distribution of fluorophore in an extended object like a neuron or a piece of animal tissue, one has to collect fluorescence from all the points with the resolution (limited by Rayleigh criterion) of the excitation volume. To take a fluorescence image, a focused excitation spot scans the extended object, and the fluorescence emitted at each point is collected through a photo detector. The plot of the fluorescence intensity vs. the position then produces the fluorescence image. This scheme is known as point scanning.

### Excitation

For setting up an MPM one needs (1) a good microscope (inverted or upright, depending on the demands of the experiment) with adequate flexibility to put external detectors, (2) scanning optics which make excitation beam scan in 2-D and which have reasonable infrared throughput (a conventional confocal scanner or simply a set of galvanometric mirrors with proper software is adequate), (3) a fs pulsed laser with high repetition rate, and (4) some ultrafast optics. The following steps are involved:

#### Coupling the Scanning Optics with the Microscope

1. Scanning optics should be coupled to the microscope such that the excitation beam entering the objective should be collimated or left a little diverging (over a 3m distance; 10% increase in the diameter is good). If the beam is converging, with a high NA, low-working-distance objective lens, the beam may get focused within the objective itself and damage it.
2. The pivot point of scanning (while scanning, the collinear beam makes an angular movement about a point, and the center of the beam always passes through it) should lie in the back focal plane of the objective lens and remain in the optical axis of the objective. This ensures homogeneous intensity distribution during spatial scanning. Otherwise as the beam moves away from the center, excitation intensity and spot size may change considerably. For achieving steps 1 and 2, one may have to use a 10× eye piece between the scanner and the microscope.

3. For good resolution, focus spot size should be the smallest possible, which is achieved by adjusting the beam diameter at the back aperture of the objective lens (beam diameter should be equal to or slightly bigger than the back aperture of the objective).
4. Beam size on the scanning mirrors should be small to avoid distortion because of the rotation about their own axes.

## Detection

### Detecting the Fluorescence

When a fluorophore is excited in an isotropic specimen it also emits isotropically. While scanning the sample, depending on the emission spectra and the demand of the experimental setup, the fluorescence can be collected mainly in three ways.

1. Epifluorescence collection
2. Non-epi collection[20]
3. $4\pi$ collection[21] (i.e., both epi and non-epi)

### Epi Collection

In an epifluorescence setup (Figure 1.4), excitation is focused with an objective and the part of the fluorescence emitted in the backward direction is collected with high efficiency by the same objective. The collected fluorescence counter propagates along the excitation path and is separated from the excitation by a dichroic mirror placed next to the objective at an angle of 45° to the beam direction.

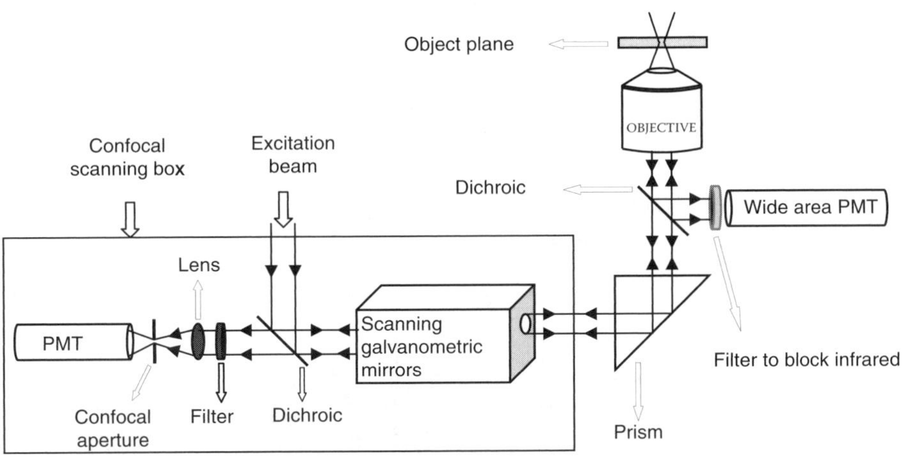


**FIGURE 1.4 Schematic diagram of the setup for epi collection.** (Adapted from Kaushalya, S.K., Balaji, J., Garai, K., and Maiti, S., Fluorescence correlation microscopy with real-time alignment readout, *Appl Opt*, 44, 3262–3265, 2005. With permission from the Optical Society of America.)

The fluorescence reflected by the dichroic passes through appropriate filters and finally falls on a wide area photo detector (usually a photo multiplier tube, PMT). A wide-area detector is required because the dichroic-reflected fluorescence is not stationary in space: it moves as the excitation beam scans the sample. No confocal pinhole is needed in front of the detector because all the emission is coming from a diffraction-limited focus spot where intensity is enough to produce MPE. This scheme is known as *nondescanned detection* and is usually the preferred (and more sensitive) way of collecting the multiphoton fluorescence signal.

In the epi setup, collection will be limited for fluorophores that emit at UV wavelengths which do not pass through the objective. Typically objectives are transparent above 350 nm (see Figure 1.6). So, for imaging a fluorophore which emits in the UV region, one should carefully check the transmission of the objective.

## Non-Epi Collection (without a Lens)

In the non-epi detection scheme, maximum possible fraction of the fluorescence emitted in the forward direction is collected onto a wide area detector placed at the nearest possible separation from the sample with suitable filters placed before it (Figure 1.5).

Non-epi detection is mainly useful for the fluorophores whose emission is deeper in the UV region and which will not pass through the objective lens. The main limitation of this scheme of detection is in blocking the excitation properly. Because excitation intensity is $10^{12} \sim 10^{13}$ times higher than the fluorescence, and it falls on the detector almost unattenuated, so excitation shielding is very crucial for obtaining

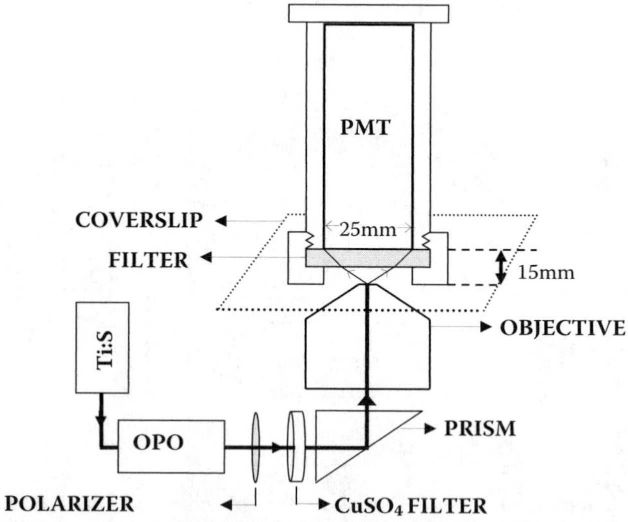


**FIGURE 1.5 Schematic diagram of the setup for non-epi collection.** Here fluorescence does not pass through any optics except for filters which block the excitation but transmit the fluorescence. (Adapted from Balaji et al.[20] With permission from the Optical Society of America.)

a good signal-to-noise ratio. This scheme[20] has been successfully used to collect signals from catecholamines such as dopamine, norepinephrine, and epinephrine, whose emission wavelengths peak around 300 nm. For serotonin, with an emission peak around 350 nm (which in vesicles is typically shifted to longer wavelengths due to interaction with other molecules), epi-collection is adequate.[2]

## Specific Optical Setup for Serotonin Imaging

Serotonin is one of the monoamines that is suitable for MPM with epi-collection because of a good overlap of its fluorescence spectra (Figure 1.6) with the detection optics. It is three-photon excitable at 740 nm, which is far less photo toxic to cells than the one-photon excitation wavelength of 270 nm. In the neuronal cells, serotonin is packed in vesicles of size ~100 nm in reasonably high concentration (~400 m*M*), which makes it easy to detect and image.

For serotonin imaging we use RN46A cells or primary culture neurons from the raphe nucleus of the 1~2-day-old rat or raphe tissue itself. For excitation, a mode-locked 100 fs, 76 MHz pulsed laser (Mira-9000, Coherent, U.S.) at 740 nm is focused with a water immersion 1.2 NA, 60× objective lens (Nikon, Japan).

A dichroic mirror (675 DCXRUV, Chroma, U.S.) is used just below the objective which reflects the emission. Emission is passed through the saturated $CuSO_4$ filter (custom-made 1-cm path length). $CuSO_4$ solution nicely serves the purpose of a filter because it blocks all the excitation and passes the emission with a wide transparent window (Figure 1.6) which allows the serotonin emission to pass through. (We note that multiphoton excitation of serotonin at 740 nm results in a second emission peak at 500 nm.) For detection, a 30 mm head-on single-photon-counting PMT (Electron Tubes, U.K.) is used. Separation between the dichroic and the filter with PMT is kept to a minimum, so that scattered fluorescence light can also be collected.

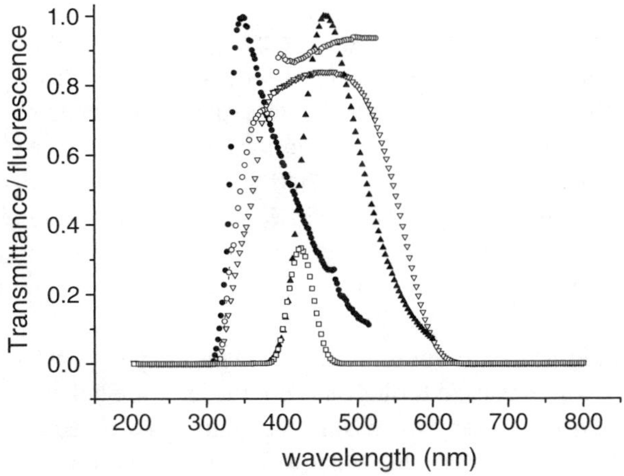


**FIGURE 1.6 Spectral characteristics of serotonin and optical elements.** Peak normalized fluorescence spectra of 400 m*M* serotonin (filled circles) and NADH (filled triangles), transmission characteristics of $CuSO_4$ (inverted empty triangles), $Cu(NH_3)_4SO_4$ (empty squares) and the objective lens (empty circles).

### Checklist before Imaging

1. To reduce optical noise and background, the whole setup, including the microscope stage and the PMT, should be made light proof. Before starting any experiment, the background leakage should be checked and minimized.
2. For calibration purposes, every time a known concentration (~100 m*M*) of serotonin solution in buffer (with a prefixed average excitation power and pulse width) should be taken, and the signal obtained should be checked with previous records.
3. In an experiment in which different excitation powers have to be used, one should be careful about the change in the pulse width, because typical means used to change the power (such as an neutral density filter) may affect the pulse width differently. One of the best ways to control power is to use a half-wave plate and a polarizer combination.
4. Polarization of the excitation beam should be fixed, as the throughput of the optical system may be strongly polarization dependent.
5. Before imaging a biological sample, one should be aware of the distance calibration of the MPI setup. From that one can easily calculate the length and the width of a pixel. For best resolved images, the settings should be such that each pixel corresponds to about half of the optical resolution in length and width. Typical pixel dimensions for good resolution would be about 120 nm in the x–y and 400 nm in the z direction.

## IMAGING SEROTONIN AND OTHER MONOAMINES IN LIVE NEURONS

### Imaging Protocol

1. Raphe neurons are isolated using standard primary culture protocols.[22] Cells cultured on homemade cover-slip-bottomed petri dishes (cover slips glued to the 15mm bottom-drilled petri dishes with Canada balsam which we find to be nontoxic) are placed in the imaging buffer (NaCl 146 m*M*, KCl 5.4 m*M*, $CaCl_2$ 1.8 m*M*, $MgSO_4$ 0.8 m*M*, $KH_2PO_4$ 0.4 m*M*, $Na_2HPO_4$ 0.3 m*M*, d-Glucose 5 m*M*, Na-HEPES 20 m*M*) after washing twice with the same buffer.
2. An appropriate field is chosen while viewing the specimen under phase contrast.
3. Cells should be viewed in transmission with a high NA water immersion objective, which is used for MPM (with the aperture before the condenser partially closed, thereby increasing the contrast by reducing the depth of field). This ensures that the object to be imaged is in the focal plane of the objective lens and reduces the time required to search for the appropriate z-plane once scanning is started.
4. The transmission lamp is switched off, and the instrument is covered so that no stray light couples to the detection system.

## Establishing the Source of the Signal

In Figure 1.7A, serotonin is seen as bright-punctate-distributed spots of varying size in the cell. What we see as the bright punctate structure is vesicular serotonin. When calibrated with known concentration of serotonin in solution, the average concentration inside the vesicle comes out to be about 400 m*M*. However, multiple components of a cell can emit autofluorescence,[23] and some of these may have spectral overlap with serotonin (this is especially true for NADH). So for any type of cell that has not been characterized previously, one has to run a battery of tests before assigning the fluorescence to serotonin. Some of the possible tests are: immunohistochemistry, inhibition of serotonin synthesis, up-regulation of serotonin synthesis, mass spectrometry of the cell extract, checking the order of excitation for its three-photon nature (many fluorescent components such as NADH would be two-photon fluorescent at 740 nm), labeling other cellular objects such as mitochondria to identify the source of the fluorescence (causing serotonin exocytosis by depolarization of the membrane), and inducing serotonin release with amphetamines. Many of these tests have been described elsewhere,[2] and we will only describe a few of these here. In the context of the applications of serotonin imaging, we will describe a few of the other ones in the subsequent section.

### Test for the Order of Excitation

At 740 nm the serotonin solution is three-photon excitable. This suggests that if the bright structures are serotonin they should also follow the same excitation order. To check this, the same field of cells was imaged with different excitation power, and the order of excitation was calculated from the fluorescence intensity of the bright structures at different excitation power. It came out to be (2.81 ± 0.29) (Figure 1.8).

### Checking Co-Localization of Bright Structures with NADH

In the cells NADH is stored inside mitochondria, and it is two-photon excitable at 740 nm.[24,25] Its fluorescence spectrum has good overlap with the serotonin spectra and the transmission window of a $CuSO_4$ filter. To check whether the bright punctuate structures

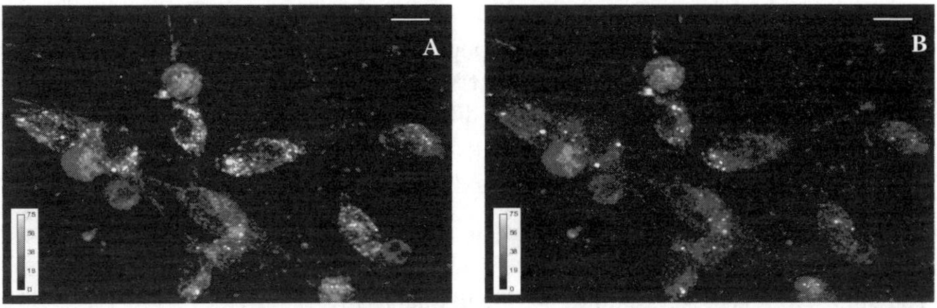

**FIGURE 1.7 Imaging serotonin in cells.** (A) a three-photon image of serotonergic cell line RN46-A. (B) image of the same field after exposure to 75 m*M* $K^+$. Serotonin vesicles are seen as bright punctate structures. Bar = 10 $\mu$m. (Adapted from Balaji et al.[2] With permission from Blackwell Publishing Ltd.)

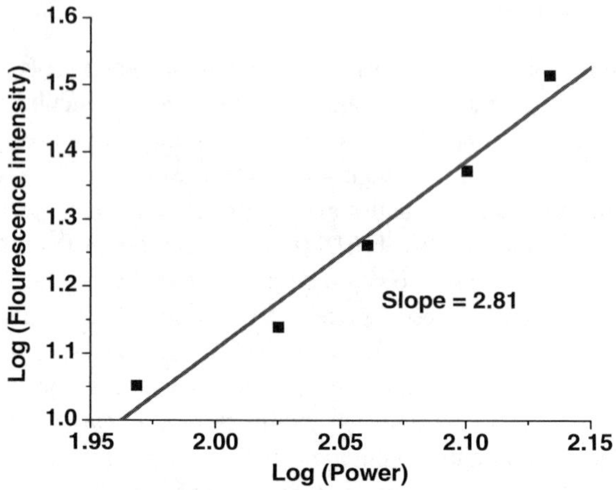


**FIGURE 1.8 The proof for three-photon excitation.** Here, log of the fluorescence signal from the bright cellular structures is plotted against the log of the corresponding excitation powers. Slope of the straight line fit is 2.81 ± 0.29, from which we conclude that the order of excitation is 3.

in the image are serotonin or NADH, one has to simultaneously image mitochondria and serotonin in two separate channels and check for colocalization of the bright structures.

In a live neuron, mitochondria can be labeled by incubating cells with 10 $\mu M$ rhodamine 1-2-3 (a fluorescent mitochondrial marker, which specifically gets packaged inside the mitochondria). To check for colocalization, simultaneous imaging for rhodamine 1-2-3 and serotonin is required. A slight modification on the detector side allows us to simultaneously image serotonin as well as rhodamine 1-2-3. The fluorescence after reflection from the dichroic (Figure 1.4) is split into two parts, using a 70/30 beam splitter; 30% of the beam is reflected for rhodamine 1-2-3 detection, whereas 70% goes straight to the serotonin channel. Rhodamine 1-2-3 fluorescence is selectively detected through a 555/50 filter (Chroma, Inc.) placed directly in front of the PMT. The serotonin channel has a 0.3 cm thick filter of saturated $Cu(NH_3)_4SO_4$ solution, which blocks the rhodamine 1-2-3 fluorescence. Filters used in either channel don't allow any cross talk and block residual infrared also. In the rhodamine 1-2-3 channel, elongated cylindrical mitochondrial structures are seen. In the serotonin channel the observed structures are mostly spherical. These two types of structures are also seen in different planes (Figure 1.9). So structures seen in the two channels do not colocalize, and the fluorescence therefore cannot originate from NADH in the mitochondria.

## Applications of Serotonin Imaging

The ability to image serotonin directly and quantitatively provides a great assay for multiple types of studies involving serotonergic signaling. We describe a few of these here: (1) depolarization induced dynamics of serotonin vesicles, (2) differentiation-induced up-regulation of serotonin synthesis, and (3) amphetamine induced nonexocytotic release of serotonin from raphe neurons.

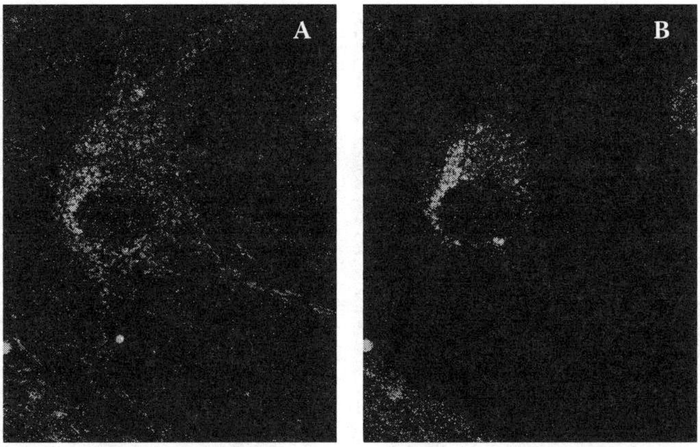


**FIGURE 1.9** **(Color figure follows p. 110.) Serotonin vesicles and the mitochondria.** (A) a pseudo-colored merged image of primary cultured cell simultaneously imaged at 6 $\mu$m above the cover slip, for serotonin (green, auto fluorescence) and mitochondria (red, labeled with rhodamine 1-2-3). (B) a pseudo-colored merged image of the same primary cultured cell simultaneously imaged at 10 $\mu$m above the cover slip. The images clearly show that emission does not colocalize with the mitochondria and therefore does not originate from NADH.

## Exocytosis with $K^+$ Depolarization

Physiologically, serotonergic vesicles in a neuron are exocytosed when the membrane is depolarized. A large number of studies have relied on electrophysiological techniques to extracellularly monitor exocytosis. Alternatively, optical techniques have also been used to monitor vesicle recycling during exocytosis[26,27] (after labeling the membrane with a fluorescent dye). However, three-photon imaging of serotonin, for the first time, allows us to directly monitor the neurotransmitter during exocytosis.[1] We provide an example here.

We use differentiated cells from a serotonergic cell line (RN46A).[2] In Figure 1.7A, we see the cells prior to any treatment. The serotonin vesicles (or vesicular clusters) are clearly visible as bright punctate structures. The cells are then treated with 50 m*M* KCl solution, with KCl replacing an equivalent amount of NaCl in the solution. The cells are then imaged repeatedly at 5-min time intervals. Figure 1.7B shows the cells after 15 min. A large number of vesicles have disappeared in the intervening time, presumably due to exocytosis. We note that most of the vesicles that remain approximately retain their original intensity, showing that the effects of photobleaching, etc., are negligible.

## Monitoring the Differentiation of Serotonergic Cells

The ability to directly monitor serotonin enables us to monitor the upregulation of serotonin synthesis and its packaging, in neurons. An interesting possibility is to use this technique as an assay for the emergence of serotonergic traits in pluripotent cells. We have followed the differentiation of a temperature-sensitive mutant cell line RN46A as the temperature is elevated from 33°C to 39°C.[2,28] The images of the

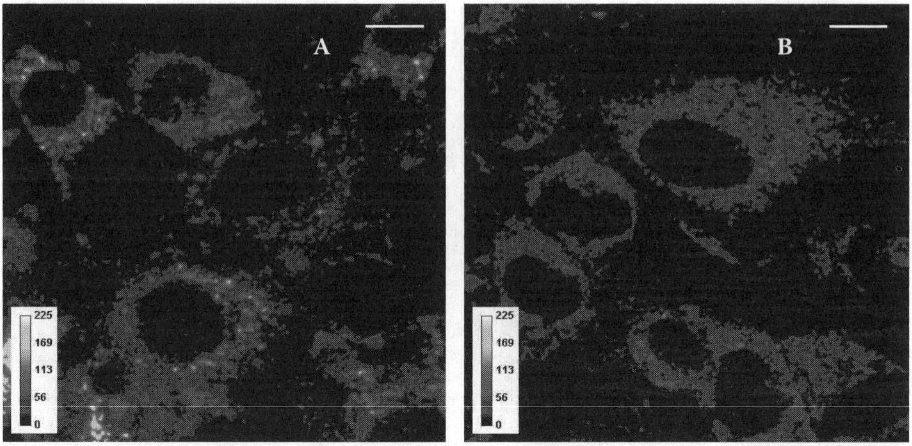


**FIGURE 1.10 Serotonin imaging as an assay for differentiation.** Differentiated RN46A cells contain more serotonin and many more vesicles (A) compared to their undifferentiated counterparts (B) Bar = 10 $\mu$m. (Adapted from Balaji et al.[2] With permission from Blackwell Publishing Ltd.)

cells in 33°C (incompletely differentiated) and those from the cells kept for 5 days at 39°C (well differentiated) are shown in Figure 1.10B and Figure 1.10A, respectively. Quantitative measurement shows that the total amount of serotonin per cell has increased by a factor of two.[2] In addition, the packaging of serotonin becomes much more prominent, and the number of serotonergic vesicles increases by a factor of four.

### Amphetamine-Induced Serotonin Release

A number of psychoactive agents work through the monoaminergic system, and serotonin imaging can provide a direct assay for the action of these agents. One of the most well-known classes of these agents is the amphetamine family. Extracellular measurements indicate that amphetamines affect the central nervous system in three main steps. They induce vesicular monoamines (serotonin and dopamine) to leak out into the cytosol, then this cytosolic monoamine is nonexocytotically expelled from the cell through reverse transport by the plasma membrane transporters. The resultant massive increase of the monoamine concentration gives rise to psychedelic effects. We can now directly image the first two steps of this process. In sequential images, we resolve the expulsion of the serotonin from the vesicles into the cytoplasm, and then observe a slower decrease of the total serotonin content of the cell (data not shown).

## Catecholamine Detection

It is important to explore whether this technique is capable of imaging other neurotransmitters. In general, if a neurotransmitter is intrinsically fluorescent, even if only under UV excitation, multiphoton microscopy should be able to image it. Neurotransmitters give us the unique opportunity of imaging them, as they are concentrated in small punctate structures in the cell and can stand out against even relatively strong but diffuse background. Catecholamines are of immediate interest in this context, as they are fluorescent in the UV, albeit at wavelengths (~300 nm) which are even shorter than that of serotonin. This is a wavelength range where the throughput of normal glass-based objective lenses and other optical elements is very poor. However, the non-epi detection scheme discussed earlier can be the perfect solution to this throughput problem. UV transparent filters do not adequately block the infrared (we need more than $10^{10}\times$ blocking of the infrared in the non-epi mode). So we need to employ a special type of fs laser, known as an optical parametric oscillator (e.g., MIRA-OPO from Coherent) for this purpose. This laser operates in the visible region and can excite the fluorophores through a two-photon route. Also, very good glass-based filters (such as UG11, Schott Glass, Germany) are available for blocking the visible wavelengths while passing through the UV. This makes the catecholamines accessible to imaging. For example, dopamine is two-photon excitable at a 550 nm wavelength. Using the scheme outlines here, we have been able to obtain sufficient signal from a dopamine solution to speculate that dopamine imaging in live neurons would also be possible in the near future, using the non-epi detection route.[20]

## ACKNOWLEDGMENTS

J. Balaji and Radha Desai have been instrumental in procuring some of the data reported here. This research is supported by a Wellcome Trust Senior Overseas Research Fellowship (no. 05995/Z/99/Z/HH/KO) to SM.

## REFERENCES

1. Maiti, S., Shear, J.B., Williams Rebacca, M., Zipfel, W.R., and Webb, W.W., Measuring Serotonin Distribution in Live Cells with Three-Photon Excitation, *Science,* 275, 1997, pp. 530–532.
2. Balaji, J., Desai, R., Kaushalya, S.K., Eaton, M.J., and Maiti, S., Quantitative measurement of serotonin synthesis and sequestration in individual live neuronal cells, *J Neurochem,* 95, 1217–1226, 2005.
3. Maria Göppert-Mayer, Über Elementarakte mit zwei Quantensprüngen, *Annalen der Physik,* 401, 273–294, 1931.
4. Singh, S. and Bradley, L.T., Three-photon absorption in napthalene crystals by laser excitation, *Phys Rev Lett,* 12, 612–614, 1964.
5. Denk, W., Strickler, J.H., and Webb, W.W., Two-Photon Laser Scanning Fluorescence Microscopy, *Science,* 248, 1990, p. 73.
6. Xu, C., Zipfel, W., Shear, J.B., Williams, R.M., and Webb, W.W., Multiphoton fluorescence excitation: new spectral windows for biological nonlinear microscopy, *Proc Natl Acad Sci USA,* 93, 10763–10768, 1996.
7. Denk, W., Piston, D.W., and Webb, W.W., in *Handbook of Confocal Microscopy,* Pawley, J., Ed., Plenum Press, New York, 1994.
8. Svoboda, K. and Yasuda, R., Principles of two-photon excitation microscopy and its applications to neuroscience, *Neuron,* 50, 823–839, 2006.
9. Xu, C. and Webb, W.W., Measurement of two photon excitation cross sections of molecular fluorophores with data from 690 to 1050 nm, *J Opt Soc Am B,* 13, 481, 1996.

10. Guild, J.B., Xu, C., and Webb, W.W., Measurement of group delay dispersion of high numerical aperture objective lenses using two-photon excited fluorescence, *Appl Opt,* 36, 397–401, 1997.
11. Moore, J.H., Davis, C.C., and Coplan, M.A., *Building Scientific Apparatus: A Practical Guide to Design and Construction,* Addison Wesley, Redwood City, Calif., 1989.
12. Born, M. and Wolf, E., *Principles in Optics: Electromagnetic Theory of Propagation, Interference and Diffraction of Light,* Cambridge University Press, 1980.
13. Sheppard, C.J.R. and Matthews, H.J., Imaging in high-aperture optical systems, *J Opt Soc Am A,* 4, 1354–1360, 1987.
14. Wilson, T. and Sheppard, C.J.R., *Theory and Practice of Scanning Optical Microscopy,* Academic, London, 1984.
15. Magde, D., Elson, E., and Webb, W.W., Thermodynamic fluctuation in a reaction system — measurement by fluorescence correlation spectroscopy, *Phys Rev Lett,* 29, 705–708, 1972.
16. Elson, E. and Magde, D., Fluorescence correlation spectroscopy: I Conceptual basics and theory, *Biopolymers,* 13, 1–27, 1974.
17. Magde, D., Elson, E., and Webb, W.W., Fluorescence correlation spectroscopy: II An experimental realization, *Biopolymers,* 13, 29–61, 1974.
18. Maiti, S., Haupts, U., and Webb, W.W., Fluorescence correlation spectroscopy: diagnostics for sparse molecules, *Proc Natl Acad Sci USA,* 94, 11753–11757, 1997.
19. Sengupta, P., Balaji, J., and Maiti, S., Measuring diffusion in cell membranes by fluorescence correlation spectroscopy, *Methods,* 27, 374–387, 2002.
20. Balaji, J., Reddy, C.S., Kaushalya, S.K., and Maiti, S., Microfluorometric detection of catecholamines with multiphoton-excited fluorescence, *Appl Opt,* 43, 2412–2417, 2004.
21. Hell, S. and Stelzer, E.H.K., Fundamental improvement of resolution with a 4Pi-confocal fluorescence microscope using two-photon excitation, *Opt Commun,* 93, 277–282, 1992.
22. Banker, G. and Goslin, K., *Culturing Nerve Cells,* The MIT Press, Cambridge, Mass., 1998.
23. Zipfel, W. R., Williams, R.M., Christie, R., Nikitin, A.Y., Hyman, B.T. and Webb, W.W., Live tissue intrinsic emission microscopy using multiphoton-excited native fluorescence and second harmonic generation, *PNAS,* 100, 7075–7080, 2003.
24. Kasischke, K.A., Vishwasrao, H.D., Fisher, P.J., Zipfel, W.R., and Webb, W.W., Neural Activity Triggers Neuronal Oxidative Metabolism Followed by Astrocytic Glycolysis, *Science,* 305, 2004, pp. 99–103.
25. Vishwasrao, H.D., Heikal, A.A., Kasischke, K.A., and Webb, W.W., Conformational dependence of intracellular NADH on metabolic state revealed by associated fluorescence anisotropy, *J Biol Chem,* 280, 25119–25126, 2005.
26. Betz, W.J. and Bewick, G.S. Optical Analysis of Synaptic Vesicle Recycling at the Frog Neuromuscular Junction, *Science,* 255, 1992, pp. 200–203.
27. Rizzoli, S.O. and Betz, W.J., The Structural Organization of the Readily Releasable Pool of Synaptic Vesicles, *Science,* 303, 2004, pp. 2037–2039.
28. White, L.A., Eaton, M.J., Castro, M.C., Klose, K.J., Globus, M.Y., Shaw, G. and Whittemore, S.R., Distinct regulatory pathways control neurofilament expression and neurotransmitter synthesis in immortalized serotonergic neurons, *J Neurosci,* 14, 6744–6753, 1994.

# 2 Monitoring Receptor-Mediated Changes of Intracellular cAMP Level by Using Ion Channels and Fluorescent Proteins as Biosensors

*Evgeni G. Ponimaskin, Martin Heine, Andre Zeug, Tatyana Voyno-Yasenetskaya, and Petrus S. Salonikidis*

## CONTENTS

Introduction ........ 20
Monitoring of Receptor Mediated cAMP Changes in Living Cells by Electrophysiological Approach ........ 21
  Experimental System, Setup, and Data Analysis ........ 21
  Examples ........ 25
  Conclusion ........ 25
Förster Resonance Energy Transfer (FRET)-Based cAMP Sensor ........ 27
  cAMP Sensors ........ 27
  FRET Principles and FRET-Based Analysis ........ 28
  Experimental System for cAMP Analysis by Using Epac1 Sensor ........ 31
    Light Sources ........ 31
    Optics ........ 31
    Camera ........ 33
  Examples ........ 33
  Outlook ........ 35
Acknowledgments ........ 36
References ........ 37

## INTRODUCTION

Five-hydroxytryptamine (5-HT or serotonin) is an important neuromodulator involved in a wide range of physiological functions. The effects of serotonin are mediated by a large family of receptors, either ionotropic or coupled to second-messenger cascades. With the exception of the 5-$HT_3$ receptor, which is a cation channel, all 5-HT receptors belong to the superfamily of 7 transmembrane-spanning receptors that are coupled to multiple heterotrimeric G-proteins.

Many of the cellular responses mediated by serotonin do not involve activation of one particular second-messenger cascade but result from the functional integration of the networks of intracellular signaling pathways. To better understand serotonergic signaling, it is therefore important to have a broad palette of methodical approaches that allow specific analysis of signaling processes with high spatial and temporal resolution. Moreover, study of receptor functions within a living cell is required to extend results obtained by biochemical and pharmacological methods. Such measurements also allow real-time analysis of signaling processes in a single cell.

Cyclic AMP (cAMP) is a key second messenger that transmits information to many different effector proteins within the cell. The cellular cAMP level depends on the activity of two groups of enzymes, the adenylyl cyclases (AC) that produce cAMP and the phosphodiesterases (PDE) that hydrolyze cAMP (Beavo, 1995; Sunahara et al., 1996). Increased cAMP levels activate a number of different effector proteins, including protein kinase A (PKA) (Francis and Corbin, 1999), hyperpolarization-activated ($I_h$) channels (DiFrancesco, 1993), the guanine–nucleotide exchange factor Epac (de Rooij et al., 1998), and cyclic nucleotide-gated (CNG) channels (Finn et al., 1996). Metabotropic serotonin receptors coupled to Gs (5HT4 and 5HT7) or Gi/o proteins (5HT1 and 5HT5) regulate AC activity, thereby changing local cAMP concentration (Barnes and Sharp, 1999).

Because biochemical methods used for cAMP measurement lack both spatial and temporal resolution, detailed understanding of how information is transduced within cAMP-regulated signaling cascades is elusive. The classical approach to analyze the receptor-mediated change in cAMP concentration includes labeling of the cells with radioactive adenine followed by the calculation of the conversion rate of [$^3$H]ATP to [$^3$H]cAMP. Although this biochemical assay is very robust and reproducible, it can not provide information about the real-time course of the cAMP level. To answer this question, cAMP signals need to be measured within the dynamic environment of the living cell.

In this chapter, we concentrate on recently established methods allowing the quantitative measurement of the intracellular cAMP concentration in living cells with good spatial and temporal resolution. These will include two different approaches: (1) electrophysiological analysis to detect electrical currents mediated by cAMP-mediated activation/inactivation of hyperpolarization-activated, cyclic nucleotide-modulated ion channels (HCN) and (2) measurement of Förster Resonance Energy Transfer (FRET) by using fluorescent-labeled guanine nucleotide exchange factor (Epac) that is activated by direct binding of cAMP (Ponsioen et al., 2004).

## MONITORING OF RECEPTOR MEDIATED cAMP CHANGES IN LIVING CELLS BY ELECTROPHYSIOLOGICAL APPROACH

Here, we will describe how to measure the intracellular cAMP concentrations using cAMP-modulated ion channels as a cAMP sensor. Because cAMP directly modulates the opening of certain ion channels such as CNG or HCN channels (Finn et al., 1996; Pape, 1996; Santoro and Tibbs, 1999; Wei et al., 1998), the measurement of their activity can be a useful tool to monitor receptor-mediated changes of intracellular cAMP concentrations.

CNG channels are fast-activating, voltage-independent nonselective cation channels. Depending on the membrane potential of the cell and the resting intracellular cAMP concentration, such channels induce a persistent ion current and can be used as a reporter of cAMP changes with high temporal and spatial resolution (Karpen and Rich, 2005). The measurement can be done using the whole cell patch clamp recordings or calcium imaging techniques. It is important to pay attention to the $Ca^{2+}$-permeability of these channels, since $Ca^{2+}$-ions can also mediate a stimulation or inhibition of several AC subtypes (Cooper, 2003). On the other hand, the $Ca^{2+}$-permeability of CNG channels can be used to monitor channel activity by fluorescent $Ca^{2+}$-indicators.

Generally, HCN channels are more sensitive to cAMP than the CNG channels (Finn et al., 1996; Santoro and Tibbs, 1999; Zagotta and Siegelbaum, 1996). For this reason, several CNG channels were genetically modified allowing for cAMP sensitivity ($K_{1/2(cAMP)} = 0.5\text{–}2\ \mu M$) similar to that obtained for natural HCN channels (Rich et al., 2001). In addition to unequal sensitivity to cAMP, CNG and HCN channels possess the different biophysical properties including channel activation and voltage dependence (Figure 2.1).

### EXPERIMENTAL SYSTEM, SETUP, AND DATA ANALYSIS

We have recently developed an experimental approach using slow activating, hyperpolarization-dependent nonspecific cation channels (HCN) as a cAMP sensor (Heine et al., 2002). The usage of the HCN channel allowed us to monitor steady-state levels and dynamic changes of cAMP. In examples described below, we used a member of the HCN-channel family, the HvCNG-channel, cloned from the antennae of the moth *Heliothis virescens* (Krieger et al., 1999) and expressed by baculovirus system in *Spodoptera frugiperda* (Sf.9) cells. Generally, we record HvCNG currents in the whole-cell patch-clamp mode, although the perforated patch configuration also can be used for such analysis (Rich and Karpen, 2002). In our experiments, we used the discontinuous single-electrode voltage-clamp amplifier SEC-05L from npi-Electronic (Tamm, Germany) connected with an ITC-16 interface from Instrutech (Greatneck, NY). This amplifier allows for a precise control over the voltage clamp of the cell independently from the access resistance. Such precise control is important, because cAMP measurement strongly depends on the voltage. For data acquisition and analysis we apply Pulse-PulseFit 8.31 software from HEKA (Lambrecht, Germany) and Igor-WaveMetrics software (Lake Oswego, OR). All measurements were performed using

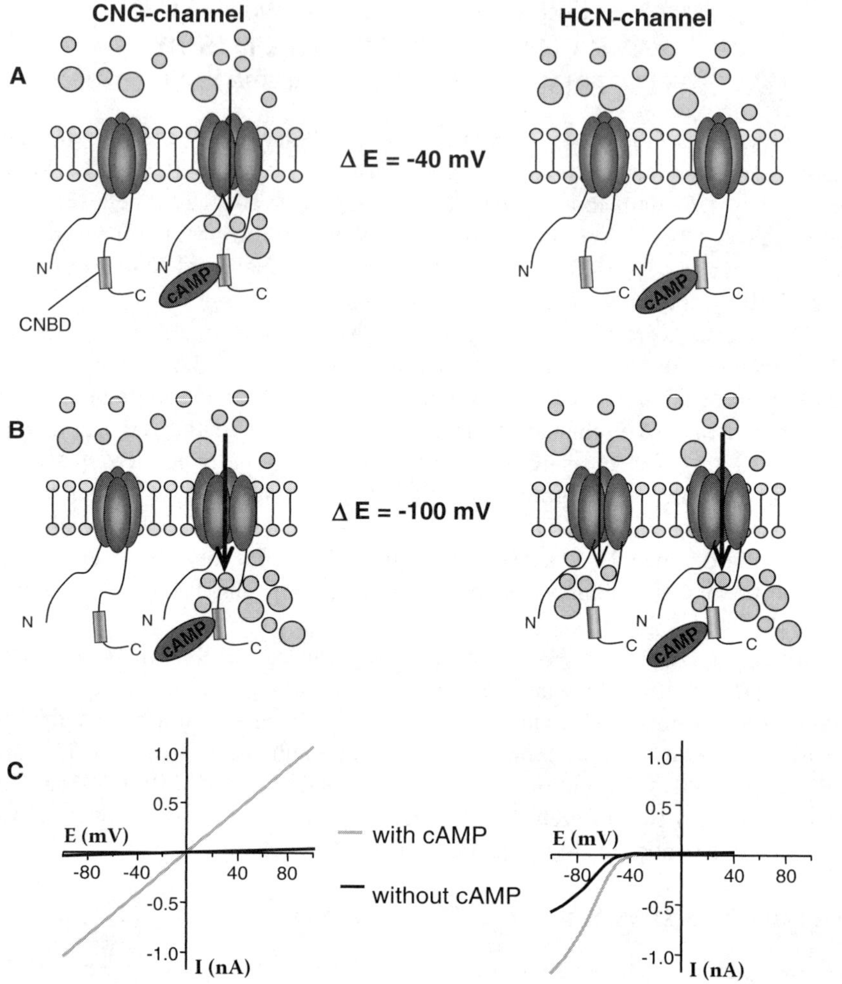


**FIGURE 2.1** Functional cyclic nucleotide gated channels (CNG) and hyperpolarization-activated cyclic nucleotide-gated channels (HCN) are oligomers composed by four subunits. The cyclic nucleotide binding domain (CNBD) is located close to the C-terminus of each subunit. (A) cAMP binding to the CNBD promotes opening of CNG but not HCN-channels at membrane potential > –40 mV. (B) cAMP binding to CNG-channels plus hyperpolarization of the cell induces a stronger inward cation current. HCN-channels are open during hyperpolarization, mid cAMP binding to the CNBD induces a faster opening of these channels. (C) Current-voltage relationship for CNG and HCN-channels in the presence (gray line) and absence (black line) of cAMP.

insect TC.100 medium as extracellular solution. In this medium it is possible to keep cells intact for 4 to 7 h at 25 ± 0.2°C. Because the possible role of $Ca^{2+}$ ions in the modulation of channel activation kinetics still remains controversial (Budde et al., 1997; Luthi and McCormick, 1998), we routinely use the highly $Ca^{2+}$-buffered (10 m*M* EGTA) pipette solution containing 15.3 n*M* free $Ca^{2+}$ concentration.

The opening kinetics of HCN channels is modulated by cAMP (Figure 2.1), so the channel activation time constant can be used as readout for the intracellular cAMP concentration. This kinetic parameter is calculated by an exponential fit of the current activation at a defined voltage step (Figure 2.2) and is independent on the expression level of the channel in the outer membrane. HCN channels are not activated at the resting potential of nonexcitable cells (around –30 to –50 m*V*) and induce no steady state calcium influx over time. These particular biophysical properties are useful to calibrate the cellular system using defined cAMP concentrations and the activation time constant as a reporter of intracellular cAMP concentration (Figure 2.2B). A second parameter that is independent on the channel expression level is half-maximal activation voltage ($V_{1/2}$). To calculate these parameters, we perform two types of fittings: an exponential fit of the time course of channel activation and a Boltzmann fit of the steady-state activation of tail currents. To fit the activation time constants we use the equation:

$$I\,(t) = I_0 + I_1 \cdot\, [1 - \exp\,(-t/\tau)], \tag{2.1}$$

where $I_0$ is the current at the beginning of activation, and $I_1$ is the maximal current at the end of the applied voltage pulses. The exponent $t/\tau$ represents the ratio between the time $t$ and the activation time constant $\tau$. For the Boltzmann fit of the steady-state activation curve of HvCNG current we used:

$$P_\infty(V_m) = 1/\{1 + \exp\,[(V_m - V_h)/k]\}, \tag{2.2}$$

where $P_\infty(V_m)$ denotes the steady-state open probability $P_\infty$ of HCN channels at a membrane potential $V_m$, $V_h$ is the half-activation voltage of the HCN current, and $k$ is the slope factor. cAMP levels were calculated from the dose–response relationship (Figure 6.2), and the data for the dose–response curve were fitted by a Hill equation with variable slope

$$\mathrm{n} = \tau_{min} + (\tau_{max} - \tau_{min})/1 + (c_{1/2}/c)^n, \tag{2.3}$$

where $\tau_{min}$ is the time constant at 0 cAMP, $\tau_{max}$ is the time constant at 1 m*M* cAMP, $c_{1/2}$ is the half-maximal cAMP concentration, and $n$ is the Hill factor. To calculate the cAMP concentration for each time constant between $\tau_{min}$ and $\tau_{max}$, we converted the Hill equation to:

$$\begin{aligned} \mathrm{c} = {} & (\tau - \tau_{min})/\ (\tau_{max} - \tau_{min}) \cdot c_{1/2}{}^{n}/ \\ & \{1 - [(\tau - \tau_{min})/\ (\tau_{max} - \tau_{min})]^{(1/n)}\} \end{aligned} \tag{2.4}$$

Based on the above strategy we determined values of $K_a$ (concentration at half-maximum activation) = 0.62 $\mu M$ for the half-maximal cAMP concentration, and $n$ = 1.51 for the Hill coefficient. The $K_a$ value was slightly lower than the $K_a$ of 0.76 $\mu M$ as previously reported by using inside–out patches (Krieger et al., 1999). On the other hand, the $K_a$ determined in our system is in good agreement with measurements obtained from other heterologously expressed HCN channels (Ludwig

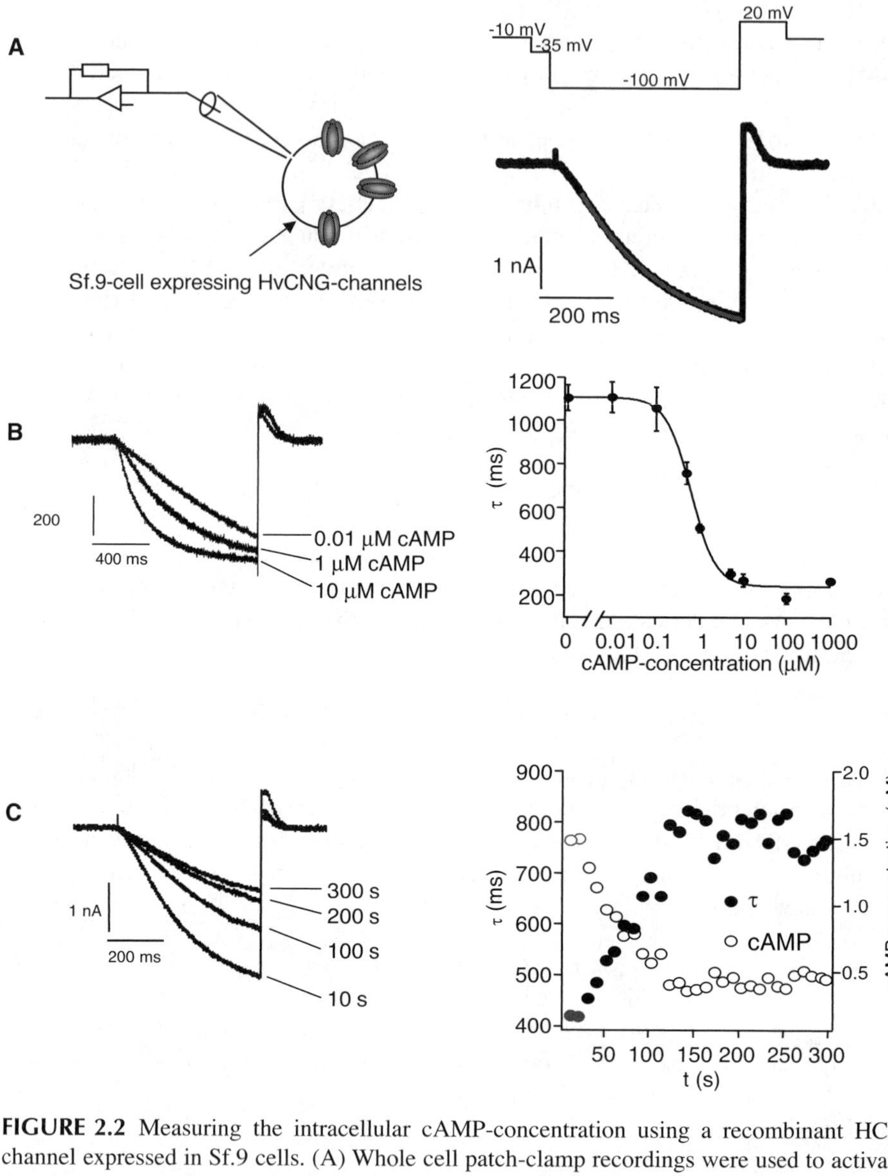


**FIGURE 2.2** Measuring the intracellular cAMP-concentration using a recombinant HCN channel expressed in Sf.9 cells. (A) Whole cell patch-clamp recordings were used to activate the channel by voltage step to –100 m*V* for 1 sec. The activation time constant (τ) was determined by the mono-exponential fit. (B) Perfusion of the cell in the whole cell patch-clamp configuration with defined cAMP-concentrations in the pipette solution was used to calibrate the system and to define the cAMP sensitivity of HCN channel. The dose-response curve is shown on the *right side* and indicates a half-maximal activation at $K_a = 0.6$ *μM* cAMP with a Hill coefficient of n = 1.5. (C) Time course of the perfusion of the cell after establishing the whole-cell configuration with a pipette solution containing 0.5 *μM* cAMP. Currents at different time points induced by 1 sec voltage steps to –100 *mV* are shown on the *right side*. The first two measuring points represent the basal cAMP-level of the cell (circles = τ, dots = cAMP concentration).

et al., 1998; Santoro et al., 2000). These results confirm that the HvCNG channel represents a sensitive cAMP sensor and encouraged us to use this channel for the quantitative analysis of cAMP level in living cells.

## Examples

The HvCNG channel expressed in Sf.9 cells was used as an endogenous sensor for cAMP changes induced by serotonin receptor activation. For that, the channel needs to be coexpressed together with appropriate receptor and G-proteins. We coexpress the HvCNG channels together with recombinant 5-HT4 receptor or 5-HT1A receptor and Gs-proteins (Gαs-, β1-, γ2-subunits) or Gi-proteins (Gαi2-, β1-, γ2-subunits), respectively.

Cells coexpressing all three components, e.g., HvCNG channels, 5-HT4 receptors, and Gs-protein revealed maximal activation of HvCNG channel-mediated currents 70 ± 4 sec after exposure to 0.1 *μM* 5-HT. The $\tau_a$ value was decreased from 769 ± 66 msec to 310 ± 38 msec, which correspond to an increase in cAMP concentration to 4.6 ± 1.45 *μM*. Noteworthy, the receptor activation could be repeated several times, although with a steadily declined response (Figure 2.3). Interestingly, $\tau_a$ values of some coinfected cells indicated that the intracellular cAMP level was elevated to 5 *μM* even under control conditions. This may indicate that overexpressed receptors reveal a basal activity that affects endogenous cAMP levels as has been previously described (Claeysen et al., 1999). On average, however, cAMP level of 1.9 ± 1.02 *μM* ($\tau_a$ = 467 ± 75 sec) was not significantly different from the basal cAMP levels of 1.5 ± 0.3 *μM* ($\tau_a$ = 571 ± 33 sec) obtained in Sf.9 cells that were only infected with the HvCNG channel.

Importantly, this system is valid not only for measuring an increase but also a decrease of cAMP. Application of serotonin to the Sf.9 cells cotransfected with HvCNG channel, 5-HT1A receptor, and Gi-protein (Gαi2-, β1-, γ2-subunits) revealed a strong inhibitory response. In this case, the $\tau_a$ value increased from 545 ± 57 msec to 805 ± 41 msec and returned to 662 ± 32 msec after wash-out, which corresponds to a decrease of the cAMP level to 0.4 ± 0.06 *μM* (basal cAMP level was 1.5 ± 0.3 *μM*). The time delay to reach stable current kinetics was 183 ± 32 sec. It is notable that in experiments with serotonin receptors we often obtained some fluctuation of the $\tau_a$ values at the start of 5-HT application and during the wash-out. We suggest that these fluctuations may reflect the activity of intracellular factors modulating the HvCNG channel activation kinetic (e.g., AC, PDE, and protein phosphorylation), which we did not control in these experiments.

## Conclusion

Taken together, the use of the HvCNG channel allows us to monitor steady-state levels and dynamic changes of cAMP. This analysis is further favored by its dual dependence on voltage and cAMP. The fact that opening of HCN channels is modulated by cAMP (Figure 2.1), enables us to use the activation time constant as readout of the intracellular cAMP concentration. The kinetic parameter is calculated by an exponential fit of the current activation at a defined voltage step (Figure 2.2)

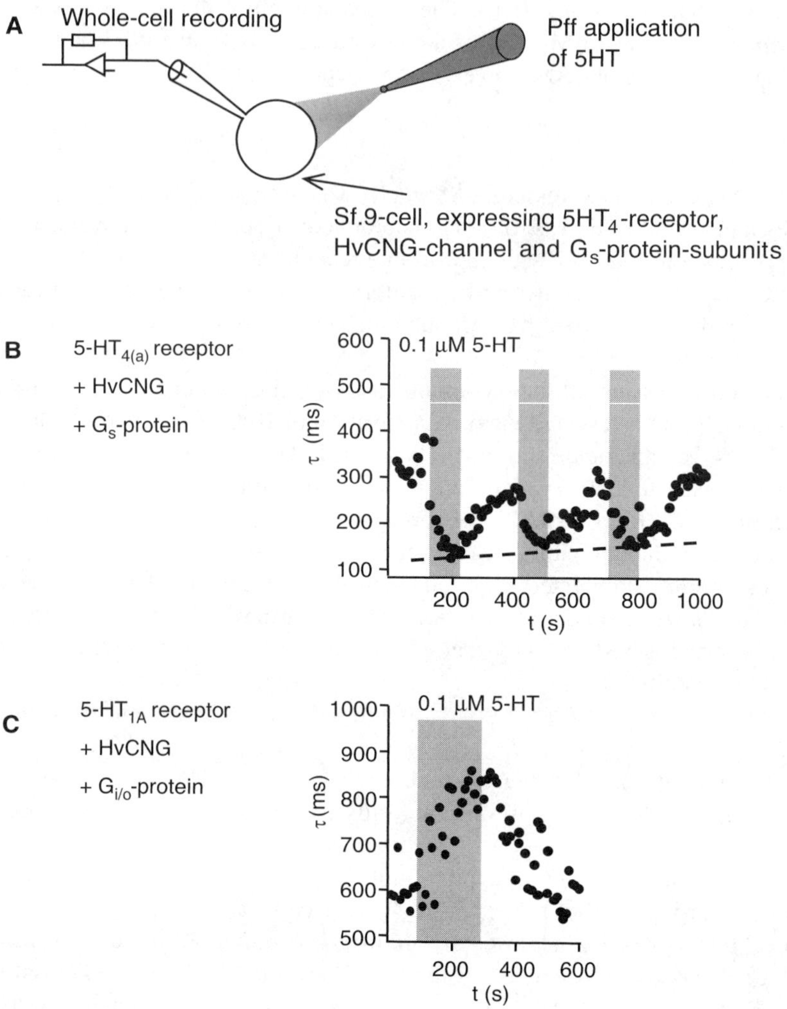


**FIGURE 2.3** Measurements of cAMP concentration after agonist-induced receptor activation. (A) Schematic drawing of the measurement configuration. (B) Coexpression of recombinant 5-HT4 receptors and *Hv*CNG channels, together with the Gs-proteins, lead to a decrease of the activation time constant for the current during 1 sec hyperpolarization to –100 m*V* in parallel with serotonin application (0.1 $\mu M$). This represents an increase in intracellular cAMP-concentration. A repetitive stimulation of the receptor results in a slow decrease of cAMP response. (C) In contrast, coexpression of recombinant 5-HT1A receptors together with the *Hv*CNG channels and Gi-proteins results in increase of the activation time constant during serotonin application (0.1 $\mu M$).

and is independent of the expression level of the channel. As the HCN channel is not activated at resting membrane potential, an overexpression of the channels will not lead to a persistent influx of $Ca^{2+}$-ions and will not change the resting membrane potential of nonexcitable cells. These particular biophysical properties are useful to calibrate the cellular system using defined cAMP concentrations and to use the activation time constant as a direct reporter of intracellular cAMP concentrations (Figure 2.2). The activation time constant is highly sensitive over a broad concentration range of cAMP concentration (0.1–5 $\mu M$), which enables measuring even small changes in cAMP level. This sensor is useful to determine endogenous cAMP levels, changes induced by constitutively active receptors, and agonist-induced changes of cAMP (Ponimaskin et al., 2005; Ponimaskin et al., 2002).

However, usage of HCN sensors has several disadvantages. This method possesses the low temporal resolution due to the slow activation kinetic (1 to 5 sec) and the absolute need to use electrophysiological techniques to induce a defined hyperpolarization of the cell as a base to measure of the activation time constant (Heine et al., 2002). In addition, the measurements derived by HCN channels do not provide the information on the subcellular organization of intracellular cAMP changes, which are accessible with different fluorescent probes (Nikolaev and Lohse, 2006) or by the use of CNG channels (Karpen and Rich, 2005; Rich and Karpen, 2002). On the other hand, CNG channels monitor cAMP or cGMP levels as a function of current amplitudes (Rich et al., 2000; Santoro et al., 2000), which strictly depends on the expression level of the channel. Therefore, the use of CNG channels demands a difficult calibration of the level of CNG channel expression within each cell.

In conclusion, the choice of a suitable sensor is mainly dependent on the answers to specific questions. HCN channels are useful for measuring constitutive activity and total cAMP levels, whereas CNG channels are more useful to investigate the temporal and spatial dynamic of receptor induced cAMP changes.

## FÖRSTER RESONANCE ENERGY TRANSFER (FRET)-BASED cAMP SENSOR

Here, we describe a more comfortable but not less sophisticated method to measure cAMP-level by using fluorescence-labeled sensors for cAMP detection. In this case, analysis of the Förster resonance energy transfer (FRET) between the fluorophores will provide information about changes in cAMP levels. To determine FRET, a fluorescence microscope, a CCD camera with computer, an image splitter, and some analyzing software are needed. The using of this approach will provide an experimenter with information about relative changes in cAMP concentrations with very good spatial and temporal resolution.

### cAMP SENSORS

Cyclic AMP is known to activate protein kinase A (PKA) and cyclic nucleotide-regulated ionchannels (CNG), as well as exchange proteins directly activated by cAMP (Epac). Therefore, these proteins have been used as a basis for the construction of fluorescence cAMP sensors (Adams et al., 1991; Zaccolo et al., 2000; Zaccolo and Pozzan, 2002). PKA was modified so that the catalytic subunit was labeled with yellow fluorescent protein (YFP) as donor, whereas the regulatory subunit was labeled with cyan fluorescent protein (CFP) as acceptor. It has been shown that the time resolution of this sensor was acceptable, but it lacked in reasonable FRET efficiency because

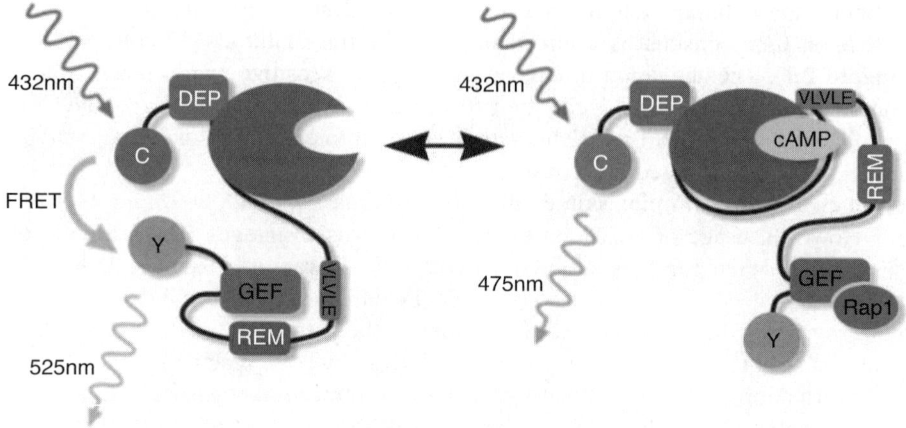


**FIGURE 2.4** The model shows a conformational change of the Epac-construct (CFP-Epac(DEP-CD)-YFP) induced by cAMP binding to the regulatory domain of Epac. (Adapted from Bos, J.L., *Nat Rev Mol Cell Biol*, 4, 738, 2003.) The distance between the two fluorophores increases after cAMP-binding, resulting in the decrease of FRET intensity. Abbreviations: DEP, domain mainly responsible for the membrane localization; GEF, guanine nucleotide exchange factor; REM, Ras-exchanger motif.

both subunits were expressed independently. Such undefined stoichiometry results in reduction of the FRET appearance. The PKA-based sensor has an affinity of ~300 n*M* for cAMP binding (Bacskai et al., 1993) and very steep dose-response relation, which rapidly reaches saturation.

Recently, an Epac1-based (de Rooij et al., 2000; de Rooij et al., 1998; Kawasaki et al., 1998) cAMP sensors has been created. These constructs are composed of a single subunit and therefore possesses a higher FRET appearance. In our experiments, we use an Epac1 sensor described by Ponsionen and colleagues (Ponsioen et al., 2004). In this construct, amino terminus of Epac1 was fused to CFP, whereas carboxy terminus was fused to YFP. In addition, the DEP domain that is responsible for membrane localization of Epac1 was deleted. Binding of cAMP to the Epac construct leads to a conformational change of the Epac1 part of the protein, resulting in a distance and/or orientation change of CFP toward YFP (Figure 2.4). By using this sensor, Ponsionen and colleagues have demonstrated that reduction of intracellular cAMP leads to an increase of the energy transfer between CFP and YFP, whereas a rise of cAMP diminishes it (Figure 2.4). In contrast to the PKA-based sensor, an Epac1-based sensor shows cAMP affinity of 14 to 50 $\mu M$ and is sensitive for an extended concentration range.

## FRET Principles and FRET-Based Analysis

The Förster resonance energy transfer (FRET) is named after Theodor Förster, who described an energy-transfer between two (not necessarily the same) fluorophores with overlapping $D_{em}/A_{ex}$ (Förster, 1948). While the donor fluorophore is excited, it can transfer its energy under certain conditions to an acceptor fluorophore. This transfer results in excitation of the acceptor. In cases of fluorescent acceptors, an induced fluorescence emission can be observed. The energy transfer, however, is nonradiative. Therefore, the often-used term *fluorescence resonance energy transfer* is misleading and should be avoided. FRET can be described by dipole–dipole coupling mechanism. The efficiency $E$ of the energy transfer depends on the distance $r$, and the orientation of the donor emission dipole moment and the acceptor absorption dipole moment. The strong distance-dependency is characterized by the Förster radius $R_o$ at the half-maximal efficiency:

$$E(r) = \frac{R_o^6}{R_o^6 + r^6} \tag{2.5}$$

In the case of the Epac construct we assume existence of two states: the first one is the unbound state *DA*, where cAMP is not bound to the construct, and the second one is the bound state *D_A* where cAMP is bound (Figure 2.4). Each state is characterized by a characteristic FRET efficiency. *DA* shows a high FRET efficiency, whereas *D_A* shows it low. In the presence of cAMP both states are populated in a certain ratio, depending on the cAMP concentration. A change in cAMP concentration will not influence the value of the FRET efficiencies but changes the population ratio between both states, resulting in a change of FRET appearance. FRET appearance of the Epac construct is therefore a function of the cAMP concentration present. Moreover it has been shown *in vitro* with the FRET analysis that a Hill kinetic could be used to explain the cAMP concentration dependent enzyme-activity of the Epac construct.

Generally, several methods are available to analyze FRET. The most sophisticated and accurate method is the measurement of the donor fluorescence lifetime. It will allow obtaining the FRET efficiency and the ratio between *DA* and *D_A* states in a quantitative manner. In contrast to many other techniques, this method is independent on the fluorophor concentration.

Other methods to analyze FRET are fluorescence intensity based. Here the appearance of FRET will result in a measurable decrease of donor intensity and an increase of acceptor intensity. This property can be used to analyze FRET and to investigate the cAMP concentration. A common intensity-based method which is applied on confocal laser scanning microscopes (cLSM) is the acceptor photobleaching method. Nevertheless, with this destructive method it is not possible to analyze changes of FRET appearance over time. Apart from the acceptor photobleaching method, laser scanning microscopes are often not ideal for kinetic measurements of FRET, because the high light intensity results in a fast bleaching of the donor as well as of the acceptor. Additionally, the scanning unit of the cLSM limits the frame rate of the microscope in a second range, resulting in a relatively poor time resolution compared to a wide field microscope with a CCD-camera (where frame rates up to 100 Hz are available).

With the aim to analyze fast changes of cAMP levels in living cells, different intensity-based methods, by which emission intensities of CFP and YFP are compared continuously are more favorable than the acceptor photobleaching method. It consists of a wide field microscope and a fast EMCCD camera allowing fast frame rates and therefore a high time resolution. It also overcomes the problem of destructive bleaching because it operates with much lower light intensities than a cLSM and is therefore more physiological. In order to obtain both emission intensities of CFP and YFP simultaneously, an image splitter has to be installed between the microscope and the camera containing two sets of filters for CFP and YFP emission.

A widely used algorithm to calculate FRET has been proposed by Gordon (Gordon et al., 1998) for measuring with only two sets of emission filters and one excitation wavelength. The ratio between the emission intensities deriving from CFP (*Ffa*) and those deriving from YFP (*Dfd*), *Ffa*/*Dfd* is used as a FRET equivalent measure, whereas FRET is inversely proportional to the ration. Because of the spectral overlap of CFP and YFP emission, these intensities cannot be measured directly. Consequently, spectral unmixing has to be applied. Taking a closer look at the emission spectra of CFP and YFP, it becomes apparent that an emission band for CFP can be chosen to exclude YFP emission. However, CFP emission cannot be excluded from YFP emission, and there is always a fraction ("cross talk") of CFP emission passing the YFP filter set. To calculate the cross talk of CFP in the emission light of the YFP filter set we applied the following considerations: with only CFP as a sample, the light intensity (*Fd*) of CFP emission in the YFP filter set is divided by the light intensity (*Dd*) of CFP emission in the CFP filter set. For our conditions:

$$Fd/Dd = 0.63 \tag{2.6}$$

With a similar procedure having only YFP as a sample, the cross-talk of YFP in the emission light of the YFP filter set can be obtained. However, as mentioned above, we chose a spectral band of our CFP filter set that no YFP emission could pass through.

Therefore we had:

$$Da/Fa = 0 \tag{2.7}$$

where *Da* is the light intensity of YFP emission in the CFP filter set and *Fa* the light intensity of YFP emission in the YFP filter set.

If the measured light intensities are *Ff* and *Df* for the YFP filter set and the CFP filter set the cross-talk corrected emission, intensities can be calculated by:

$$Ff - (Fd / Dd)Df \text{ and } Df - (Da / Fa)Ff \tag{2.8}$$

resulting in the following equation:

$$\frac{Ffa}{Dfd} = \frac{Ff - (Fd / Dd)Df}{Df - (Da / Fa)Ff} \tag{2.9}$$

and with the condition of Equation 2.6 and Equation 2.7 we gained:

$$\frac{Ffa}{Dfd} = \frac{Ff - Df \cdot 0.63}{Df} \tag{2.10}$$

One has to be aware that, although this measure of FRET is corrected for cross talk, it will not separate FRET and non-FRET signals. Consequently, it can not be used to analyze absolute values of FRET and absolute values of cAMP concentration, but it can be used to describe time-dependent changes.

## Experimental System for cAMP Analysis by Using Epac1 Sensor

### Light Sources

The main aspect of a suitable light source for donor excitation is an adequate light intensity. On the one hand, the intensity needs to be strong enough. This depends on the concentration of the fluorophores and the sensitivity of the camera, which limits the exposure time and frame rate, respectively. It must be assured that enough photons are collected during the exposure time by the CCD chip in order to receive a respectable signal-to-noise ratio. On the other hand, high light intensity results in bleaching of the fluorophores and even destruction of the specimen. Another important aspect of the light source is its spectrum. In order to gain a maximum of light intensity, it is useful to choose a light source which possesses a high intensity at the wavelength needed. This guarantees having less autofluorescence and a high signal-to-noise ratio. The highest intensity at a certain wavelength would give a laser or a high efficient LED. A xenon lamp emits its light through a wide range of wavelengths (~400–800 nm) with a more or less same intensity, whereas the emission of a mercury lamp will show several intensity bands along its spectrum. One of the mercury bands is at 435 nm, which makes mercury lamps reasonable for exiting the fluorophore CFP but not for GFP, because no band is present at 488 nm. For a xenon of mercury lamps one needs to select the bandwidth of wavelengths for excitation. This can be realized by using a monochromator, or much easier, with optical filters. We use a very light-sensitive camera with an EMCCD chip (iXon, Andor), allowing use of a 100 W xenon lamp connected to a monochromator (Optoscan, Kinetic Imaging) and light fiber (Figure 2.5).

Because the FRET efficiency is calculated by a ratio analysis, an inhomogeneous illumination of the specimen will not disturb the measurements as long as the inhomogeneity will not change during the time.

### Optics

A high numeric aperture of the objective is needed in order to gain a high light intensity. We used an epifluorescent (reflected light) microscope, where the excitation light as well as the emitting light passes the same objective (Figure 2.5). In this case, the intensity of the detected light is proportional to the fourth power of numerical aperture (Inoué, 1986). It is recommended to use the whole area of the camera chip to obtain the highest possible resolution of the camera. The objective should magnify the image

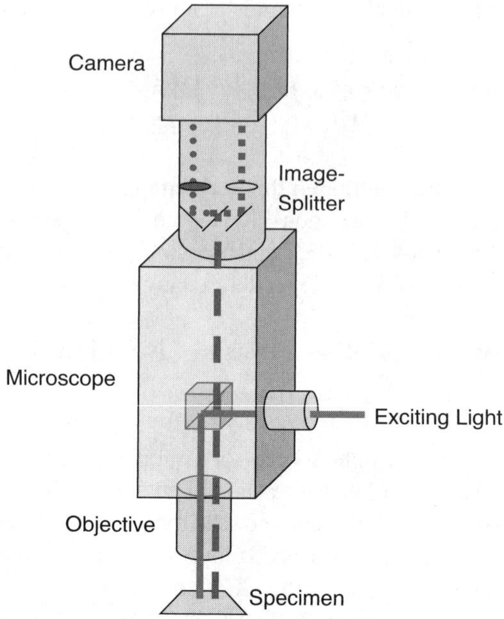


**FIGURE 2.5** Pathway of excitation as well as emission light through the setup is shown schematically. Excitation light originating from a monochromator enters the wide-field microscope through a light fiber. It is reflected by a dichotic mirror (DM) into the objective exiting the specimen. The emission from the specimen is collected by the objective and passes the dichotic mirror without being reflected. The image splitter splits the fluorescence light with a set of dichroic mirror and bandpath filters into to a CFP and YFP channel. This results in the formation of two spatially identical, but spectrally different images of the specimen on both half sides of the camera CCD chip.

of the relevant cell or cell area to half of the area of the chip. We used a water immersion objective from Olympus (LUMFI) with 40× magnification and NA = 1.1.

The excitation light for CFP was filtered at 436/20 nm. Because we used an epifluorescent microscope, the excitation light had to be reflected with a 455 nm dichroic mirror into the objective pathway in order to excite the specimen (Figure 2.5). The 455 nm dichroic mirror passes the emission light of CFP and YFP collected by the objective to an image splitter. This image splitter represents the heart of the FRET setup, and it is placed between the microscope and the camera. It allows capturing two pictures simultaneously by using a single camera. A set of dichroic filters and mirrors splits the emitted and collimated light from the specimen in two spectrally separated light channels (Figure 2.5). Both channels are equipped with different filters. This results in a formation of two spatially identical but spectrally different images of the specimen on the camera (each image is positioned such that it uses one half of the CCD total area). Because we are analyzing FRET between CFP and YFP, we chose a 470/30 nm filter for the CFP channel, a 535/30 nm filter for the YFP channel and 515 nm for the dichroic mirror. In an ensuing image processing for further analysis, the two pictures have to be overlaid exactly, pixel over pixel. Therefore, it is very important to accurately adjust the two images on the CCD chip using micro-grids provided by the manufacturer.

### Camera

Sensitivity, speed, and number of pixels are major parameters necessary for FRET experiments. The camera sensitivity will mainly influence the time resolution of the system. Higher light sensitivity allows shortening exposure time, resulting in a higher frame rate. However, the frame rate is limited due to relative slow translocation of charge into read-out areas of the CCD chip. We used the iXon camera from AndorTechnology with an electron-multiplying CCD chip (EMCCD) having a maximal frame rate of 34 Hz. By using this camera, we were able to work with exposure times of 100 msec and less with a good signal-to-noise ratio. Even if fast frame rates are not necessary for the experiment, one should consider using shorter exposure times with longer delays between two frames because this would reduce bleaching effects of the samples.

The spatial resolution of the image is determined not only by the number of pixels on the chip but also by the optical resolution of the microscope. The intensity resolution, however, is determined by the bit depth of each pixel. The CCD chip in the iXon camera we used consisted of 512 × 512 pixels. Each pixel had a bit depth of 14 bit. Thus, signals are digitalized using up to 16,384 gray scale levels, which give a dynamic range of 84 dB.

## Examples

In the experiments described next we used an Epac sensor and FRET analysis to examine effect of serotonin receptors on intracellular cAMP level. For that, neuroblastoma glioma cells N1E-115 were transfected with the Epac construct and the serotonin receptor 5-HT7, which is known to activate the adenylyl cyclase (AC) via Gs-protein upon stimulation with an agonist. The measurement consisted in a series of 400 images taken with a frame rate of 1/sec. In order to prevent strong bleaching of the fluorophores by excitation, light exposure time was preferably short. We found we could still have a good signal-to-noise ratio using 100 msec as an exposure time. For the rest of the 900 msec excitation light was turned off by a shutter. After 200 msec, the cells were treated with 5-carboxamidotryptamine (5-CT), which is a selective 5-HT7 receptor agonist. An agonist was applied with an inline solution application MPRE8 from Cell Micro Controls directly on top of the respective cell, which allows a solution exchange in few seconds. The disadvantage of the inline solution application system MPRE8 is that it is vulnerable to have air bubbles emitted from the opening. The example of such an artifact produced by the air bubble is shown in Figure 2.6. It is important to note that this artifact will not appear in the analysis of the CFP to YFP ratio, which proves the analysis to be insensitive towards non-FRET-related intensity fluctuations. From the monochromic series of images, a two-color series was created by the IQ-program. From each image one half-image (Figure 2.7, left) was overlapped with the other half-image (Figure 2.7, right). The two colors of the series represented the emission light of the two channels of the image splitter.

An area of interest was defined covering the cell and from the mean intensity of this area, the background intensity was subtracted. The background intensity was gained

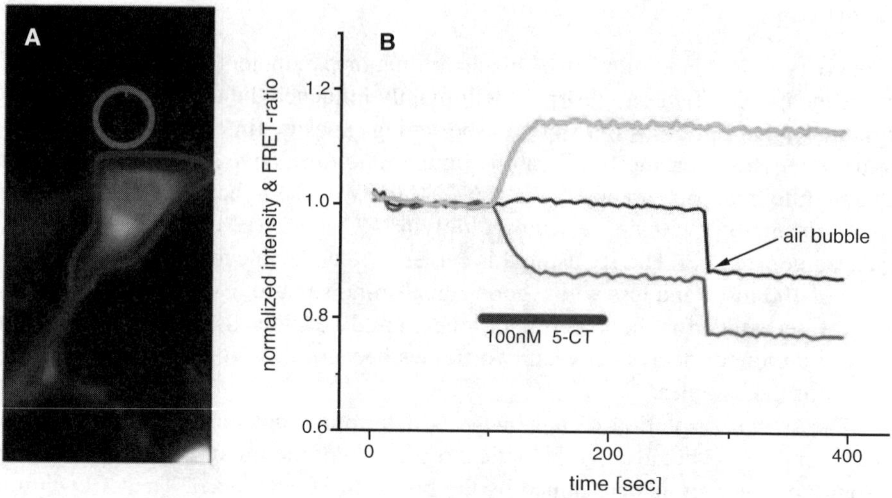


**FIGURE 2.6 (Color figure follows p. 110.)** FRET analysis results from the experiments introduced in this chapter. CFP and YFP intensities were measured from a neuroblastoma cell N1E-115 transfected with Epac and 5-HT7 receptor. (A) Region of interest (blue) drawn around the cell as well as the background region (red) used for background correction. (B) Normalized intensity traces of CFP (black) and YFP (red) resulting from Equation 2.12. The respective CFP to YFP ratio calculated with Equation 2.13 is shown as a green trace. A dramatic effect due to intracellular cAMP increase is visible after application of the 5-HT7 receptor agonist 5-CT. An air bubble arriving from the solution application system and resting underneath the objective reduces the intensity of the CFP and YFP traces to the same fraction. Therefore, it has no influence in the trace showing CFP to YFP ratio.

from the region of interest beside the cell. The intensity traces resulting from the subtraction represent the variables *Ff* and *Df*, respectively, from Equation 2.8. Using this equation we calculated the cross talk corrected intensity $Ff_{corr}$ from the YFP channel:

$$Ff^{corr} = Ff - Df \cdot 0.63 \tag{2.11}$$

Compared to Equation 2.10 we did additional normalization of the traces shown in Figure 2.6 in order to receive a ratio value of one before the agonist application. The traces were normalized to the value 0 at 50 msec:

$$Ff_{norm}^{corr} = \frac{Ff^{corr}(t)}{Ff^{corr}(50\text{ms})} \text{ and } Df_{norm} = \frac{Df(t)}{Df(50\text{ms})} \tag{2.12}$$

The equation for the ration trace, which is inversely proportional to FRET, has therefore following form related to Equation 2.10:

$$\left(\frac{Ffa}{Dfd}\right)_{norm} = \frac{Ff_{norm}^{corr}}{Df_{norm}} \tag{2.13}$$

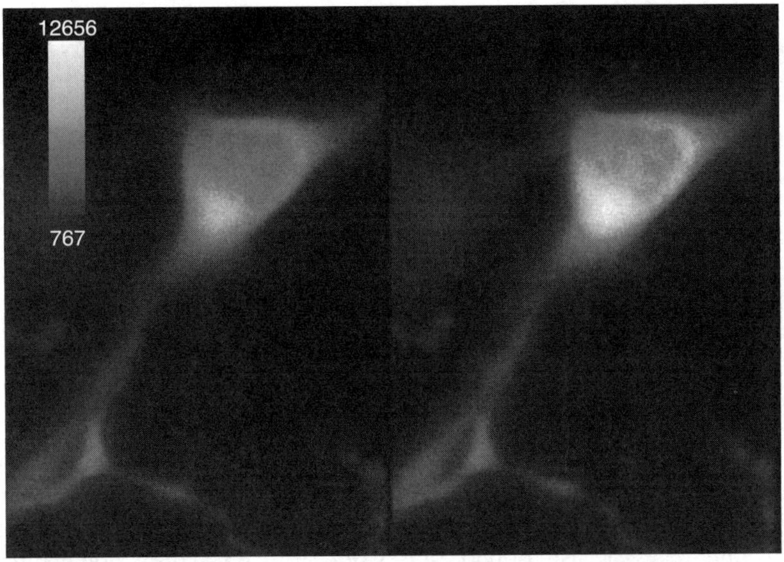


**FIGURE 2.7** Half-images originating from the same specimen. The left side represents the CPF channel and the right side does the YFP channel from the image splitter. The image shown is one of a whole image series taken during the experiment introduced in this chapter. A gray scale table was used, ranging from 767 to 12,650 arbitrary units (a.u). The cell in the YFP channel shows a higher intensity than in the CFP channel. This is caused by the cross talk between CFP intensity in the YFP channel.

It has to be mentioned that non-FRET signals are still included in the normalized FRET signal. Therefore, it is questionable to compare normalized CFP to YFP ration as readout for FRET from different cells with the intention to compare amplitudes. But nothing argues against a comparison of time-depended characteristics. With respect to this argumentation, a high accuracy in selecting the area of interest is needless as long as no other cells are located in close proximity.

## Outlook

So far we have discussed the relative changes of the FRET signal only. With this technique it is hardly possible to compare measurements between individual cells and to acquire the absolute values of cAMP concentration. A more quantitative analysis strategy has been recently proposed by Hoppe and colleagues (Hoppe et al., 2002). In this work, a stoichiometric method is described that uses two excitation wavelengths and three filter sets to measure the FRET efficiency and the relative concentrations of donor and acceptor, as well as the fractions of donor and acceptor in complex. The fluorescence images obtained with the filter sets are $F_{ex_D,em_D}$ with the donor filter set, $F_{ex_A,em_A}$ with the acceptor filter set, and $F_{ex_D,em_A}$ with the FRET filter set, where the subscriptions $_{ex_{D/A}}$ stand for the excitation wavelength for donor and acceptor, respectively, and the subscriptions $_{em_{D/A}}$ stand for the emission wavelength range for donor and acceptor, respectively. The filter sets must be chosen so that (1) no acceptor emission is detected at the donor emission wavelength $F^{A}_{exD,emD} = 0$, and (2) no donor excitation occurs at the acceptor excitation wavelength $F^{D}_{exA,emA} = 0$. Using the theory derived by Lacowicz (Lakowicz, 1999) and taking into account the bleed through or the "cross talk" of the donor emission to the acceptor emission band and the acceptor emission excited at the donor excitation wavelength, the characteristic efficiency of energy transfer can be calculated by the following equation:

$$E_C = \gamma\left(\frac{F_{exD,emA} - \beta F^{DA}_{exD,emD} - \alpha F^{AD}_{exA,emA}}{\alpha F^{AD}_{exA,emA}}\right)\frac{1}{f_A}$$
$$\text{with } \alpha = \frac{F^{A}_{exD,emA}}{F^{A}_{exA,emA}} = \frac{\varepsilon^{A}_{exD}}{\varepsilon^{A}_{exA}},\ \beta = \frac{F^{D}_{exD,emA}}{F^{D}_{exD,emD}} \text{ and } \gamma = \frac{\varepsilon^{A}_{exD}}{\varepsilon^{D}_{exD}}. \tag{2.14}$$

where $f_A$ is the fractional labeling of the acceptor with donor or the fraction of acceptor in complex with the donor. The correction terms α and β can be obtained by separate calibration measurements using acceptor only with the FRET and the acceptor filter sets, and using donor only with the FRET and the donor filter sets. γ could be calculated from literature values. If $E_C$ is known, $f_A$ can be directly obtained by the use of the three filter sets.

$$f_A = \frac{[DA]}{[A_{tot}]} = \gamma\left(\frac{F^{DA}_{exD,emA} - \beta F^{DA}_{exD,emD} - \alpha F^{DA}_{exA,emA}}{\alpha F^{DA}_{exA,emA}}\right)\frac{1}{E_C} \tag{2.15}$$

If the extinction coefficients of the donor and acceptor at the donor excitation wavelength are not available, γ can be obtained by a tandem construct where $f_A$ is equal to one. However, in both cases $E_C$ must be acquired in a separate measurement.

If $E_C$ can not be obtained, an apparent efficiency $E_A$ of transfer to the acceptor

$$E_A = E_C f_A = \gamma\left(\frac{F_{exD,emA} - \beta F^{DA}_{exD,emD} - \alpha F^{AD}_{exA,emA}}{\alpha F^{AD}_{exA,emA}}\right) \tag{2.16}$$

can be discussed. This efficiency is still quantitative in that changes in $E_A$ reflect real changes in the number of acceptor-labeled molecules in complex.

By using this strategy, the absolute measure of the cAMP concentration is possible after calibration of cAMP concentration as a function of $f_A$. Due to its quantitative nature, this strategy also allows a comparison between individual experiments. In this way the specimen can be investigated under physiological conditions at a sufficient time resolution.

## ACKNOWLEDGMENTS

We thank Dr. Kees Jalink from the Department of Cellular Biophysics, The Netherlands Cancer Institute, who kindly provided us with cDNA encoding for the

CFP-Epac-YFP fusion construct. This work was supported by the Deutsche Forschungsgemeinschaft through the Centre of Molecular Physiology of the Brain and Grant PO 732/1-2 to E.G.P. Additional support was provided by grants from the American Heart Association to TVY. TVY is an "Established Investigator" of the American Heart Association.

## REFERENCES

Adams, S.R., Harootunian, A.T., Buechler, Y.J., Taylor, S.S., and Tsien, R.Y., Fluorescence ratio imaging of cyclic AMP in single cells, *Nature*, 349, 697, 1991.

Bacskai, B.J., Hochner, B., Mahaut-Smith, M., Adams, S.R., Kaang, B.K., Kandel, E.R., and Tsien, R.Y., Spatially Resolved Dynamics of cAMP and Protein Kinase A Subunits in Aplysia Sensory Neurons, *Science*, 260, 1993, p. 226.

Barnes, N.M. and Sharp, T., A review of central 5-HT receptors and their function, *Neuropharmacology*, 38, 1083–1152, 1999.

Beavo, J.A., Cyclic nucleotide phosphodiesterases: functional implications of multiple isoforms, *Physiol Rev*, 75, 725–748, 1995.

Bos, J.L., Epac: a new cAMP target and new avenues in cAMP research, *Nat Rev Mol Cell Biol*, 4, 738, 2003.

Budde, T., Biella, G., Munsch, T., and Pape, H.C., Lack of regulation by intracellular $Ca^{2+}$ of the hyperpolarization-activated cation current in rat thalamic neurones, *J Physiol*, 503(Pt. 1), 79–85, 1997.

Claeysen, S., Sebben, M., Becamel, C., Bockaert, J., and Dumuis, A., Novel brain-specific 5-HT4 receptor splice variants show marked constitutive activity: role of the C-terminal intracellular domain, *Mol Pharmacol*, 55, 910–920, 1999.

Cooper, D.M., Molecular and cellular requirements for the regulation of adenylate cyclases by calcium, *Biochem Soc Trans*, 31, 912–915, 2003.

de Rooij, J., Rehmann, H., van Triest, M., Cool, R.H., Wittinghofer, A. and Bos, J.L., Mechanism of regulation of the Epac family of cAMP-dependent RapGEFs, *J Biol Chem*, 275, 20836, 2000.

de Rooij, J., Zwartkruis, F.J., Verheijen, M.H., Cool, R.H., Nijman, S.M., Wittinghofer, A., and Bos, J.L., Epac is a Rap1 guanine-nucleotide-exchange factor directly activated by cyclic AMP, *Nature*, 396, 474–477, 1998.

DiFrancesco, D., Pacemaker mechanisms in cardiac tissue, *Annu Rev Physiol*, 55, 455–472, 1993.

Finn, J.T., Grunwald, M.E., and Yau, K.W., Cyclic nucleotide-gated ion channels: an extended family with diverse functions, *Annu Rev Physiol*, 58, 395–426, 1996.

Förster, T., Zwischenmolekulare Energiewanderung und Fluoreszenz, *Annalen der Physik*, 437, 75, 1948.

Francis, S.H. and Corbin, J.D., Cyclic nucleotide-dependent protein kinases: intracellular receptors for cAMP and cGMP action, *Crit Rev Clin Lab Sci*, 36, 275–328, 1999.

Gordon, G.W., Berry, G., Liang, X.H., Levine, B., and Herman, B., Quantitative fluorescence resonance energy transfer measurements using fluorescence microscopy, *Biophys J*, 74, 2713, 1998.

Heine, M., Ponimaskin, E., Bickmeyer, U., and Richter, D.W., 5-HT-receptor-induced changes of the intracellular cAMP level monitored by a hyperpolarization-activated cation channel, *Pflugers Arch*, 443, 418–426, 2002.

Hoppe, A., Christensen, K., and Swanson, J.A., Fluorescence resonance energy transfer-based stoichiometry in living cells, *Biophys J*, 83, 3664, 2002.

Inoué, S., *Video Microscopy: The Fundamentals*, Plenum Press, New York, 1986, p. 128.

Karpen, J.W. and Rich, T.C., High-resolution measurements of cyclic adenosine monophosphate signals in 3D microdomains, *Methods Mol Biol*, 307, 15–26, 2005.

Kawasaki, H., Springett, G.M., Mochizuki, N., Toki, S., Nakaya, M., Matsuda, M., Housman, D.E., and Graybiel, A.M., A Family of cAMP-Binding Proteins that Directly Activate Rap1, *Science*, 282, 1998, p. 2279.

Krieger, J., Strobel, J., Vogl, A., Hanke, W. and Breer, H., Identification of a cyclic nucleotide- and voltage-activated ion channel from insect antennae, *Insect Biochem Mol Biol*, 29, 255–267, 1999.

Lakowicz, J., *Principles of Fluorescence Spectroscopy*, Kluwer, Plenum Press, New York, 1999.

Ludwig, A., Zong, X., Jeglitsch, M., Hofmann, F., and Biel, M., A family of hyperpolarization-activated mammalian cation channels, *Nature*, 393, 587–591, 1998.

Luthi, A. and McCormick, D.A., Periodicity of thalamic synchronized oscillations: the role of $Ca^{2+}$-mediated upregulation of Ih, *Neuron*, 20, 553–563, 1998.

Nikolaev, V.O. and Lohse, M.J., Monitoring of cAMP synthesis and degradation in living cells, *Physiology (Bethesda)*, 21, 86–92, 2006.

Pape, H.C., Queer current and pacemaker: the hyperpolarization-activated cation current in neurons, *Annu Rev Physiol*, 58, 299–327, 1996.

Ponimaskin, E., Dumuis, A., Gaven, F., Barthet, G., Heine, M., Glebov, K., Richter, D.W., and Oppermann, M., Palmitoylation of the 5-hydroxytryptamine4a receptor regulates receptor phosphorylation, desensitization, and beta-arrestin-mediated endocytosis, *Mol Pharmacol*, 67, 1434–1443, 2005.

Ponimaskin, E.G., Heine, M., Joubert, L., Sebben, M., Bickmeyer, U., Richter, D.W., and Dumuis, A., The 5-hydroxytryptamine(4a) receptor is palmitoylated at two different sites, and acylation is critically involved in regulation of receptor constitutive activity, *J Biol Chem*, 277, 2534–2546, 2002.

Ponsioen, B., Zhao, J., Riedl, J., Zwartkruis, F., van der Krogt, G., Zaccolo, M., Moolenaar, W.H., Bos, J.L., and Jalink, K., Detecting cAMP-induced Epac activation by fluorescence resonance energy transfer: Epac as a novel cAMP indicator, *EMBO Rep*, 5, 1176–1180, 2004.

Rich, T.C., Fagan, K.A., Nakata, H., Schaack, J., Cooper, D.M., and Karpen, J.W., Cyclic nucleotide-gated channels colocalize with adenylyl cyclase in regions of restricted cAMP diffusion, *J Gen Physiol*, 116, 147–161, 2000.

Rich, T.C. and Karpen, J.W., Review article: cyclic AMP sensors in living cells: what signals can they actually measure?, *Ann Biomed Eng*, 30, 1088–1099, 2002.

Rich, T.C., Tse, T.E., Rohan, J.G., Schaack, J., and Karpen, J.W., *In vivo* assessment of local phosphodiesterase activity using tailored cyclic nucleotide-gated channels as cAMP sensors, *J Gen Physiol*, 118, 63–78, 2001.

Santoro, B., Chen, S., Luthi, A., Pavlidis, P., Shumyatsky, G.P., Tibbs, G.R., and Siegelbaum, S.A., Molecular and functional heterogeneity of hyperpolarization-activated pacemaker channels in the mouse CNS, *J Neurosci*, 20, 5264–5275, 2000.

Santoro, B. and Tibbs, G.R., The HCN gene family: molecular basis of the hyperpolarization-activated pacemaker channels, *Ann NY Acad Sci*, 868, 741–764, 1999.

Sunahara, R.K., Dessauer, C.W., and Gilman, A.G., Complexity and diversity of mammalian adenylyl cyclases, *Annu Rev Pharmacol Toxicol*, 36, 461–480, 1996.

Wei, J.Y., Roy, D.S., Leconte, L., and Barnstable, C.J., Molecular and pharmacological analysis of cyclic nucleotide-gated channel function in the central nervous system, *Prog Neurobiol*, 56, 37–64, 1998.

Zaccolo, M., De Giorgi, F., Cho, C.Y., Feng, L., Knapp, T., Negulescu, P.A., Taylor, S.S., Tsien, R.Y., and Pozzan, T., A genetically encoded, fluorescent indicator for cyclic AMP in living cells, *Nat Cell Biol*, 2, 29, 2000.

Zaccolo, M. and Pozzan, T., Discrete Microdomains with High Concentration of cAMP in Stimulated Rat Neonatal Cardiac Myocytes, *Science*, 295, 2002, p. 1715.

Zagotta, W.N. and Siegelbaum, S.A., Structure and function of cyclic nucleotide-gated channels, *Annu Rev Neurosci*, 19, 235–263, 1996.

# 3 Membrane Organization and Dynamics of the Serotonin$_{1A}$ Receptor Monitored Using Fluorescence Microscopic Approaches

*Shanti Kalipatnapu, Thomas J. Pucadyil, and Amitabha Chattopadhyay*

**CONTENTS**

Relevance of Membrane Organization and Dynamics to Protein Function ........ 42
The Serotonin$_{1A}$ Receptor ........ 43
Membrane Organization of the Serotonin$_{1A}$ Receptor Monitored Using Detergent Insolubility ........ 44
Rationale ........ 44
Experimental Methodology ........ 45
Detergent Insolubility of the Serotonin$_{1A}$ Receptor ........ 47
Membrane Dynamics of the Serotonin$_{1A}$ Receptor ........ 49
Rationale ........ 49
Experimental Methodology ........ 50
G-protein Dependent Cell Surface Dynamics of the Serotonin$_{1A}$ Receptor ........ 53
Conclusion ........ 55
Acknowledgments ........ 55
References ........ 56

## RELEVANCE OF MEMBRANE ORGANIZATION AND DYNAMICS TO PROTEIN FUNCTION

Biological membranes are complex noncovalent assemblies of a diverse variety of lipids and proteins. From the original proposal of membranes containing lipids by Overton in 1895 to Gorter and Grendel's concept of a membrane bilayer, to Singer and Nicolson's fluid mosaic model, it has been a long journey of evolution in ideas on the organization of biological membranes (see [1] for a historical perspective). Subsequent developments have greatly refined this model. The current understanding of the structure of biological membranes will be described in this section in the context of its implications in the functioning of membrane proteins.

Because a significant portion of integral membrane proteins remains in contact with the membrane [2] and reaction centers in them are often buried within the membrane, the function of membrane proteins depends on the surrounding membrane environment. Lipid–protein interaction in membranes has attracted much attention in relation to its role in assembly, stability, and function of membrane proteins [2–4]. These effects have been attributed either to specific interactions of lipids with residues in proteins or to bulk properties of membranes. Considering the diverse array of membrane lipids, and a large repertoire of membrane proteins, it is believed that physiologically relevant processes occurring in membranes involve an intense coordination of multiple lipid–protein interactions. Since the organization and dynamics of membranes have considerable impact on membrane protein structure and function [5,6], the development and characterization of experimental tools to analyze these aspects of membranes assume significance.

The fluid mosaic model for cell membranes [7] envisages a largely fluid membrane bilayer in which proteins are embedded. This model proposes a dynamic bilayer with free translational diffusion of lipids and proteins and possible interactions between them, and a restricted movement of the membrane components across the bilayer which would preserve asymmetry of the bilayer. Some of the tenets put forward by this model were soon modified to accommodate experimental observations emerging from several laboratories, both in model membrane systems and in biological membranes, which favored a nonrandom organization of lipids and proteins in the form of domains [1,8]. The current understanding of membrane organization incorporates the idea of membrane domains, which are enriched in specific lipids and proteins, and which facilitate processes such as trafficking, sorting, and signal transduction [9–11]. Several forms of membrane domains such as caveolae, lipid rafts, and glycolipid-enriched domains, which could have overlapping characteristics in terms of composition and physical properties, have been proposed [9,11,12].

The implication of membrane organization on the signaling functions of membrane proteins in general, and on G-protein coupled receptors (GPCR) in particular, is an interesting and emerging area. The relevance of membrane organization to GPCRs arises from the fact that they represent the largest class of molecules involved in signal transduction across the plasma membrane [13]. Indeed, genomewide analysis of integral membrane proteins indicates a larger representation of proteins with seven transmembrane domains than others in the human genome [14]. GPCRs are prototypical members of the family of seven transmembrane domain proteins and include >800 members which together constitute ~2% of the human genome [15]. They respond to a variety of ligands and mediate multiple physiological processes and have therefore emerged as major targets for the development of novel drug candidates in all clinical areas [16,17]. The classical view of receptor-G-protein function in cells proposes free diffusion of molecules on the cell surface such that the probability of such interaction would depend on random collisions [18]. However, the specific and rapid signaling responses, characteristic of GPCR activation, cannot be explained solely based on a uniform distribution of receptors, G-proteins and effectors, one or more of which could even be in low abundance, on the cell surface [19,20]. This leads to the possibility that receptor-G-protein interactions may be dependent on their organization in membranes and not solely on the binding sites present on the interacting proteins. Spatiotemporal regulation of interactions between receptor, G-proteins, and effectors on the cell surface by the restriction imposed on their mobility along with selectivity of receptors to specific G-protein subunits and effectors is now believed to be an important determinant in GPCR signaling [18–21].

New technologies to analyze GPCR function in an intact cellular environment are predicted to have a major impact on GPCR research [16]. Such technologies currently involve fluorescence-based approaches to gain insight into GPCR functions such as receptor–receptor and receptor–ligand interactions, real-time assessment of signal transduction, receptor organization and dynamics in the plasma membrane, and intracellular trafficking of receptors. Fluorescence-based approaches in general are considered superior over other existing molecular detection technologies due to their enhanced sensitivity, minimal perturbation, multiplicity of measurable parameters, and suitable time scales that allow the analysis of several biologically relevant molecular processes [22,23]. This chapter aims to provide experimental guidelines to the successful application of fluorescence-based approaches such as quantitative fluorescence imaging and fluorescence recovery after photobleaching (FRAP) to yield novel information regarding the organization and dynamics of the serotonin$_{1A}$ receptor, a representative member of the GPCR superfamily.

## THE SEROTONIN$_{1A}$ RECEPTOR

The serotonin$_{1A}$ receptor binds the neurotransmitter serotonin and signals across the membrane through its interaction with heterotrimeric G-proteins which are membrane-associated signaling molecules on the cytoplasmic side of the membrane. Among ~14 different subtypes of serotonin receptors, the serotonin$_{1A}$ receptor is one of the most extensively studied for a number of reasons. These include its important role in neuronal physiology and the availability of a selective ligand 8-OH-DPAT (8-hydroxy-2-(di-*N*-propylamino)tetralin) allowing extensive biochemical, physiological, and pharmacological characterization of the receptor [24-26]. The serotonin$_{1A}$ receptor has been shown to have a role in neural development [27], and protection of stressed neuronal cells undergoing degeneration and apoptosis [28]. Treatment using agonists for the serotonin$_{1A}$ receptor constitutes a potentially useful approach in case of children with developmental disorders [29]. The serotonin$_{1A}$ receptor agonists and antagonists represent a major class of molecules with potential therapeutic effects in anxiety- or stress-related disorders [25,26]. As a result, the serotonin$_{1A}$ receptor serves as an important target in the development of therapeutic agents for neuropsychiatric disorders such as anxiety and depression. Interestingly, mutant (knockout) mice lacking the serotonin$_{1A}$ receptor exhibit enhanced anxiety-related behavior [30], and represent an excellent model system to understand anxiety-related behavior in higher animals [31]. On the clinical front, serotonin$_{1A}$ receptor levels have been shown to be altered in schizophrenia, and in patients suffering from major depression [25]. Interestingly, a recent observation has associated genetic polymorphisms at the upstream repressor region of the serotonin$_{1A}$ receptor gene to major depression and suicide in humans [32] linking its expression status to these clinical syndromes. The selective serotonin$_{1A}$ receptor agonist 8-OH-DPAT has recently been shown to inhibit growth of *Plasmodium falciparum* (reviewed in [33]) opening novel possibilities in antimalarial drug research. Besides, serotonin$_{1A}$ receptors are implicated in feeding, regulation of blood pressure, temperature, and working memory [25]. Taken together, these reports highlight the key role played by the serotonin$_{1A}$ receptor in a multitude of physiological processes, and point toward the significance of serotonin$_{1A}$ receptors in human health and disease.

## MEMBRANE ORGANIZATION OF THE SEROTONIN$_{1A}$ RECEPTOR MONITORED USING DETERGENT INSOLUBILITY

### RATIONALE

In the context of the membrane environment being an important modulator of protein function, membrane domains could serve as platforms for signaling by concentrating certain lipids (such as cholesterol and sphingolipids) and proteins while excluding others [20,34,35]. Such an organization of membranes assumes importance due to its potential role in a number of processes such as membrane trafficking, sorting, signal transduction, and pathogen entry [9,36–38]. Insolubility of membrane components in non-ionic detergents such as Triton X-100 at low temperature has been a widely utilized biochemical tool to identify and characterize membrane domains [12,39]. Evidence from model membrane studies shows that enrichment with lipids such as sphingolipids (with high melting temperature) and cholesterol serves as an important determinant for the phenomenon of detergent resistance [40,41]. The tight acyl chain packing in cholesterol–sphingolipid-rich membrane regions is thought to confer detergent resistance to membrane regions enriched in these lipids and to the proteins which reside in them. Several GPI-anchored proteins, a few transmembrane proteins and certain G-proteins have been found to reside in detergent resistant membrane domains, popularly referred to as DRMs [34,39,42]. In spite of concerns on the possibility of membrane perturbation due to the use of detergents, resistance to detergent extraction continues to be a principal tool to study membrane domains since the need for relatively simple and straightforward biochemical methods for detecting membrane domains persists. Further, information obtained from this extensively used biochemical approach can often form the basis for a more detailed analysis of membrane domains utilizing other specialized techniques [43,44].

The pharmacological and signaling functions of the serotonin$_{1A}$ receptor have been well addressed since the receptor has been cloned and heterologously expressed [reviewed in 25]. However, molecular details of the membrane organization and dynamics of the serotonin$_{1A}$ receptor have been relatively unexplored and are only beginning to be addressed now. The application of detergent insolubility as a biochemical means to understand membrane organization of the serotonin$_{1A}$ receptor will be described in the following section.

## Experimental Methodology

Detection of proteins in detergent resistant membranes (DRMs) is usually performed either by immunoblotting or ligand binding studies. However, these methods are not suitable in detecting proteins (e.g., in the case of the serotonin$_{1A}$ receptor) when the availability of antibodies with high specificity is limited [45] and/or ligand binding is compromised in presence of the detergent [46]. Membrane proteins tagged with the green fluorescent protein (GFP) provide an alternative which can overcome these difficulties. GFP from the jellyfish *Aequoria victoria* and its variants have become popular reporter molecules for monitoring protein expression, localization, and dynamics of various membrane and cytoplasmic proteins [47]. More specifically, tagging of GPCRs with GFP has allowed direct visualization of signaling and their real-time trafficking in living cells [48]. The use of fluorescent reporter proteins has its advantages over fluorescently labeled ligands to visualize receptors because (1) the stoichiometry of the receptor and fluorescent protein is well defined as the latter is covalently attached to the receptor at the DNA level, (2) complications encountered while using fluorescent ligands such as ligand dissociation are avoided, (3) this approach allows analysis of the unliganded states of the receptor (not possible with fluorescently labeled ligands), (4) the possibility of perturbation induced by bulky fluorescent groups to small endogenous ligands such as biogenic amines is eliminated, and (5) cellular biosynthesis ensures the presence of receptors attached to fluorescent proteins in cells and eliminates the necessity of labeling receptors with fluorescent ligands before each experiment.

The serotonin$_{1A}$ receptor has been fused to the enhanced yellow fluorescent protein (EYFP), a variant of the GFP, in order to visualize the receptor [49]. The EYFP in particular displays enhanced brightness and a more red-shifted fluorescence emission compared to the GFP. A schematic diagram indicating the site of the EYFP tag on the serotonin$_{1A}$ receptor, and its typical fluorescence distribution when stably expressed in Chinese hamster ovary (CHO) cells are shown in Figure 3.1. The serotonin$_{1A}$-EYFP receptor was found to be essentially similar to the native receptor in pharmacological assays and therefore can be used to reliably explore aspects such as cellular distribution and dynamics on account of its intrinsic fluorescence [49]. We have employed this fusion protein to directly determine detergent insolubility of the serotonin$_{1A}$ receptor by a GFP-fluorescence-based approach [50]. This method is based on quantitating fluorescence of the membrane protein in cells before and after detergent treatment. It is important to compare the fluorescence of the same group of cells before and after detergent treatment at 4°C in order to obtain an unambiguous estimate of the fraction of receptors which are detergent insoluble.

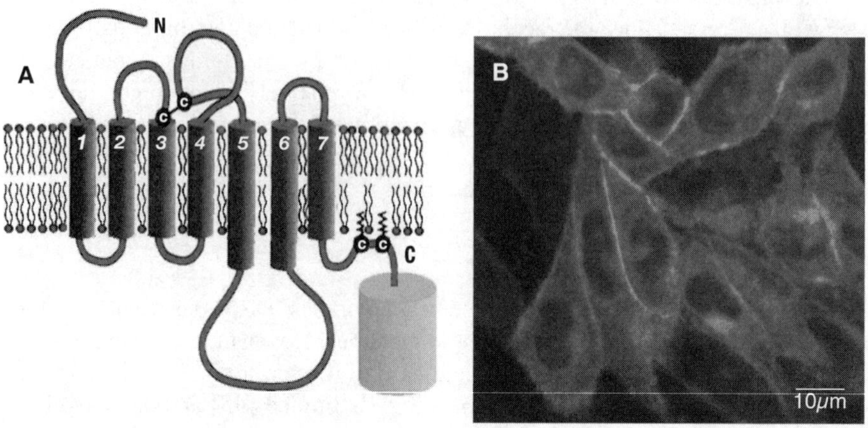


**FIGURE 3.1 (Color figure follows p. 110.)** Cellular distribution of serotonin$_{1A}$-EYFP receptors. (A) Overall topology of the serotonin$_{1A}$ receptor with the site of the EYFP tag on the receptor. (B) Typical fluorescence distribution of the serotonin$_{1A}$-EYFP receptor stably expressed in Chinese hamster ovary (CHO) cells. The image represents a midplane section of this group of cells acquired on a Zeiss LSM 510 Meta confocal microscope. The scale bar represents 10 μm.

Custom-built cell culture dishes with their base replaced with photoetched coverslips can be used for this purpose. One can identify groups of cells grown in such a dish and record their fluorescence and their position with the help of the numbers on the photoetched grids of the coverslip. The dish can be taken out of the microscope stage for the detergent extraction to be performed at 4°C. The same group of cells whose fluorescence has been recorded before detergent treatment can be identified with the help of the numbered grids on the coverslip of the glass bottom dish. The details of the experiment are given below.

1. CHO-K1 cells expressing the serotonin$_{1A}$-EYFP receptors are plated on a glass bottom dish custom built by replacing the bottoms of 35 mm plastic tissue culture dishes with photoetched grid coverslips (Bellco, Vineland, NJ) as previously described [51], and grown for 2 days. The cells are plated such that ~70–80% confluency is reached on the second day.
2. The medium is washed off and cells are imaged in HEPES-Hanks, pH 7.4 buffer to record the fluorescence intensity before detergent extraction on a Meridian Ultima 570 confocal laser scanning microscope system attached to an inverted Olympus fluorescence microscope. Optical sections of the cells are recorded using a 60×, 1.4 NA, oil-immersion objective using the 514 nm line of an Ar laser at a z-slice thickness of 0.5 $\mu$m. Fluorescence emission is collected using the 505–535 nm bandpass filter.
3. Cells are then incubated with a small volume (~60 μl) of cold Triton X-100 (0.05%, w/v) prepared in the same buffer for 10 min. on a bed of ice and water.
4. The detergent solution is then removed and cells are carefully washed using cold buffer before imaging the same group of cells whose fluorescence intensity was recorded earlier. Because cells tend to detach from

the coverslip during this wash, it is advisable to perform the wash gently by allowing small volumes of cold buffer to pass over the coverslip. This should be repeated a few times to ensure substantial removal of the detergent.

5. Fluorescence images of the same group of cells acquired before and after extraction with detergent are analyzed using the Meridian DASY Master Program v4.19. Sections of cells largely representing the plasma membrane are selected and projected together resulting in a single combined image of the chosen sections. Outlines of each cell (or a small group of cells) are drawn, and integrated fluorescence intensities within these outlines are determined using the Meridian DASY Master Program. The extent of detergent insolubility of the receptor is estimated by comparing the integrated fluorescence intensities before and after detergent extraction of cells.
6. In order to validate this fluorescence-based approach, lipid and protein markers, whose insolubility in nonionic detergents such as Triton X-100 has been well described in literature, have been employed. For this, CHO-K1 cells labeled with membrane domain-specific fluorescent lipid probes, DiIC$_{16}$ and *FAST* DiI, or fluorescently labeled transferrin (see below) are used. Stock solutions of the DiI probes are made in absolute ethanol and diluted in HEPES-Hanks buffer to prepare the labeling solutions while ensuring that the residual ethanol concentration was always low (<1%, v/v). Stock solutions of Texas-red-labeled transferrin are prepared in PBS. Cells grown for 2 days are washed twice in cold HEPES-Hanks buffer (pH 7.4) before labeling them with either of these reagents. Cells are labeled either with DiIC$_{16}$ (8 $\mu$M for 75 min), or *FAST* DiI (14 μM for 35 min) at 4°C, or with Texas-red-labeled transferrin (100 μg/ml for 30 min) at 37°C, in HEPES-Hanks buffer. Labeled cells are washed three times in cold HEPES-Hanks buffer before performing detergent extractions as described above.

## Detergent Insolubility of the Serotonin$_{1A}$ Receptor

A typical fluorescence distribution of the serotonin$_{1A}$-EYFP receptor upon detergent extraction is shown in Figure 3.2. Utilizing the approach described above, ~26% of fluorescence of the serotonin$_{1A}$-EYFP receptor is found to be retained upon extraction with 0.05% (w/v) Triton X-100 [50]. This represents the fraction of serotonin$_{1A}$ receptors which are resistant to detergent treatment under these conditions. In order to validate this fluorescence microscopic approach toward determination of detergent insolubility of membrane components, specific lipid (DiIC$_{16}$ and *FAST* DiI) and protein (transferrin receptor) markers were utilized, whose organization in membranes and ability to be extracted by cold, nonionic detergents have been well documented. The dialkylindocarbocyanine (DiI) series of lipid analogues have been shown to exhibit preferential phase partitioning into biological and model membranes of varying degrees of order (fluidity) depending on the relative headgroup to tail cross-sectional areas and the chain length [52–54]. For example, DiIC$_{16}$ with its two 16-carbon saturated alkyl chains preferentially partitions into relatively rigid (highly ordered) domains, whereas *FAST* DiI which has two 18-carbon chains with

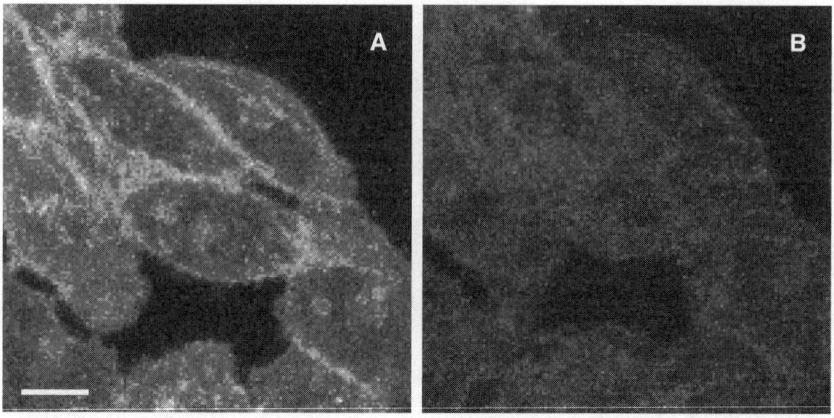


**FIGURE 3.2** Detergent insolubility of serotonin$_{1A}$-EYFP receptors. Cells expressing serotonin$_{1A}$-EYFP receptors are shown (A) before and (B) after treatment with cold Triton X-100 (0.05%, w/v) for 10 min. The images represent combined mid-plane confocal sections of the same group of cells before and after detergent extraction. The scale bar represents 10 μm. Reproduced from Ref. [50].

two *cis* double bonds in each chain preferentially partitions into fluid domains in membranes [54]. Accordingly, when cells labeled with either of these probes are extracted with Triton X-100, DiIC$_{16}$ was found to be insoluble in detergent to a greater extent than *FAST* DiI [50]. This fluorescence-based approach for determining detergent insolubility of membrane components has been further tested by monitoring detergent insolubility of a protein marker, the transferrin receptor, a transmembrane protein. Several reports have earlier shown this receptor to be soluble in Triton X-100, so it is often employed as a control in detergent insolubility experiments [51,55]. In agreement with these reports, transferrin receptor is found to be largely soluble in Triton X-100 by the GFP-based fluorescence approach to monitor detergent insolubility of the serotonin$_{1A}$ receptor [50]. Taken together, results obtained using lipid (DiI) and protein (transferrin receptor) markers validate the observation of detergent insolubility of the serotonin$_{1A}$ receptor in particular, and the GFP fluorescence-based approach in general.

Using this approach, the membrane organization of the serotonin$_{1A}$ receptor tagged to EYFP stably expressed in cells has been critically analyzed under conditions of reduced membrane cholesterol and agonist stimulation [56]. This method of analysis of detergent insolubility could be potentially useful in exploring membrane localization of other G-protein coupled receptors. In addition, this approach is free from possible artifacts induced by antibodies in immunoblotting experiments. Thus, this GFP fluorescence-based approach represents a useful application of GPCR-GFP fusion proteins to explore membrane organization of G-protein coupled receptors. This approach has the potential to be used in large-scale screenings of detergent insolubility of membrane proteins with their fusion to fluorescent proteins and in testing for insolubility by an automated fluorescence imaging system capable of handling multiple samples.

## MEMBRANE DYNAMICS OF THE SEROTONIN$_{1A}$ RECEPTOR

### RATIONALE

As mentioned earlier, the lateral organization and dynamics of GPCRs in membranes have significant implications in the manner in which cellular signaling processes involving GPCRs are regulated. The significance of receptor lateral diffusion on the plasma membrane in the signaling functions of GPCRs forms the basis of the mobile receptor hypothesis [57]. This model proposes that receptor–effector interactions at the plasma membrane are controlled by lateral mobility of the interacting components. Evidence for this comes from reports that correlate receptor signaling to membrane dynamics of individual components involved in such signaling. These include experimental evidences such as (1) the dependence of the vasopressin $V_2$ receptor to activate adenylyl cyclase through G-proteins on the fraction of receptors that are mobile on the cell surface [58], (2) dependence of the agonist-stimulated adenylyl cyclase signal transduction process on the mobile fractions of proteins in reticulocyte plasma membranes [59], (3) the correlation between lateral diffusion of rhodopsin in the membrane and light-stimulated G-protein activation [60], and (4) theoretical calculations in simulated models where the efficacy of cellular signaling could be modeled more accurately based on the diffusion limited collisional encounter of receptors and G-proteins rather than mere density of receptors and G-protein in a given membrane [61]. This model has evolved taking into consideration more recent observations on the nature and specificity of GPCR signal transduction events along with the current understanding of the organization of cell membranes. Recent evidence indicates a spatiotemporally organized system of receptors and G-proteins in membrane domains (such as caveolae or nonionic detergent-insoluble membrane microdomains such as lipid rafts, see previous text) rather than a freely diffusible system that is responsible for rapid and specific propagation of extracellular stimuli to intracellular signaling molecules [18,21]. For example, the efficient interaction of $\beta_1$- and $\beta_2$-adrenergic receptors with adenylyl cyclase (compared to prostaglandin E2 receptors) appears to correlate with the localization of $\beta_1$- and $\beta_2$-adrenergic receptors and adenylyl cyclase (and absence of prostaglandin E2 receptors) together in caveolae [62]. Overexpression of adenylyl cyclase selectively enhances $\beta$-adrenergic receptor-mediated stimulation of adenylyl cyclase activity, but not that stimulation mediated by prostaglandin E2 receptors. Furthermore, $\beta_1$-adrenergic receptors are found to stimulate adenylyl cyclase more efficiently than $\beta_2$-adrenergic receptors. Although both $\beta_1$- and $\beta_2$-adrenergic receptors are initially localized in caveolae along with adenylyl cyclase, the latter signal to adenylyl cyclase with lower efficiency due to their translocation out of caveolae upon agonist-stimulation [63]. In addition, constitutive localization of the gonadotrophin-releasing hormone (GnRH) receptor into low-density, non-ionic detergent-insoluble membrane fractions appears to be necessary for its signaling functions, namely activation of the extracellular signal-related kinase (ERK) [64]. Interestingly, stimulation of the GnRH receptor by its agonist has earlier been reported to reduce its lateral diffusion in the membrane [65], and induce homodimerization [66]. On the other hand, targeting of the oxytocin receptor, which is predominantly excluded from caveolae [67], to such membrane microdomains by its fusion with caveolin can turn the receptor-mediated inhibition of cell growth into a proliferative response [68]. Taken together, the spatiotemporal segregation of GPCRs and their effectors into microdomains has given rise to new challenges and complexities in receptor signaling since signaling now has to be understood in context of the three dimensional organization of various signaling components which include receptors and G-proteins.

Fluorescence recovery after photobleaching (FRAP) is a widely used approach to quantitatively estimate diffusion properties of molecules in cells. This approach provides information on the diffusion behavior of an ensemble of molecules, as the area monitored is large (in the order of micrometers) [69,70]. Fluorescence recovery after photobleaching involves generating a concentration gradient of fluorescent molecules by irreversibly photobleaching a fraction of fluorophores in the sample region. The dissipation of this gradient with time owing to diffusion of fluorophores into the bleached region from unbleached regions in the membrane is an indicator of mobility of fluorophores in the membrane. The recovery of fluorescence into the bleached spot in FRAP experiments is described by two parameters, an apparent diffusion coefficient (D) and mobile fraction ($M_f$). Thus, the rate of fluorescence recovery provides an estimate of the apparent D of molecules, whereas the extent of fluorescence recovery provides an estimate of $M_f$ of diffusing molecules. It must be kept in mind that $M_f$ is only an estimate of the fraction of molecules mobile in the time scale of the FRAP experiment.

We have analyzed the diffusion characteristics of serotonin$_{1A}$-EYFP receptors using FRAP to monitor the role of receptor-G-protein interaction in determining its membrane dynamics [49,71]. The following sections describe experimental details involved in a typical FRAP experiment on cells stably expressing the serotonin$_{1A}$-EYFP receptor.

## Experimental Methodology

FRAP experiments are performed on an inverted Zeiss LSM 510 Meta confocal microscope (Jena, Germany), with a 63×, 1.2 NA water-immersion objective using the 514 nm line of an argon laser. Fluorescence emission is collected using the 535–590 nm band pass filter. Images are recorded using a 225 $\mu$m pinhole thereby providing a resolution in the z-axis of 1.7 $\mu$m. Since temperature has a direct influence on the diffusion of molecules, FRAP experiments described here are performed on cells which have been acclimatized to a particular temperature by incubating them in a temperature-controlled chamber. Before imaging, glass coverslips with attached cells on them are washed with HEPES-Hanks buffer and mounted on an FCS2 closed temperature-controlled Bioptechs chamber (Butler, PA). The chamber is gently perfused with buffer and allowed to attain the required temperature. Importantly, cells are maintained under these conditions for a period of ~10 min before performing FRAP experiments.

1. The rule of thumb for any FRAP experiment is to achieve significant photobleaching within the shortest time possible. Thus, a short bleach event with high laser strength is preferable to a long event with low laser strength. This is because the bleach duration should in principle be

infinitesimally small compared to the characteristic diffusion time ($\tau_d$), or the time it takes for fluorescence recovery after photobleaching to reach its half-maximal value. A bleach duration comparable to or longer than the characteristic diffusion time of the fluorophore in a particular system can induce formation of a "corona" around the bleached region. This occurs due to repeated photobleaching of fluorophores adjacent to the bleach region on account of their diffusion into the bleach region effectively increasing the bleached spot size [72,73]. A consequence of this is an underestimation of the diffusion coefficient. To avoid this, our experiments are typically performed under conditions where the illuminating light intensity (laser strength) is set to its maximum to achieve the shortest possible bleach period. Consequently, optimization of fluorescence imaging parameters is necessary to avoid photobleaching of the sample during routine monitoring of fluorescence.

2. Optimization involves tuning imaging parameters (excitation wavelength, beam splitter configuration, emission filters, pinhole diameters) to achieve optimal collection of fluorescence from the sample thereby improving signal to noise. In addition, it is advisable to determine the maximum number of times a particular sample can be scanned (at the decided imaging parameters) before significant bleaching (>5–10%) becomes apparent. For this, it is important to scan a spatially disconnected region in the sample (or scan an entire cell). This is due to the fact that optimization performed on a spatially connected region in the sample may provide an overestimated number of scans before the sample starts to bleach due to diffusion of fluorophores from adjacent regions in the sample that are not exposed to the laser.
3. The area to be monitored and photobleached in the sample (preferably a circular region of interest (ROI) of ~1–2 $\mu$m in radius) is then selected. We frequently perform FRAP on the basal surface of well spread cells in contact with the glass coverslip. This is because analysis of FRAP data based on the theoretical framework described below assumes that fluorescence recovery into the bleached spot is isotropic in the plane of the membrane, which would be more true when monitoring the uniformly fluorescent bottom surface of cells attached to the coverslip. Furthermore, the planar geometry of the uniformly fluorescent bottom surface of cells ensures that the dimensions of the circular ROI used for bleaching are not distorted in the actual sample.
4. Present-day confocal microscopes use an area bleaching protocol where a focused laser beam is used to bleach and monitor an ROI, as opposed to a point bleach protocol where the laser bleaches and monitors a single spot in the sample. The number of times the laser scans the ROI to achieve a significant extent of bleach then needs to be set. This number depends on intrinsic factors such as photostability and mobility of the fluorophore in a given system, and on extrinsic factors such as laser power and scan speed of the laser beam, and should typically reduce fluorescence in the ROI area by 70–80% of the initial value in order to observe appreciable

fluorescence recovery. Higher extents of bleach can induce artifacts in FRAP experiments since it could damage the sample or induce crosslinking of fluorophores.

5. The total number of scans to be acquired for the entire experiment using the time series dialog is then set. This refers to the number of scans determined in the optimization procedure before significant photobleaching of the sample becomes apparent. At the same time, the time interval between successive scans is specified. This value is dependent on the dynamics of the particular system being monitored. For a fast process, the time interval would be small whereas for a slow process, the time interval would be large. To ensure complete recovery of fluorescence after bleach, it is advisable to monitor fluorescence recovery for a period of ~$4\tau_d$.
6. After performing a FRAP experiment based on the guidelines specified above, data representing the mean fluorescence intensity in the ROI as a function of time is background subtracted. Data for background fluorescence can be collected by performing a mock FRAP experiment on an area without cells (or fluorescence). This data is normalized to both the prebleach fluorescence intensity in the ROI and the time of bleach. The latter is achieved by subtracting the midpoint of the bleach duration from each data point on the time axis. This results in the first postbleach time point starting from a time $t > 0$.
7. Data on the change in the mean background-subtracted fluorescence intensity in the ROI (F(t)) vs. normalized time (t) is analyzed based on the uniform-disk illumination model [74] (according to Equation 3.1) which describes fluorescence recovery into a spot that is uniformly photobleached or which has a steplike intensity profile across the bleached spot (as is nearly the case in the present experiments).

$$F(t) = [F(\infty) - F(0)][\exp(-2\tau_d/t)(I_0(2\tau_d/t) + I_1(2\tau_d/t))] + F(0) \tag{3.1}$$

where $F(\infty)$ is the recovered fluorescence at time $t \rightarrow \infty$, F(0) is the bleached fluorescence intensity in the ROI immediately after bleach, and $\tau_d$ is the characteristic diffusion time. $I_0$ and $I_1$ are modified Bessel functions. We routinely perform nonlinear curve fitting of the recovery data to Equation 3.1 using Graphpad Prism software version 4.00 (San Diego, CA). Diffusion coefficient (D) is determined from the equation:

$$D = \omega^2/4\tau_d \tag{3.2}$$

where $\omega$ is the radius of the ROI. Mobile fraction ($M_f$) is determined from the equation:

$$M_f = [F(\infty) - F(0)]/[F(p) - F(0)] \tag{3.3}$$

where F(p) is the mean background corrected and normalized prebleach fluorescence intensity.

8. It should be kept in mind that precise determination of the diffusion coefficient from a FRAP experiment depends to a large extent on how similar the dimensions of the set ROI ($\omega$) are to the actual bleached spot in the sample (see Equation 3.2). For routine experiments, one assumes that the actual size of the bleach spot is the same as the dimensions of the ROI. However, this may not be true for all experiments. Since the bleach duration is never very small compared to the characteristic diffusion time ($\tau_d$), especially for present day confocal microscopes with relatively low power lasers, the effective size of the bleach spot would depend on the duration of bleach. A long bleach duration in a sample with high mobility can broaden the bleach spot leading to an underestimation in D. These estimates can be corrected to a significant extent by analyzing the effective bleached spot size obtained in such FRAP experiments by relatively simple image analysis procedures previously described by us [75]. Obtaining consistent and reliable quantitative estimates of mobility of molecules using FRAP can be hindered by the lack of appropriate standards for calibrating the FRAP set-up (microscope configuration and data fitting algorithms) used in a given experiment [75]. We have validated our FRAP experiments performed on serotonin$_{1A}$-EYFP receptors in cells by independent measurement of the mobility of a standard solution of EGFP in a viscous solution using a similar experimental procedure described above [75]. Our experimentally determined diffusion coefficient of EGFP under such conditions is in excellent agreement with the value predicted for GFP in a solution of comparable viscosity as calculated using the Stokes-Einstein equation.

## G-Protein Dependent Cell Surface Dynamics of the Serotonin$_{1A}$ Receptor

In light of the proposed significance of a spatiotemporally restricted environment in modulating receptor and G-protein interaction, we have analyzed the effect of serotonin$_{1A}$ receptor activation on its cell surface dynamics (diffusion characteristics) using FRAP [49,71]. These experiments indicate that the mobility of the receptor is dependent on its interaction with G-proteins. For example, pre-incubation with agents that activate G-proteins through receptor-dependent and -independent means increases receptor mobility on the plasma membrane. A typical FRAP experiment performed on cells expressing serotonin$_{1A}$-EYFP receptors under optimized imaging conditions is shown in Figure 3.3. The figure also shows the increase in fluorescence recovery kinetics of serotonin$_{1A}$-EYFP receptors in presence of aluminum fluoride ($AlF_4^-$), a receptor-independent activator of G-proteins in cells.

The G-protein heterotrimer is a large protein complex with an average molecular mass of ~88 kDa [76], which would be ~1.2 times the mass of the receptor tagged to EYFP. It is therefore possible that their association with the receptor would reduce its mobility. Receptor-dependent and -independent activation of G-proteins stimulates the exchange of a GTP for the existing GDP molecule at the G$\alpha$ subunit of G-proteins, resulting in the dissociation of G-protein heterotrimer complex from the

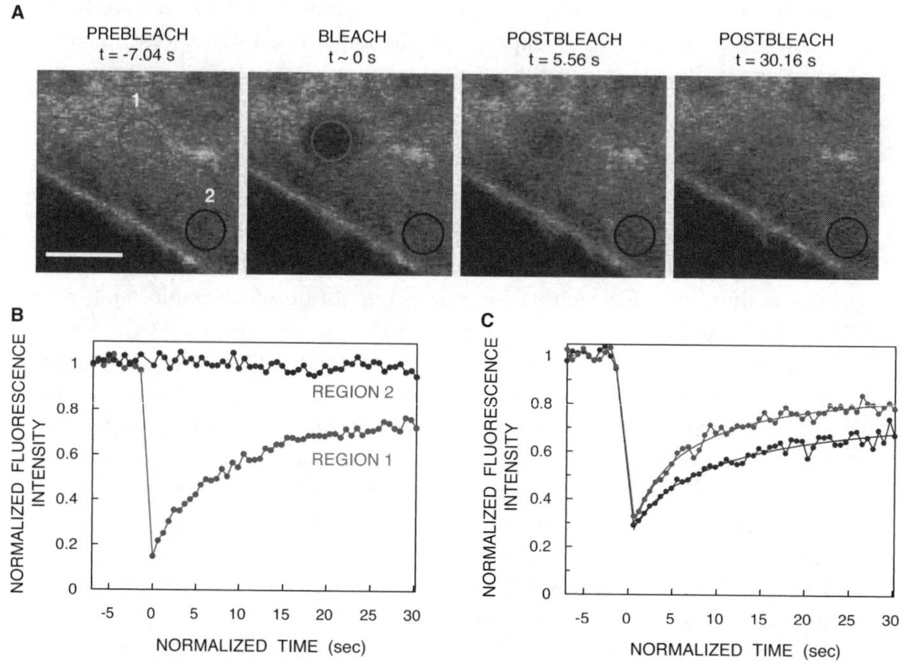


**FIGURE 3.3 (Color figure follows p. 110.)** FRAP of serotonin$_{1A}$-EYFP receptors in CHO cells. (A) Confocal fluorescence images corresponding to the base of the same cell are shown before and after photobleaching for the indicated duration of time. The prebleach image is shown at time t < 0 and the bleach event is shown at time t ~ 0. (B) Normalized fluorescence intensity in regions 1 (bleach region, red) and 2 (control region, blue) of the images in panel A are shown for the entire duration of the FRAP experiment. The constant fluorescence intensity in region 2 indicates no significant photobleaching of the field due to repeated imaging. The prebleach intensities are shown at time, t < 0. (C) Typical fluorescence recovery plots of the serotonin$_{1A}$-EYFP receptor in cells in the absence (blue) or presence (red) of aluminum fluoride ($AlF_4^-$), a receptor-independent activator of G-proteins. The curves are nonlinear regression fits to the model describing fluorescence recovery under uniform disk illumination condition. The scale bar represents 5 μm. Adapted and modified from Ref. [49].

receptor. The proposal that the association of G-proteins to the receptor reduces its mobility is further validated by the observation that treatment of cells with pertussis toxin that reduces receptor and G-protein interaction also causes an increase in receptor mobility. Importantly, these results for the first time provide convincing evidence that the cell surface dynamics of a GPCR is dependent of its interaction with G-proteins.

Diffusion behavior of several integral membrane proteins indicates that the cytoskeleton underlying the plasma membrane can act as a barrier to free diffusion of these proteins [77]. This is thought to occur due to the steric hindrance imposed by the cytoskeleton on the cytoplasmic domains of these proteins. Treatment of cells with agents that disrupt the cytoskeleton [78], truncation of the cytoplasmic domains of transmembrane proteins [79], or a lack of interaction of membrane proteins with cytoplasmic effector molecules [80] tends to increase their mobility on the cell surface. Likewise, the presence of the bulky heterotrimeric G-protein complex associated with the receptor (since G-proteins, when bound to membrane receptors, could be considered as equivalent to cytoplasmic domains of membrane proteins) could further reduce (over the differences arising due to molecular mass of G-proteins) receptor diffusion, which would be partially relieved when the G-protein dissociates from the receptor. Another possibility could be that the increase in receptor diffusion could reflect changes in the oligomeric state of the receptor, as has been shown for the δ-opioid receptor [81] and the cholecystokinin receptor [82], or their partitioning into or out of domains proposed to exist on the cell surface [20]. Incidentally, there is growing evidence on the compartmentalized localization of G-proteins in detergent-insoluble, cholesterol-rich membrane domains [19], which have been reported to diffuse as separate entities on the membrane [77,83]. Whether differences in the diffusion properties of the receptor upon agonist activation result from its movement into or out of such domains enriched in cholesterol represents an interesting possibility. In this regard, our recent report on the detergent insolubility of serotonin$_{1A}$-EYFP receptors under conditions of agonist stimulation suggests that this may not be the case [56]. The extent of detergent insolubility of serotonin$_{1A}$-EYFP receptors remains unchanged in the presence or absence of the agonist, although agonist treatment is found to enhance the diffusion coefficient of the receptor [49].

Our results on the G-protein dependent cell surface dynamics of the serotonin$_{1A}$ receptor provide novel insight into signal transduction involving this receptor. Due to similarity in the initial events of signal transduction involving GPCRs, it is possible that the increase in receptor mobility upon G-protein activation could be occurring for other GPCRs as well. Analysis of GPCR mobility, therefore, could be a sensitive and powerful approach to assess receptor/G-protein interaction in intact cells.

## CONCLUSION

The fluorescence-based approaches and their application to the serotonin$_{1A}$ receptor described in this chapter provide novel insights into the membrane biology of the serotonin$_{1A}$ receptor, and represent promising strategies to understand membrane protein function under nativelike conditions. In conjunction with recent advances on the role of the membrane lipid environment (especially cholesterol) on the ligand binding, G-protein coupling and membrane organization of the serotonin$_{1A}$ receptor (reviewed in 6,25,26), the methodologies described here provide significant insight into developing a comprehensive understanding of receptor function. This area of research assumes relevance considering the fact that membrane protein function and its dependence on the lipid environment are found to have enormous implications in health and disease [6].

## ACKNOWLEDGMENTS

Work in A.C.'s laboratory is supported by the Council of Scientific and Industrial Research, Department of Biotechnology, Life Sciences Research Board, and the

International Society for Neurochemistry. A.C. is an honorary faculty member of the Jawaharlal Nehru Centre for Advanced Scientific Research, Bangalore (India). S.K. thanks Jawaharlal Nehru Centre for Advanced Scientific Research, Bangalore (India) for a Research Associate Fellowship. We thank members of our laboratory for critically reading the manuscript.

## REFERENCES

1. Edidin, M., Lipids on the frontier: a century of cell-membrane bilayers, *Nat. Rev. Mol. Cell Biol.*, 4, 414, 2003.
2. Lee, A.G., Lipid-protein interactions in biological membranes: a structural perspective, *Biochim. Biophys. Acta*, 1612, 1, 2003.
3. Lee, A.G., How lipids affect the activities of integral membrane proteins, *Biochim. Biophys. Acta*, 1666, 62, 2004.
4. Palsdottir, H. and Hunte, C., Lipids in membrane protein structures, *Biochim. Biophys. Acta,* 1666, 2, 2004.
5. Burger, K., Gimpl, G., and Fahrenholz, F., Regulation of receptor function by cholesterol, *Cell. Mol. Life Sci.*, 57, 1577, 2000.
6. Pucadyil, T.J. and Chattopadhyay, A., Role of cholesterol in the function and organization of G-protein coupled receptors, *Prog. Lipid Res.*, 45, 295, 2006.
7. Singer, S.J. and Nicolson, G.L., The fluid mosaic model of the structure of cell membranes, *Science*, 175, 1972, p. 720.
8. Jacobson, K., Sheets, E.D., and Simson, R., Revisiting the fluid mosaic model of membranes, *Science*, 268, 1995, p. 1441.
9. Simons, K. and Ikonen, E., Functional rafts in cell membranes, *Nature*, 387, 569, 1997.
10. Mukherjee, S. and Maxfield, F.R., Membrane domains, *Annu. Rev. Cell Dev. Biol.*, 20, 839, 2004.
11. Pike, L.J., Lipid rafts: heterogeneity on the high seas, *Biochem. J.*, 378, 281, 2004.
12. Hooper, N.M., Detergent-insoluble glycosphingolipid/cholesterol-rich membrane domains, lipid rafts and caveolae, *Mol. Membr. Biol.*, 16, 145, 1999.
13. Pierce, K.L., Premont, R.T., and Lefkowitz R.J., Seven-transmembrane receptors, *Nat. Rev. Mol. Cell Biol.*, 3, 639, 2002.
14. Wallin, E. and von Heijne, G., Genome-wide analysis of integral membrane proteins from eubacterial, archaean, and eukaryotic organisms, *Protein Sci.*, 7, 1029, 1998.
15. Fredriksson, R., Lagerström, M.C., Lundin, L.-G., and Schiöth, H.B., The G-protein-coupled receptors in the human genome form five main families. Phylogenetic analysis, paralogon groups, and fingerprints, *Mol. Pharmacol.*, 63, 1256, 2003.
16. Nature reviews drug discovery GPCR questionnaire participants, The state of GPCR research in 2004, *Nat. Rev. Drug Discov.*, 3, 577, 2004.
17. Hill, S.J., G-protein-coupled receptors: past, present and future, *Br. J. Pharmacol.*, 147, S27, 2006.
18. Neubig, R.R., Membrane organization in G-protein mechanisms, *FASEB J.*, 8, 939, 1994.
19. Huang, C., Hepler, J.R., Chen, L.T., Gilman, A.G., Anderson, R.G., and Mumby, S.M., Organization of G proteins and adenylyl cyclase at the plasma membrane, *Mol. Biol. Cell*, 8, 2365, 1997.
20. Ostrom, R.S. and Insel, P.A., The evolving role of lipid rafts and caveolae in G protein-coupled receptor signaling: implications for molecular pharmacology, *Br. J. Pharmacol.*, 143, 235, 2004.

21. Hur, E.-M. and Kim, K.-T., G protein-coupled receptor signalling and cross-talk: achieving rapidity and specificity, *Cell. Signal.*, 14, 397, 2002.
22. Chattopadhyay, A., Exploring membrane organization and dynamics by the wavelength-selective fluorescence approach, *Chem. Phys. Lipids*, 122, 3, 2003.
23. Raghuraman, H., Kelkar, D.A., and Chattopadhyay, A., in *Reviews in Fluorescence 2005*, Geddes, C.D. and Lakowicz, J.R. Eds., Springer, New York, 2005, p. 199.
24. Raymond, J.R., Mukhin, Y.V., Gettys, T.W., and Garnovskaya, M.N., The recombinant 5-HT$_{1A}$ receptor: G protein coupling and signalling pathways, *Br. J. Pharmacol.*, 127, 1751, 1999.
25. Pucadyil, T.J., Kalipatnapu, S., and Chattopadhyay, A., The serotonin$_{1A}$ receptor: a representative member of the serotonin receptor family, *Cell. Mol. Neurobiol.*, 25, 553, 2005.
26. Kalipatnapu, S. and Chattopadhyay, A., The serotonin$_{1A}$ receptor and its interaction with the membrane environment, *Proc. Indian Natl. Sci. Acad.*, B71, 61, 2006.
27. del Olmo, E., López-Giménez, J.F., Vilaró, M.T., Mengod, G., Palacios, J.M., and Pazos, A., Early localization of mRNA coding for 5-HT$_{1A}$ receptors in human brain during development, *Mol. Brain Res.*, 60, 123, 1998.
28. Singh, J.K., Chromy, B.A., Boyers, M.J., Dawson, G., and Banerjee, P., Induction of the serotonin$_{1A}$ receptor in neuronal cells during prolonged stress and degeneration, *J. Neurochem.*, 66, 2361, 1996.
29. Azmitia, E.C., Neuronal instability: implications for Rett's syndrome, *Brain and Dev.*, 23, S1(Suppl.1), 2001.
30. Julius, D., Serotonin receptor knockouts: a moody subject, *Proc. Natl. Acad. Sci. U.S.A.*, 95, 15153, 1998.
31. Tóth, M., 5-HT$_{1A}$ receptor knockout mouse as a genetic model of anxiety, *Eur. J. Pharmacol.*, 463, 177, 2003.
32. Lemonde, S., Turecki, G., Bakish, D., Du, L., Hrdina, P.D., Bown, C.D., Sequeira, A., Kushwaha, N., Morris, S.J., Basak, A., Ou, X.-M., and Albert, P.R., Impaired repression at a 5-hydroxytryptamine$_{1A}$ receptor gene polymorphism associated with major depression and suicide, *J. Neurosci.*, 23, 8788, 2003.
33. Chattopadhyay, A. and Kalipatnapu, S., Serotonin$_{1A}$ receptor agonist acquires an antimalarial connection, *J. Biosci.*, 29, 1, 2004.
34. Brown, D.A. and London, E., Structure and origin of ordered lipid domains in biological membranes, *J. Membr. Biol.*, 164, 103, 1998.
35. Edidin, M., Shrinking patches and slippery rafts: scales of domains in the plasma membrane, *Trends Cell Biol.*, 11, 492, 2001.
36. Simons, K. and Toomre, D., Lipid rafts and signal transduction, *Nat. Rev. Mol. Cell Biol.*, 1, 31, 2000.
37. van der Goot, F.G. and Harder, T., Raft membrane domains: from a liquid-ordered membrane phase to a site of pathogen attack, *Semin. Immunol.*, 13, 89, 2001.
38. Pucadyil, T.J., Tewary, P., Madhubala, R., and Chattopadhyay, A., Cholesterol is required for *Leishmania donovani* infection: implications in leishmaniasis, *Mol. Biochem. Parasitol.*, 133, 145, 2004.
39. Brown, D.A. and Rose, J.K., Sorting of GPI-anchored proteins to glycolipid-enriched membrane subdomains during transport to the apical cell surface, *Cell*, 68, 533, 1992.
40. Schroeder, R.J., Ahmed, S.N., Zhu, Y., London, E., and Brown, D.A., Cholesterol and sphingolipid enhance the Triton X-100 insolubility of glycosylphosphatidylinositol-anchored proteins by promoting the formation of detergent-insoluble ordered membrane domains, *J. Biol. Chem.*, 273, 1150, 1998.
41. Masserini, M. and Ravasi, D., Role of sphingolipids in the biogenesis of membrane domains, *Biochim. Biophys. Acta*, 1532, 149, 2001.

42. Chamberlain, L.H., Detergents as tools for the purification and classification of lipid rafts, *FEBS Lett.*, 559, 1, 2004.
43. Laude, A.J. and Prior, I.A., Plasma membrane microdomains: organization, function and trafficking, *Mol. Membr. Biol.*, 21, 193, 2004.
44. Kenworthy, A.K., Fleeting glimpses of lipid rafts: how biophysics is being used to track them, *J. Investig. Med.*, 53, 312, 2005.
45. Zhou, F.C., Patel, T.D., Swartz, D., Xu, Y., and Kelley, M.R., Production and characterization of an anti-serotonin 1A receptor antibody which detects functional 5-$HT_{1A}$ binding sites, *Brain Res. Mol. Brain Res.*, 69, 186, 1999.
46. Banerjee, P., Joo, J.B., Buse, J.T., and Dawson, G., Differential solubilization of lipids along with membrane proteins by different classes of detergents, *Chem. Phys. Lipids*, 77, 65, 1995.
47. Tsien, R.Y., The green fluorescent protein, *Annu. Rev. Biochem.*, 67, 509, 1998.
48. Kallal, L. and Benovic, J.L., Using green fluorescent proteins to study G-protein-coupled receptor localization and trafficking, *Trends Pharmacol. Sci.*, 21, 175, 2000.
49. Pucadyil, T.J., Kalipatnapu, S., Harikumar, K.G., Rangaraj, N., Karnik, S.S., and Chattopadhyay, A., G-protein-dependent cell surface dynamics of the human $serotonin_{1A}$ receptor tagged to yellow fluorescent protein, *Biochemistry*, 43, 15852, 2004.
50. Kalipatnapu, S. and Chattopadhyay, A., A GFP fluorescence-based approach to determine detergent insolubility of the human $serotonin_{1A}$ receptor, *FEBS Lett.*, 576, 455, 2004.
51. Mayor, S. and Maxfield, F.R., Insolubility and redistribution of GPI-anchored proteins at the cell surface after detergent treatment, *Mol. Biol. Cell*, 6, 929, 1995.
52. Klausner, R.D. and Wolf, D.E., Selectivity of fluorescent lipid analogues for lipid domains, *Biochemistry*, 19, 6199, 1980.
53. Spink, C.H., Yeager, M.D. and Feigenson, G.W., Partitioning behavior of indocarbocyanine probes between coexisting gel and fluid phases in model membranes, *Biochim. Biophys. Acta*, 1023, 25, 1990.
54. Mukherjee, S., Soe, T.T., and Maxfield, F.R., Endocytic sorting of lipid analogues differing solely in the chemistry of their hydrophobic tails, *J. Cell Biol.*, 144, 1271, 1998.
55. Shogomori, H. and Brown, D.A., Use of detergents to study membrane rafts: the good, the bad, and the ugly, *Biol. Chem.*, 384, 1259, 2003.
56. Kalipatnapu, S. and Chattopadhyay, A., Membrane organization of the human $serotonin_{1A}$ receptor monitored by detergent insolubility using GFP fluorescence, *Mol. Membr. Biol.*, 22, 539, 2005.
57. Peters, R., Lateral mobility of proteins and lipids in the red cell membrane and the activation of adenylate cyclase by beta-adrenergic receptors, *FEBS Lett.*, 234, 1, 1988.
58. Jans, D.A., Peters, R., Zsigo, J., and Fahrenholz, F., The adenylate cyclase-coupled vasopressin $V_2$-receptor is highly laterally mobile in membranes of LLC-PK1 renal epithelial cells at physiological temperature, *EMBO J.*, 8, 2481, 1989.
59. Zakharova, O.M., Rosenkranz, A.A., and Sobolev, A.S., Modification of fluid lipid and mobile protein fractions of reticulocyte plasma membranes affects agonist-stimulated adenylate cyclase. Application of the percolation theory, *Biochim. Biophys. Acta*, 1236, 177, 1995.
60. Calvert, P.D., Govardovskii, V.I., Krasnoperova, N., Anderson, R.E., Lem, J., and Makino, C.L., Membrane protein diffusion sets the speed of rod phototransduction, *Nature*, 411, 90, 2001.
61. Stickle, D. and Barber, R., Collisions and encounters in simulations of receptor/GTP-binding protein interactions via simple diffusion, *Biochim. Biophys. Acta*, 1310, 242, 1996.

62. Ostrom, R.S., Gregorian, C., Drenan, R.M., Xiang, Y., Regan, J.W., and Insel, P.A., Receptor number and caveolar co-localization determine receptor coupling efficiency to adenylyl cyclase, *J. Biol. Chem.*, 276, 42063, 2001.
63. Rybin, V.O., Xu, X., Lisanti, M.P., and Steinberg, S.F., Differential targeting of β-adrenergic receptor subtypes and adenylyl cyclase to cardiomyocyte caveolae. A mechanism to functionally regulate the cAMP signaling pathway, *J. Biol. Chem.*, 275, 41447, 2000.
64. Navratil, A.M., Bliss, S.P., Berghorn, K.A., Haughian, J.M., Farmerie, T.A., Graham, J.K., Clay, C.M., and Roberson, M.S., Constitutive localization of the gonadotropin-releasing hormone (GnRH) receptor to low density membrane microdomains is necessary for GnRH signaling to ERK, *J. Biol. Chem.*, 278, 31593, 2003.
65. Nelson, S., Horvat, R.D., Malvey, J., Roess, D.A., Barisas, B.G., and Clay, C.M., Characterization of an intrinsically fluorescent gonadotropin-releasing hormone receptor and effects of ligand binding on receptor lateral diffusion, *Endocrinology*, 140, 950, 1999.
66. Horvat, R.D., Roess, D.A., Nelson, S.E., Barisas, B.G., and Clay, C.M., Binding of agonist but not antagonist leads to fluorescence resonance energy transfer between intrinsically fluorescent gonadotropin-releasing hormone receptors, *Mol. Endocrinol.*, 15, 695, 2001.
67. Gimpl, G. and Fahrenholz, F., Human oxytocin receptors in cholesterol-rich vs. cholesterol-poor microdomains of the plasma membrane, *Eur. J. Biochem.*, 267, 2483, 2000.
68. Reversi, A., Rimoldi, V., Brambillasca, S., and Chini, B., Effects of cholesterol manipulation on the signaling of the human oxytocin receptor, *Am. J. Physiol. Regul. Integr. Comp. Physiol.*, 291, R861, 2006.
69. Edidin, M., Patches, posts and fences: proteins and plasma membrane domains, *Trends Cell Biol.*, 2, 376, 1992.
70. Lagerholm, B.C., Weinreb, G.E., Jacobson, K., and Thompson, N.L., Detecting microdomains in intact cell membranes, *Annu. Rev. Phys. Chem.*, 56, 309, 2005.
71. Pucadyil, T.J. and Chattopadhyay, A., The human serotonin$_{1A}$ receptor exhibits G-protein-dependent cell surface dynamics, *Glycoconj. J.*, 24, 25, 2007.
72. Weiss, M., Challenges and artifacts in quantitative photobleaching experiments, *Traffic*, 5, 662, 2004.
73. Braga, J., Desterro, J.M.P., and Carmo-Fonseca, M., Intracellular macromolecular mobility measured by fluorescence recovery after photobleaching with confocal laser microscopes, *Mol. Biol. Cell*, 15, 4749, 2004.
74. Soumpasis, D.M., Theoretical analysis of fluorescence photobleaching recovery experiments, *Biophys. J.*, 41, 95, 1983.
75. Pucadyil, T.J. and Chattopadhyay, A., Confocal fluorescence recovery after photobleaching of green fluorescent protein in solution, *J. Fluoresc.*, 16, 87, 2006.
76. Casey, P.J. and Gilman, A.G., G-protein involvement in receptor–effector coupling, *J. Biol. Chem.*, 263, 2577, 1988.
77. Lenne, P.-F., Wawrezinieck, L., Conchonaud, F., Wurtz, O., Boned, A., Guo, X.-J., Rigneault, H., He, H.-T., and Marguet, D., Dynamic molecular confinement in the plasma membrane by microdomains and the cytoskeleton meshwork, *EMBO J.*, 25, 3245, 2006.
78. Fujiwara, T., Ritchie, K., Murakoshi, H., Jacobson, K., and Kusumi, A., Phospholipids undergo hop diffusion in compartmentalized cell membrane, *J. Cell Biol.*, 157, 1071, 2002.

79. Edidin, M., Zuniga, M.C., and Sheetz, M.P., Truncation mutants define and locate cytoplasmic barriers to lateral mobility of membrane glycoproteins, *Proc. Natl. Acad. Sci. U.S.A.*, 91, 3378, 1994.
80. Zhang, F., Yang, B., Odin, J.A., Shen, Z., Lin, C.T., Unkeless, J.C. and Jacobson, K., Lateral mobility of FcγRIIa is reduced by protein kinase C activation, *FEBS Lett.*, 376, 77, 1995.
81. Cvejic, S. and Devi, L.A., Dimerization of the δ opioid receptor: implication for a role in receptor internalization, *J. Biol. Chem.*, 272, 26959, 1997.
82. Cheng, Z.-J. and Miller, L.J., Agonist-dependent dissociation of oligomeric complexes of G protein-coupled cholecystokinin receptors demonstrated in living cells using bioluminescence resonance energy transfer, *J. Biol. Chem.*, 276, 48040, 2001.
83. Pralle, A., Keller, P., Florin, E.-L., Simons, K., and Hörber, J.K.H., Sphingolipid-cholesterol rafts diffuse as small entities in the plasma membrane of mammalian cells, *J. Cell Biol.*, 148, 997, 2000.

# 4 Calmodulin Is a 5-HT Receptor-Interacting and Regulatory Protein

*Justin N. Turner, Sonya D. Coaxum, Andrew K. Gelasco, Maria N. Garnovskaya, and John R. Raymond*

## CONTENTS

Abstract ..... 61
Introduction ..... 62
5-HT Receptors ..... 62
G Protein-Coupled Receptor-Interacting Proteins ..... 62
Calmodulin ..... 64
Role of CaM in 5-HT Receptor Signaling ..... 65
Interaction of CaM with GPCRs ..... 66
5-HT Receptors and Calmodulin ..... 67
5-$HT_{1A}$ Receptor ..... 67
Other 5-$HT_1$ Receptors ..... 70
5-$HT_2$ Receptors ..... 70
Other 5-HT Receptors (5-$HT_4$, 5-$HT_5$, 5-$HT_6$, 5-$HT_7$) ..... 72
Conclusions ..... 74
Acknowledgments ..... 74
References ..... 74

## ABSTRACT

This chapter explores emerging roles for calmodulin as a regulator of 5-HT receptor function and describes recent work suggesting that calmodulin may be a so-called RIP (receptor-interacting protein) that binds to and modifies the functions of G protein-coupled 5-HT receptors.

# INTRODUCTION

## 5-HT RECEPTORS

5-hydroxytryptamine (5-HT, serotonin) is a monoamine that serves as a neurotransmitter, mitogen and hormone. 5-HT influences cells in the brain, nervous system and peripheral tissues by binding to, and activating, a diverse array of cell surface receptors. 5-HT receptors have been divided into seven families based on their molecular structures, pharmacology, and signal transduction linkages. One of the families of 5-HT receptors (5-$HT_3$) is comprised of ligand-gated ion channels (termed *5-$HT_{3A}$–5-$HT_{3E}$*), whereas six of the families (5-$HT_1$, 5-$HT_2$, 5-$HT_4$, 5-$HT_5$, 5-$HT_6$, 5-$HT_7$) are comprised of heptahelical receptors that signal primarily through coupling to heterotrimeric guanine nucleotide binding and regulatory proteins (G proteins). The G protein-coupled 5-HT receptor families have been characterized by pharmacological properties, amino acid sequences, gene organization, and second messenger coupling pathways. These receptors are integral membrane proteins with seven putative hydrophobic transmembrane domains connected by three intracellular loops (termed *iL1–iL3*) and three extracellular loops (termed *eL1–eL3*). The amino terminus of each receptor is oriented toward the extracellular space, whereas the carboxyl terminus (CT) is oriented toward the cytoplasm. These receptors possess conserved or common sites for posttranslational modifications. The extracellular domains are typically glycosylated and possess cysteine residues that may participate in disulfide bonds, which provide structural constraints on the conformation of the receptors. The intracellular domains variably contain phosphorylation sites for a wide array of kinases, palmitoylation sites in the CT for lipid anchoring, and various potential sites for protein–protein interactions.

Recent work has demonstrated an unanticipated diversity of signals linked to the various G protein-coupled receptors (GPCRs), including 5-HT receptors. The work also has led to a growing awareness that GPCR signaling diversity can be modulated by diverse regulatory proteins, many of which are likely to bind directly to the receptor. Those proteins that bind directly to specific motifs on GPCRs are termed *receptor-interacting proteins* (RIPs). The topic of this chapter focuses on the interaction of $Ca^{2+}$–calmodulin (CaM), which is potential 5-HT receptor RIP, with G protein-coupled 5-HT receptors.

## G PROTEIN-COUPLED RECEPTOR-INTERACTING PROTEINS

A growing body of work suggests that GPCRs can bind to a variety of proteins. This should not be entirely surprising in that, very early on, it was recognized that GPCRs could bind to G proteins. Subsequently, molecular methods allowed the identification of other proteins (arrestins, protein kinase A, protein kinase C (PKC), and G protein-coupled receptor kinases) that bind to GPCRs and modify their functions and interactions with G proteins (48,63,64). Within the last several years, novel protein–protein interactions with GPCRs have been reported, including physical interactions with transmembrane proteins such as other GPCRs (47,76), and RAMPs (receptor associated modifying proteins) (17,19,58).

The most striking and well-characterized examples of GPCR RIPs are the RAMPs. The human calcitonin receptorlike receptor (hCRLR) has drastically altered the ligand-binding properties depending upon the RAMP with which it is expressed. It becomes a functional calcitonin gene-related peptide receptor when cotransported with human RAMP1 to the cell surface, whereas it becomes a functional adrenomedullin receptor when cotransported with RAMP2 (53). Another excellent example is provided by the angiotensin II $AT_{1A}$ receptor, which has been shown to physically interact with ARAP1 ($AT_1$ receptor associated protein 1), thereby promoting recycling of the receptor to the plasma membrane, enhancing its signaling (37). The importance of this relationship was highlighted by the recent finding that transgenic mice overexpressing ARAP1 develop hypertension and renal hypertrophy (36). Although the interaction of RAMPS with the hCRLR, and ARAP1 with the $AT_{1A}$ receptor, are quite remarkable, other GPCRs have more recently been demonstrated in a preliminary manner to interact with nonmembrane-spanning proteins such as 14-3-3 proteins (18,26,62,65,90), CaM (16,94), PDZ (PSD-95 discs-large ZO-1) proteins (8,75), and cytoskeletal elements (75,83,86). Most of these reports are preliminary, and the functional consequences of the interactions of GPCRs with these putative RIPs have not been completely elucidated. Indeed, unequivocal verification that these proteins actually interact with GPCRs in intact cells has generally not been completed.

For example, cytoskeletal elements (microtubules and spinophilin) have been shown to physically bind to peptides derived from the $\alpha_2$-adrenergic receptor (75,82,86). The dopamine $D_2$ receptor i3L binds spinophilin (but not to its related protein, neurabin) in yeast two-hybrid and fusion protein interaction assays, although the functional significance of this interaction remains undefined. Spinophilin is expressed ubiquitously and contains multiple protein interaction domains (81). The $\alpha_2$-adrenergic receptor i3L was shown to interact with amino acids 169–255 of spinophilin, in a region between its F-actin-binding and phosphatase 1 regulatory domains. Because the interaction between the $\alpha_2$-adrenergic receptor and spinophilin was increased by agonist treatment, it would seem to serve a dynamic function (24). The same group showed that the disruption of microtubules with colchicine or nocodazole could alter the apical steady state distribution and delivery of $A_1$-adenosine receptors and increase the binding capacity of $\alpha_{2B}$-adrenergic receptors expressed in MDCK-II cells (31). The authors suggested that the cytoskeleton and G proteins interact with distinct domains of the i3L of the $\alpha_{2B}$-adrenergic receptor and that the cytoskeletal effects may be mediated, independent of G proteins. These studies show that abundantly expressed cellular proteins, such as components of the actin cytoskeleton (or CaM), might play important roles in GPCR function, possibly independent of G proteins.

At least five well-characterized examples of GPCR RIPs involve 5-HT receptors. These include the interactions of caveolin-1 and PSD-95 with the $5\text{-HT}_{2A}$ receptor, MUPP1 (multi-PD2 domain protein 1) with the $5\text{-HT}_{2C}$ receptor, and EBP50 (ezrin/radixin/moesin-binding phosphoprotein 50) and SNX27 (new sorting nexin) with the $5\text{-HT}_{4a}$ receptor. MUPP1 interacts with a PDZ domain of the $5\text{-HT}_{2B}$ receptor. PDZ recognition motifs are contained in many signaling proteins, and these domains mediate key protein–protein interactions. For GPCRs, the best characterized example of a functionally significant GPCR interaction with a PDZ protein is the $\beta_2$-adrenergic receptor, which interacts with NHERF ($Na^+/H^+$ exchanger regulatory factor)/EBP50 (32) *via* agonist-dependent binding of the first PDZ domain of NHERF to the CT of the $\beta_2$-adrenergic receptor, resulting in regulation of the type 3 sodium-proton exchanger, NHE-3 (33). PDZ interactions of the $\beta_2$-adrenergic receptor with NHERF have also been shown to control recycling of internalized receptors through an endocytic sorting pathway that leads to lysosomal degradation (34).

In yeast two-hybrid screens, 5-$HT_2$ receptors interact with a multivalent PDZ protein called *MUPP1*. Coimmunoprecipitations and mutagenesis studies have shown that an SXV sequence at the extreme CT of the 5-$HT_{2C}$ receptor selectively interacts with the 10th PDZ domain of MUPP1 (23). This interaction is functionally important in that it induces a conformational change in MUPP1 (23), attenuates receptor phosphorylation on serine 453 ($S^{453}$), and attenuates desensitization (35).

Bhatnagar and colleagues surprisingly demonstrated that caveolin-1 binds to 5-$HT_{2A}$ receptors in rat synaptic membranes, C6 glioma cells, and transfected HEK-293 cells (14). They demonstrated that this interaction is significant in that caveolin-1 (but not caveloin-2) facilitated increases in intracellular $Ca^{2+}$ through increased coupling of the receptor with $G_{q\alpha}$. This specificity of the interaction between caveolin-1 and 5-$HT_{2A}$ receptor was further supported by experiments showing that caveolin-1 knockdown greatly diminished both 5-$HT_{2A}$ and $P_{2Y}$ purinergic receptor signaling without altering PAR-1 (protease activated receptor-1 thrombin receptor) signaling. The same group also demonstrated that interaction of PSD-95 with the 5-$HT_{2A}$ receptor increases coupling of the receptor to $G_{q\alpha/11}$ and blunts receptor internalization without altering its desensitization (9,95,96).

5-$HT_2$ receptors are not the only 5-HT receptors that have been shown to interact functionally with RIPs. The subcellular localization of the 5-$HT_{4a}$ receptor is regulated by its binding to two RIPs. EBP50 also called *NHERF*) and SNX27a (a new sorting nexin, also called *Mrt1a*) have been shown to interact directly with the 5-$HT_{4a}$ receptor and to modify its subcellular localization. SNX27a redirects the 5-$HT_{4a}$ receptor into early endosomes, whereas EBP50/NHERF recruits the receptor into ezrin-containing microvilli (41). In the remainder of this chapter, we will describe findings that support the idea that CaM is an important 5-HT receptor RIP.

## Calmodulin

CaM is a member of the superfamily of EF-hand proteins. CaM has four EF-hand motifs, each of which is composed of two α-helices connected by a 12-amino acid loop. When intracellular $Ca^{2+}$ levels rise to the low micromolar range, all four EF-hands bind $Ca^{2+}$, inducing a conformational change that results in binding to various target proteins (4,5,27,97). In CaM's non$Ca^{2+}$-bound state, the α-helices of the EF hands are aligned in a nearly parallel fashion, which is termed the *closed* conformation, whereas in its $Ca^{2+}$-bound state, the EF hands are oriented in a nearly perpendicular fashion termed the *open conformation.* We liken this to two outstretched arms with hands ready to grasp a target. CaM is shaped like a barbell, and when it binds to a target α-helical peptide, it folds over the peptide (27,97), assuming a globular configuration. The crystal structure of CaM and other biochemical evidence have allowed for the construction of algorithms to predict potential CaM-binding proteins (97). Those algorithms identify α-helical peptides that have groupings of hydrophobic amino acids interspersed with positively charged amino acids. In helical wheel diagrams, the positively charged amino acids are often located on the opposite "side" of the peptide from the hydrophobic amino acids. Interestingly, many GPCRs (including 5-HT receptors) possess similar motifs that could serve as CaM-binding domains.

CaM is classically activated by increases in intracellular $Ca^{2+}$, resulting in conformational changes in CaM and activation of target proteins. However, other mechanisms of regulating CaM are possible, although poorly understood. Primary among the alternate mechanisms of activating CaM is phosphorylation of CaM on serine–threonine or tyrosine residues. CaM can be phosphorylated by receptor and nonreceptor tyrosine kinases and serine–threonine kinases (25,30,38,68,77,80). Casein kinase II phosphorylates CaM *in vitro* on threonine and serine residues ($T^{79}$, $S^{81}$, $S^{101}$, and $T^{117}$) (78), whereas $Ca^{2+}$-calmodulin-dependent protein kinase IV phosphorylates CaM primarily on $T^{44}$ (40). In contrast, the EGF (epidermal growth factor) receptor phosphorylates $Y^{99}$ of bovine brain CaM (10,28) with a stoichiometry of 1:1 (11), and the insulin receptor phosphorylates $Y^{99}$ and $Y^{138}$ of CaM in CHO-IR (Chinese hamster ovary-insulin receptor) cells (42,79). Nonreceptor tyrosine kinases Src (30) and Jak2 also phosphorylate CaM in response to a number of stimuli (31,32,49,59), but the identity of the tyrosine residues that are phosphorylated by these kinases have not been established. In any case, because nearly all of the G protein-coupled 5-HT receptors either increase intracellular $Ca^{2+}$ or activate tyrosine kinases, any number of 5-HT receptors could reasonably be expected to activate CaM.

## Role of CaM in 5-HT Receptor Signaling

There is a small but growing literature suggesting that CaM and CaM-dependent enzymes and effectors play key roles in signaling pathways initiated by various G protein-coupled 5-HT receptors. In that regard, several vascular and neuronal effects of 5-HT have been attributed to CaM. For example, an undefined 5-HT receptor subtype induces contractions in human umbilical artery that are mediated *via* CaM (54); an undefined 5-HT receptor subtype mediates contractions of bovine middle cerebral artery through CaM and CaM-dependent myosin light chain kinase (60); 5-HT evokes outward currents in dissociated rat hippocampal pyramidal neurons, and increased membrane conductance are sensitive to CaM inhibitors (93). CaM dysfunction has been suggested to be involved in the mechanism of enhanced intracellular $Ca^{2+}$ response to 5-HT in bipolar disorder (89).

$G_{i/o}$-coupled 5-$HT_{1A}$ receptors exert tonic inhibition of AMPA receptors in excitatory hippocampal neurons through $Ca^{2+}$-calmodulin-dependent protein kinase II (CaMK-II) (57,84,85). Similarly, 5-$HT_{1A}$ receptors inhibit the NR2B subunit involved in NMDA receptor-mediated ionic and synaptic currents in prefrontal cortex pyramidal neurons, and the NMDA receptor through CaMK-II (21,99). The 5-$HT_{1A}$ receptor transfected into CHO cells manifests CaM-dependent internalization and Erk (extracellular receptor kinase) activation (29).

CaM is involved in various functions of the $G_{q/11}$-coupled 5-$HT_2$ receptors, including their desensitization (43,71). 5-$HT_{2A}$ (but not 5-$HT_{2C}$) receptor desensitization requires CaM and CaMK-II (13). For example, 5-$HT_{2A}$ receptors regulate cyclic AMP accumulation in embryonic cortex A1A1 neuronal cells by protein kinase C-dependent and CaM-dependent mechanisms (12). 5-$HT_{2A}$-mediated BDNF release in C6 glioma cells is mediated by CaMK-II (55). 5-$HT_{2A}$ receptors in PC12 cells activate extracellular signal-regulated kinase and tyrosine phosphorylation of a number of proteins by a CaM and tyrosine kinase-dependent pathway (69). 5-$HT_{2A}$ receptors in failing human cardiac ventricles mediate activation of $Ca^{2+}$-calmodulin-dependent myosin light chain kinase (70). 5-$HT_{2A}$ receptors in mesangial cells increase COX-2 (cyclooxygenase 2) expression *via* CaM and CaMK-II (34,88). 5-$HT_{2B}$ and 5-$HT_{2C}$ receptors also couple to CaM-dependent signals. 5-$HT_{2B}$ receptors significantly potentiate NMDA-induced depolarizations in frog spinal cord motoneurones *via* calmodulin but not CaMK-II (39). 5-$HT_{2A/2B}$ receptors cause cardiac hypertrophy *via* calcineurin, which is a CaM-dependent phosphatase (20). 5-$HT_{2C}$ receptors expressed in A9 cells evoke $Ca^{2+}$-dependent outward currents, which rapidly desensitize. This desensitization involves CaM (15).

$G_s$-coupled 5-$HT_7$ receptors, but not 5-$HT_6$ receptors, activate CaM-sensitive adenylyl cyclases, AC1 and AC8 through mobilization of $Ca^{2+}$ (7). Thus, it is clear that a number of 5-HT receptors couple to diverse signaling pathways through $G_{i/o}$, $G_{q/11}$, and $G_s$. However, it is not at all clear whether the CaM-dependent effects are due to increased intracellular $Ca^{2+}$, phosphorylation of CaM, or direct binding to the G protein-coupled 5-HT receptors. The notion that CaM can interact with 5-HT receptors is explored in the section titled "Interaction of CaM with GPCRs," and in the subsections of "5-HT Receptors and Calmodulin."

## Interaction of CaM with GPCRs

CaM has been demonstrated to bind to peptides derived from at least six GPCRs, including glutamate $mGlu_7$ receptor (58), $\mu$-opioid $OP_3$ receptor (21), dopamine $D_2$ receptor (22), 5-$HT_{1A}$ receptor (91), 5-$HT_{2A}$ receptor (92), and angiotensin $AT_{1A}$ receptor (65). CaM is a ubiquitous $Ca^{2+}$ sensor that can regulate numerous enzymes involved in signaling of GPCRs. Although the best characterized interactions between GPCRs and CaM occur downstream of the receptor, Sadee's group detected a CaM-binding motif in i3L of the $OP_3$ receptor. They showed that peptides derived from the $OP_3$ receptor i3L strongly bound to CaM and could diminish binding between CaM and immuno-purified $OP_3$ receptors. Agonist stimulation of the $OP_3$ receptor resulted in release of CaM from the plasma membrane. CaM also had some effects on receptor functions in that CaM reduced basal and agonist-stimulated $^{35}S$-labeled guanosine 5′-3-*O*-(thio)triphosphate [$^{35}$S] GTPβS incorporation (a measure of G protein activity). Overexpression of CaM or antisense to CaM inversely affected basal and agonist-induced G protein activity. They interpreted the results to mean that CaM competes with G proteins for binding to opioid receptors and that CaM could serve as an independent second messenger that is released from the membrane upon receptor stimulation (21).

CaM has been proposed to fulfill two major functions by binding to GPCRs, including blunting the G protein coupling and blunting the receptor phosphorylation. Bofill-Cardona et al. used a variety of methods to demonstrate that a peptide fragment of the dopamine $D_2$ receptors binds to CaM, and that CaM suppresses $D_2$ receptor activation of G proteins. Moreover, the effects of CaM were relatively specific as there were no effects on $A_1$ adenosine and $Mel_{1A}$ melatonin receptors (22). For the $mGlu_7$ receptor, there appears to be a complex interaction between serine–threonine phosphorylation of the CT of the receptor, binding of G protein $\beta\gamma$ subunits, and CaM (58–61). Thus, there is tantalizing evidence that CaM can directly bind to, and regulate, various GPCRs. The evidence for direct interaction of CaM with $5\text{-}HT_{1A}$ and $5\text{-}HT_{2A}$ receptors is described in the sections titled "$5\text{-}HT_{1a}$ Receptor" and "5-$HT_2$ Receptors," respectively.

# 5-HT RECEPTORS AND CALMODULIN

## $5\text{-}HT_{1A}$ RECEPTOR

Because the $5\text{-}HT_{1A}$ receptor has been linked to CaM in a few preliminary reports (21,29,57,84,85,99), it is a logical candidate to study the possible interaction between the receptor and CaM. The $5\text{-}HT_{1A}$ receptor i3L contains a number of sequences that could potentially interact with RIPs. In that regard, we identified two putative CaM-binding domains in the proximal and distal juxtamembrane regions of the i3L of the $5\text{-}HT_{1A}$ receptor, using a computer search algorithm (http://calcium.uhnres.utoronto.ca/ctdb/pub_pages/general/index.htm), for which scoring is based on evaluation criteria including hydropathy, α-helical propensity, residue charge, helical class, residue weight, and hydrophobic residue content (98). As a general rule, CaM-binding regions are characterized by the presence of several hydrophobic residues interspersed with several positively charged residues, often forming amphipathic α-helices. CaM-binding regions described to date have been divided into several motifs based on the distance between key hydrophobic residues. The putative proximal i3L CaM-binding region of the $5\text{-}HT_{1A}$ receptor ($Y^{215}GRIFRAARFRIRKTVKKVEKTG^{237}$) was identified as a 1–12 motif, with key hydrophobic residues at positions 1 and 12. The more proximal sequence is also a putative G protein contact site. The more distal sequence ($A^{330}KRKMALARERKTVKTLGIIMG^{352}$) conforms to a standard CaM-recognition motif of hydrophobic residues at positions 1, 8, and 14 interspersed with positively charged amino acids. This distal peptide is in an intriguing location, as it overlaps with a G protein-contact site and a PKC phosphorylation site in the $5\text{-}HT_{1A}$ receptor (45,46,50,51,72). The predicted locations of the two CaM-binding regions of the 5-$HT_{1A}$ receptor (as well as other 5-HT receptors) are depicted in Figure 4.1.

In order to illustrate the amphipathic, α-helical nature of the two putative CaM-binding regions from the $5\text{-}HT_{1A}$ receptor, we modeled them with an α-helical wheel algorithm (http://marqusee9.berkeley.edu/kael/helical.htm), which showed that each peptide is predicted to form an α-helix with hydrophobic amino acids (λ) and charged amino acids (+) located on opposite sides of each helix (Figure 4.2), typical of the amphipathic nature of CaM-binding sites.

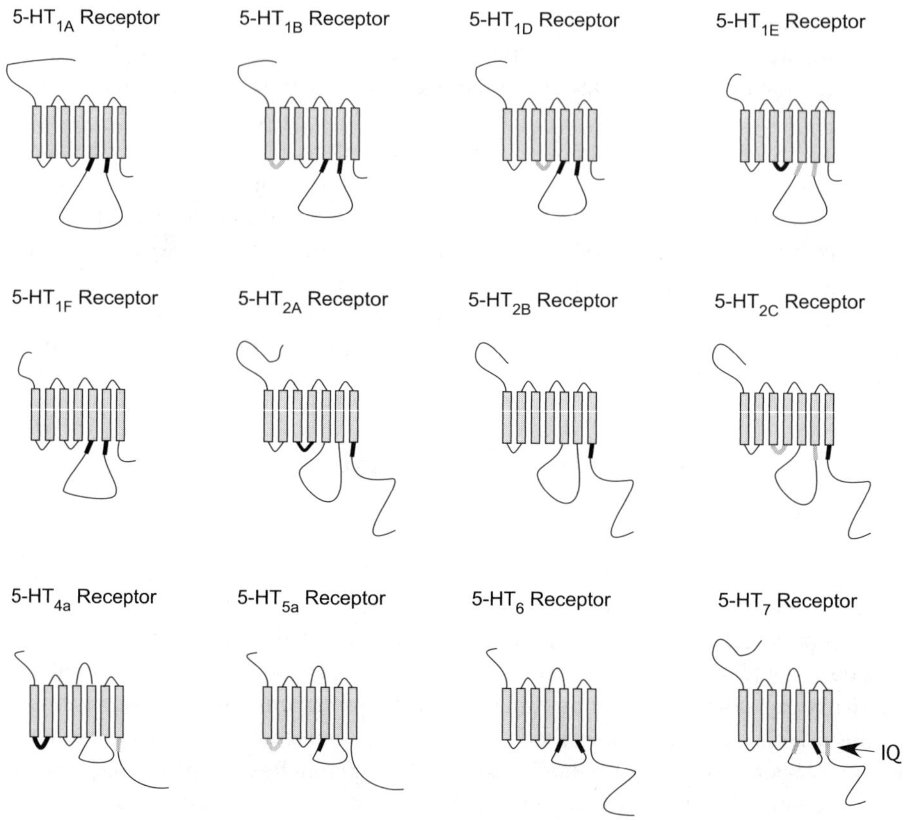


**FIGURE 4.1** Schematic illustration of putative CaM-binding domains of various G protein-coupled 5-HT receptors. The upper face of each receptor represents the extracellular domains, and the lower face represents the intracellular domains. The grey bars represent the seven transmembrane spanning domains of each receptor. Moderate-high or high-likelihood-CaM-binding domains, as predicted by a computer algorithm (http://calcium.uhnres.utoronto.ca/ctdb/pub_pages/general/index.htm) (98), are depicted with thick black lines, whereas moderate-likelihood-CaM-binding domains are depicted with thick grey lines. The arrow indicates a potential IQ (non$Ca^{2+}$-dependent) CaM-binding domain.

We used a number of methods (coimmunoprecipitation, gel-shift analysis, surface plasmon resonance spectroscopy [SPRS], slot blot, dansyl chloride spectroscopy, and bioluminescence resonance energy transfer [BRET]) to demonstrate that the 5-$HT_{1A}$ receptor interacts with CaM *in vitro* and in live cells and to calculate the affinities of the two sites for CaM (91). The apparent affinity of the proximal i3L putative CaM site was 87 ± 23 n*M*, and the apparent affinity of the distal i3L site was 1.70 ± 0.16 *μM.* Regarding the functional effects of the CaM sites, we have discovered that CaM-binding and phosphorylation of the 5-$HT_{1A}$ receptor i3L peptides by (PKC) or protein kinase C are mutually antagonistic processes *in vitro*, suggesting a possible role for CaM in the regulation of 5-$HT_{1A}$ receptor phosphorylation and desensitization (91). Furthermore, we have found that binding of CaM to

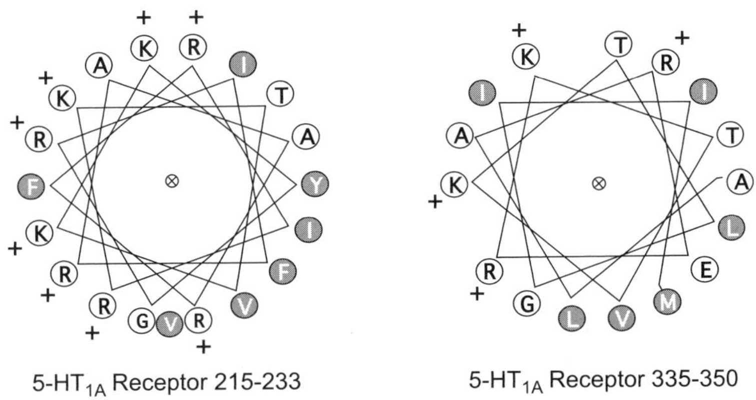


**FIGURE 4.2** Helical wheel projections of putative CaM-binding domains of the 5-HT$_{1A}$ receptor. Projections were made using a computer algorithm (http://marqusee9.berkeley.edu/kael/helical.htm). Each circle represents an amino acid residue. Grey circles indicate hydrophobic residues, whereas "+" signs indicate positively charged residues. The analyzed sequences encompassed amino acids 215–233 and 335–350 of the human 5-HT$_{1A}$ receptor.

the 5-HT$_{1A}$ receptor decreases coupling to $G_{i/o}$ G proteins as assayed by GTPgS binding to crude membrane preparations (Turner et al., unpublished data).

These data support the idea that the 5-HT$_{1A}$ receptor contains high- and moderate-affinity CaM-binding regions that regulate G protein coupling, receptor phosphorylation, and (perhaps) desensitization.

The idea that CaM interferes with receptor phosphorylation is not limited to 5-HT receptors. Indeed, G protein-coupled receptor kinase (GRK)-mediated phosphorylation of rhodposin is inhibited by CaM (66). Additionally, phosphorylation of a conserved serine residue in the CT of group III metabotropic glutamate receptors by PKC inhibits CaMbinding (1). Thus, there seems to be a mutually antagonistic relationship between phosphorylation and CaM binding for GPCRs and for other signaling proteins. For example, phosphorylation of GRK2 (G protein-coupled receptor kinase 2) by PKC abolishes the ability of CaM to inhibit GRK2 (44), and phosphorylation of the MARCKS proteins (myristoylated alanine rich C-kinase substrate) inhibits CaM binding (35,52). However, this effect is not universal in that phosphorylation of a calcineurin peptide by CaMK-II does not significantly alter the binding of CaM when compared to the nonphosphorylated peptide (22).

Why would phosphorylation and CaM binding be mutually antagonistic? Although the structural basis for this antagonism is not known, it should be intuitive that phosphorylation, by adding a negative charge to a target protein, could disrupt the CaM recognition motif that relies on positive charges. Phosphorylation might also induce conformational changes in the target protein, perhaps by disrupting electrostatic tethering of the positively charged face of the peptide to acidic (negatively charged) phospholipids that are resident on the inner leaflet of the plasma membrane. The inner leaflet of mammalian plasma membranes typically contains between 15–30% of acidic lipids, primarily phosphatidylserine. There is increasing awareness that peptides can interact with plasma membrane lipids through positive charge domains and, especially, with negatively charged lipids in the inner leaflet of the plasma membrane (87). Indeed, electrostatic interactions between positive charges in the transmembrane domains of the S4 family of voltage-sensitive ion channels and negative charges on the inner plasma membrane induce a conformational change in the channels that results in pore opening and closing (56). The electrostatic interaction of positive charge domains with negatively charged lipids would be expected to occur selectively on the plasma membrane, rather than to other internal membranes, in that an inner leaflet of the plasma membrane has far more electrostatic charge than other cellular membranes (24,61).

Conversely, CaM could attenuate phosphorylation by competing with or blocking the access of various kinases to phosphorylation sites within or near to the CaM-binding domains. Alternatively, CaM could mask the positive charges in its target-binding domains, thereby disrupting the electrostatic interaction of the CaM-binding domains with the plasma membrane, resulting in conformational changes that are less favorable to phosphorylation.

## Other 5-HT$_1$ Receptors

Virtually nothing is known regarding the interaction of CaM with other 5-HT$_1$ receptors, although a computer algorithm predicts that each 5-HT$_1$ receptor might have several CaM sites (http://calcium.uhnres.utoronto.ca/ctdb/pub_pages/general/index.htm) (98). The 5-HT$_{1B}$ receptor has three potential CaM-binding domains, including a moderate- to high-likelihood site in the proximal juxtamembrane region of the i3L (L$^{227}$YGRIYVEARSRILK$^{241}$), an extended high-likelihood site in the distal juxtamembrane region of the i3L (N$^{288}$QVKVRVSDALLEKKKLMAARERKATKTLGIILG$^{321}$), and a moderate-likelihood site in the i1L (N$^{67}$AFVIATVYRTRK$^{79}$) (Figure 4.2).

The 5-HT$_{1D}$ receptor has three potential CaM-binding domains, including a moderate-likelihood site in the proximal juxtamembrane region of the i3L (L$^{212}$LIILYGRIYRAARNRI$^{228}$), an extended high-likelihood site in the distal juxtamembrane region of the I3L (N$^{275}$HVKIKLADSALERKRISAARERKATKILGIIL$^{307}$), and a moderate-likelihood site in the i2L K$^{148}$RRTAGHAATMIAIV$^{162}$).

The 5-HT$_{1E}$ receptor has three potential CaM-binding domains, including a moderate- to high-likelihood site in the i2L (a$^{123}$itnaieyarkrtakraalmiltv$^{146}$), and two moderate-likelihood sites in the proximal and distal juxtamembrane regions of the i3L (y$^{203}$riyhaakslyqkr$^{216}$ and s$^{282}$strerkaarilglilg$^{298}$).

The 5-HT$_{1F}$ receptor has two potential CaM-binding domains, including a high-likelihood site in the proximal and distal juxtamembrane regions of i3L (l$^{198}$ilyykiyraaktlyhkrqas$^{218}$ and i$^{283}$sgtrerkaattlglilg$^{300}$). Figure 9.2 clearly shows that all of the putative 5-HT-receptor CaM-binding domains reside in juxtamembrane regions that are also thought to be critical for G protein coupling, supporting the idea that CaM-binding to 5-HT receptors either shields the receptor from G proteins and/or induces conformational changes that could alter G protein coupling.

## 5-HT$_2$ Receptors

5-HT$_{2A}$ receptors in various neuronal and peripheral cells mediate numerous effects through the intermediate actions of CaM (12,23,31,33,43,69,71,88) or CaM-dependent enzymes such as CaMK-II (3,13,55) and myosin light chain kinase (70). We therefore assessed the sequence of the human 5-$HT_{2A}$ receptor for possible CaM-binding sites. As a receptor that couples to $G_{q/11}$-type G proteins, the 5-$HT_{2A}$ receptor is capable of stimulating phosphoinositide turnover and subsequent increases in intracellular $Ca^{2+}$ (73). Accordingly, we postulated that the 5-$HT_{2A}$ receptor might exert some of its $Ca^{2+}$-sensitive intracellular effects by directly interacting with CaM. A search of the amino acid sequence using a computer algorithm (http://calcium.uhnres.utoronto.ca/ctdb/pub_pages/general/index.htm) (98) revealed two novel potential CaM-binding motifs, located in the i2L ($S^{184}$RFNSRTKAFLKIIAVWTI$^{202}$) and the juxtamembrane region of the CT ($P^{367}$LVY TLFNKTYRSAFSRYIQ$^{396}$) of the 5-$HT_{2A}$ receptor. Both sequences contain consensus phosphorylation sites and are potentially important for G protein coupling, suggesting that interaction of CaM with those sites could play roles in regulating receptor function. The CaM-binding region in the i2L of the 5-$HT_{2A}$ receptor conforms to a 1-8-14 motif, which is characterized by hydrophobic residues at positions 1, 8, and 14. The putative CaM-binding domain in the CT of the 5-$HT_{2A}$ receptor conforms to a 1-10 motif, with critical hydrophobic residues separated by 8 amino acids.

In order to illustrate the amphipathic nature of the putative CaM-binding sequences of the 5-$HT_{2A}$ receptor, we modeled them with an $\alpha$-helical wheel algorithm (http://marqusee9.berkeley.edu/kael/helical.htm). Figure 4.3 shows that both putative CaM-binding domains of the 5-$HT_{2A}$ receptor contain clusters of positively charged amino acids on one side of the $\alpha$-helix, with mostly hydrophobic amino acids concentrated on the opposite side.

We used a variety of techniques (coimmunoprecipitation, gel-shift analysis, slot blot, dansyl chloride spectroscopy, and BRET) to demonstrate that the 5-$HT_{2A}$ receptor interacts with CaM *in vitro* and in live cells, and to calculate the affinities of each of the two sites for CaM. Peptides from each of the sites bound to CaM in a $Ca^{2+}$-dependent

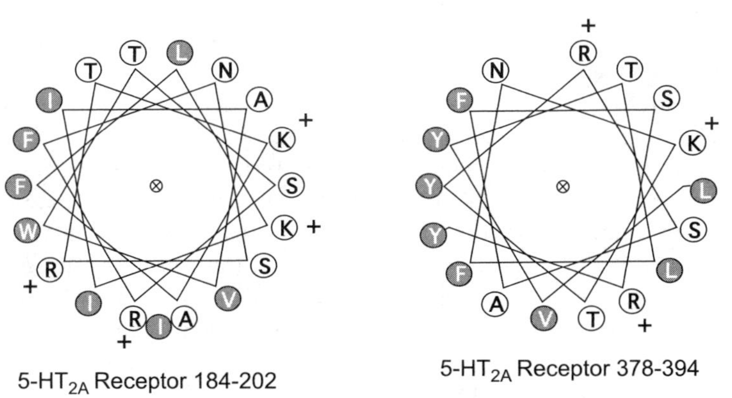


**FIGURE 4.3** Helical wheel projections of putative CaM-binding domains of the 5-$HT_{2A}$ receptor. Projections were made using a computer algorithm (http://marqusee9.berkeley.edu/kael/helical.htm). Each circle represents an amino acid residue. Grey circles indicate hydrophobic residues, whereas "+" signs indicate positively charged residues. The analyzed sequences encompassed amino acids 184–202 and 378–394 of the human 5-$HT_{2A}$ receptor.

fashion, with the i2L peptide binding with an apparent higher affinity than that of the CT peptide, based on mobility shifting of CaM in a nondenaturing gel shift assay. We used fluorescence emission spectral analyses of dansyl-CaM to calculate the apparent $K_D$ values of 65 ± 30 n*M* for the i2L peptide and 168 ± 38 n*M* for the CT peptide (92).

Because the putative CT CaM-binding domain overlaps with a putative PKC site, we tested whether CaM and PKC exerted mutually antagonistic effects on this peptide sequence. We demonstrated that the CT peptide was readily phosphorylated by PKC *in vitro* and that CaM-binding and phosphorylation of this peptide were antagonistic. These results suggest that there is a potential role for CaM in the regulation of 5-$HT_{2A}$ receptor phosphorylation and desensitization (similar to what was shown for the 5-$HT_{1A}$ receptor in the section titled "5-$HT_{1a}$ Receptor"). We also tested whether CaM had an effect on G protein coupling of the 5-$HT_{2A}$ receptor by measuring [$^{35}$S]GTPγS binding in the presence and absence of 5-HT. Those experiments showed that CaM decreases 5-$HT_{2A}$ receptor-mediated [$^{35}$S]GTPγS binding to NIH-3T3 cell membranes, supporting a possible role for CaM in modulating receptor-G protein coupling. The i2L peptide (but not the CT peptide) was able to stimulate [$^{35}$S]GTPγS binding, and this effect was blocked by CaM, suggesting that the i2L is critical for G protein activation. In contrast, the CT peptide (but not the i2L peptide) was a strong substrate for PKC-induced phosphorylation. These results strongly support two regulatory roles for CaM in 5-$HT_{2A}$ receptor function. First, CaM appears to blunt receptor coupling to G proteins through interaction with the i2L. Second, CaM blunts phosphorylation (and possibly desensitization) of the CT peptide of the 5-$HT_{2A}$ receptor (92). Thus, the 5-$HT_{2A}$ receptor contains two high-affinity-CaM-binding domains that appear to play important roles in both the initiation and the termination of receptor signaling.

The evidence for important roles of CaM in regulating other 5-$HT_2$ receptors is less clear than for the 5-$HT_{2A}$ receptors. 5-$HT_{2B}$ receptors potentiate NMDA-induced depolarizations in frog spinal cord motor neurons through CaM, but not through elevations of intracellular $Ca^{2+}$ or activation of CaMK-II (39). The human 5-$HT_{2B}$ receptor has one high-likelihood CaM-binding domain, ($L^{382}$FNKTFRDAFGRYITCNYRATKSVKTLR$^{409}$), which is located at the juxtamembrane region of the CT of the receptor.

There are two reports showing that calcineurin, which is a CaM-dependent phosphatase, inhibits the desensitization of the 5-$HT_{2C}$ receptor (2,15). The 5-$HT_{2C}$ receptor has a high-likelihood CaM-binding domain ($V^{367}$YTLFNKIYRRAFSNYLRCNYK$^{388}$) at the juxtamembrane region of the CT of the receptor, as well as two moderate-likelihood CaM-binding domains in the i2L ($F^{165}$NSRTKAIMKIAIVW$^{179}$) and distal i3L ($A^{302}$INNERKASKVLGIVF$^{317}$). Thus, it is possible that all three of the major subtypes of 5-$HT_2$ receptors might be regulated through direct interactions with CaM, although that possibility has not been rigorously examined for the 5-$HT_{2B}$ and 5-$HT_{2C}$ receptors.

## Other 5-HT Receptors (5-$HT_4$, 5-$HT_5$, 5-$HT_6$, 5-$HT_7$)

We were unable to find references mentioning the coupling of 5-$HT_4$, 5-$HT_5$, or 5-$HT_6$ receptors to CaM. However, each of those receptors has potential CaM-binding domains. The human 5-$HT_{4A}$ receptor has two putative CaM-binding domains, including a high-likelihood site in iL1 ($V^{44}$CWDRQLRKIKTNYFIVSLA$^{63}$), and a moderate-likelihood site in the proximal juxtamembrane region of the CT (N$^{308}$PFLYAFLNKSFRRAFLI$^{325}$). The human 5-$HT_{5a}$ receptor has a moderate-likelihood site in the proximal juxtamembrane region of i3L (N$^{206}$PFLYAFLNKSFRRAFLI$^{224}$) and a high-likelihood site at the distal juxtamembrane region of i3L (D$^{252}$SRRLATKHSRKALKASLTLGIL$^{273}$). The human 5-$HT_6$ receptor has two putative CaM-binding sites in the juxtamembrane regions of the i3L, including a potential moderate-likelihood site in the proximal i3L (D$^{259}$SRRLATKHSRKALKASLTLGIL$^{275}$) and a high-likelihood site in the distal i3L (H$^{312}$ERKNISIFK REQKAATTLG IIVGA$^{335}$).

In contrast to the human 5-$HT_4$, 5-$HT_5$, and 5-$HT_6$ receptors, 5-$HT_{7a}$ receptors have been functionally linked to CaM in that they have been shown to activate CaM-sensitive adenylyl cyclases, AC1 and AC8, whereas the 5-$HT_6$ receptor do not activate them (7). In their study, Baker and colleagues showed that the 5-$HT_6$ receptor behaved in a typical manner for $G_s$-coupled receptors in that it stimulated AC5, a $G_s$-sensitive adenylyl cyclase, but not AC1 or AC8. AC1 and AC8 typically are not activated by $G_s$-coupled receptors *in vivo*. In contrast, the 5-$HT_{7a}$ receptor stimulated AC1 and AC8 by increasing intracellular $Ca^{2+}$ (7).

The 5-$HT_{7a}$ receptor has three potential CaM-binding domains, two of which are in the juxtamembrane regions of i3L, including a proximal moderate-likelihood site (Y$^{259}$YQIYKAARKSAAKHKF$^{275}$) and a distal high-likelihood site (R$^{313}$KNISIFK REQKAATTLG IIVGA$^{335}$). The 5-$HT_{7a}$ receptor also has an unusual CaM-binding motif in the proximal juxtamembrane region of the CT (F$^{381}$IYAFFNRDLRTTYRS$^{397}$). This site contains a sequence that is termed an *ilimaquinone* or *IQ motif*, which is a consensus site for Ca $^{2+}$-independent CaM binding (74). The IQ motif is widely distributed in a variety of CaM-binding proteins, often in association with classical $Ca^{2+}$-dependent binding motifs. IQ motif-containing proteins possess a dazzling array of biological functions including signal transduction, phosphorylation, cytoskeletal regulation, trafficking, cell cycle control, and polarized growth (74). The presence of the IQ motif in the 5-$HT_{7a}$ receptor is intriguing in that IQ motifs play critical roles in neuronal growth and plasticity, and in assembling neuronal signal transduction complexes (6).

The presence of the IQ motif in the human 5-$HT_{7a}$ receptor is also intriguing at the molecular level. One group has shown that the presence of the IQ motif can influence the binding of CaM to classical $Ca^{2+}$-dependent CaM-binding motifs. They showed that the IQ motif in an abundant neuronal protein, PEP-19, accelerates the slow association and dissociation of $Ca^{2+}$ from the C-domain of free CaM 40–50-fold, and also increases the rate of dissociation of $Ca^{2+}$ from CaM when CaM is bound to CaMK-II. Thus, the IQ motif could regulate the dynamics of $Ca^{2+}$-binding to both free CaM and CaM bound to target proteins. The authors of this study pointed out that this relationship is critical in that the binding of $Ca^{2+}$ to the C-domain of CaM is a rate-limiting step for CaM-dependent enzyme activation (67). Scenarios in which the 5-$HT_{7a}$ receptor IQ motif enhances the dynamic interaction of CaM with the two putative juxtamembrane CaM-binding domains in the i3L of the 5-$HT_{7a}$ receptor, or with CaM target enzymes colocalized within a signaling complex with the 5-$HT_{7a}$ receptor, could be easily envisioned. Indeed, we speculate that the IQ motif could facilitate CaM-binding to GPCRs that heterodimerize with the 5-$HT_{7a}$ receptor. Those possibilities would be interesting to pursue through further experimentation.

## CONCLUSIONS

There is a growing awareness of the potential importance of the interplay between G protein-coupled 5-HT receptors, CaM, and CaM-dependent enzymes in critical functions of neuronal and nonneuronal cells. The fact that all known G protein-coupled 5-HT receptors possess putative CaM-binding motifs, as described in this chapter using a computer algorithm (98), raises the possibility that some of the CaM-dependent effects of 5-HT receptors could be mediated by, or facilitated by, direct binding of CaM to the receptors of interest. Indeed, both 5-$HT_{1A}$ and 5-$HT_{2A}$ receptors possess *bona fide* CaM-binding domains that appear to regulate G protein coupling and receptor phosphorylation (91,92). Thus, CaM is a potentially important 5-HT receptor RIP. These insights open up new avenues of investigation for all G protein-coupled 5-HT receptors. In particular, it will be important to develop mutant 5-HT receptors that maintain normal G protein coupling, but which have impaired CaM-binding in order to fully elucidate the roles of direct binding of CaM in 5-HT receptor signal transduction, trafficking, and desensitization.

## ACKNOWLEDGMENTS

The authors gratefully acknowledge support from the Medical and Research Services of the Department of Veterans Affairs (Merit Awards and a REAP award to JRR and MNG, and a Veterans Integrated Service Network-7 Career Development Award to AKG), the National Institutes of Health (DK52448 and GM63909 to JRR, DK59950 to AKG, and GM08716 to JHT), the American Heart Association Mid-Atlantic Affiliate (a predoctoral fellowship to JHT), and a joint endowment sponsored by the Medical University of South Carolina and Dialysis Clinics, Incorporated (to JRR). The majority of the work by the authors described herein was performed in VA space leased from the Medical University of South Carolina. Shared equipment grants from the Department of Veterans Affairs and the National Institutes of Health (S10 RR13005) made portions of this work possible.

## REFERENCES

1. Airas, J.M., Betz, H., and El Far, O., PKC phosphorylation of a conserved serine residue in the C-terminus of group III metabotropic glutamate receptors inhibits calmodulin binding, *FEBS Lett,* 494, 60, 2001.
2. Aiyar, J., Grissmer, S., and Chandy, K.G., Full-length and truncated Kv1.3 K+ channels are modulated by 5-HT1c receptor activation and independently by PKC, *Am J Physiol,* 265, C1571, 1993.
3. Arvanov, V.L., Liang, X., Russo, A., and Wang, R.Y., LSD and DOB: interaction with 5-$HT_{2A}$ receptors to inhibit NMDA receptor-mediated transmission in the rat prefrontal cortex, *Eur J Neurosci,* 11, 3064, 1999.
4. Babu, Y.S., Bugg, C.E., and Cook, W.J., Structure of calmodulin refined at 2.2 Å resolution, *J Mol Biol,* 204, 191, 1988.
5. Babu, Y.S., Sack, J.S., Greenhough, T.J., Bugg, C.E., Means, A.R., and Cook, W.J., Three-dimensional structure of calmodulin, *Nature,* 315, 37, 1985.

6. Bahler, M. and Rhoads, A., Calmodulin signaling via the IQ motif, *FEBS Lett,* 513, 107, 2002.
7. Baker, L.P., Nielsen, M., Impey, S., Metcalf, M.A., Poser, S.W., Chan, G., Obrietan, K., Hamblin, M.W., and Storm, D.R., Stimulation of type 1 and type 8 $Ca^{2+}$/calmodulin-sensitive adenylyl cyclases by the $G_s$-coupled 5-hydroxytryptamine subtype 5-$HT_{7A}$ receptor, *J Biol Chem,* 273, 17469, 1998.
8. Becamel, C., Figge, A., Poliak, S., Dumuis, A., Peles, E., Bockaert, J., Lubbert, H., and Ullmer, C., Interaction of serotonin 5-hydroxytryptamine type 2C receptors with PDZ10 of the multi-PDZ domain protein MUPP1, *J Biol Chem,* 276, 12974, 2001.
9. Becamel, C., Gavarini, S., Chanrion, B., Alonso, G., Galeotti, N., Dumuis, A., Bockaert, J., and Marin, P., The serotonin 5-$HT_{2A}$ and 5-$HT_{2C}$ receptors interact with specific sets of PDZ proteins, *J Biol Chem,* 279, 20257, 2004.
10. Benaim, G., Cervino, V., and Villalobo, A., Comparative phosphorylation of calmodulin from trypanosomatids and bovine brain by calmodulin-binding protein kinases, *Comp Biochem Physiol C Pharmacol Toxicol Endocrinol,* 120, 57, 1998.
11. Benguria, A., Hernandez-Perera, O., Martinez-Pastor, M.T., Sacks, D.B., and Villalobo, A., Phosphorylation of calmodulin by the epidermal-growth-factor-receptor tyrosine kinase, *Eur J Biochem,* 224, 909, 1994.
12. Berg, K.A., Clarke, W.P., Chen, Y., Ebersole, B.J., McKay, R.D.G., and Maayani, S., 5-Hydroxytryptamine type 2A receptors regulate cyclic AMP accumulation in a neuronal cell line by protein kinase C-dependent and calcium/calmodulin-dependent mechanisms, *Mol Pharmacol,* 45, 826, 1994.
13. Berg, K.A., Stout, B.D., Maayani, S., and Clarke, W.P., Differences in rapid desensitization of 5-hydroxytryptamine$_{2A}$ and 5-hydroxytryptamine$_{2C}$ receptor-mediated phospholipase C activation, *J Pharmacol Exp Ther,* 299, 593, 2001.
14. Bhatnagar, A., Sheffler, D.J., Kroeze, W.K., Compton-Toth, B.A., and Roth, B.L., Caveolin-1 interacts with 5-$HT_{2A}$ serotonin receptors and profoundly modulates the signaling of selected $G_{aq}$-coupled protein receptors, *J Biol Chem,* 279, 34614, 2004.
15. Boddeke, H.W., Hoffman, B.J., Palacios, J.M., and Hoyer, D., Calcineurin inhibits desensitization of cloned rat 5-$HT_{1C}$ receptors, *Naunyn Schmiedebergs Arch Pharmacol,* 348, 221, 1993.
16. Bofill-Cardona, E., Kudlacek, O., Yang, Q., Ahorn, H., Freissmuth, M., and Nanoff, C., Binding of calmodulin to the $D_2$-dopamine receptor reduces receptor signaling by arresting the G protein activation switch, *J Biol Chem,* 275, 32672, 2000.
17. Bomberger, J.M., Spielman, W.S., Hall, C.S., Weinman, E.J., and Parameswaran, N., Receptor activity-modifying protein (RAMP) isoform-specific regulation of adrenomedullin receptor trafficking by NHERF-1, *J Biol Chem,* 280, 23926, 2005.
18. Brock, C., Boudier, L., Maurel, D., Blahos, J., and Pin, J.P., Assembly-dependent surface targeting of the heterodimeric $GABA_B$ Receptor is controlled by COPI but not 14-3-3, *Mol Biol Cell,* 16, 5572, 2005.
19. Buhlmann, N., Leuthauser, K., Muff, R., Fischer, J.A., and Born, W., A receptor activity modifying protein (RAMP)2-dependent adrenomedullin receptor is a calcitonin gene-related peptide receptor when coexpressed with human RAMP1, *Endocrinology,* 140, 2883, 1999.
20. Bush, E., Fielitz, J., Melvin, L., Martinez-Arnold, M., McKinsey, T.A., Plichta, R., and Olson, E.N., A small molecular activator of cardiac hypertrophy uncovered in a chemical screen for modifiers of the calcineurin signaling pathway, *Proc Natl Acad Sci USA,* 101, 2870, 2004.

21. Cai, X., Gu, Z., Zhong, P., Ren, Y., and Yan, Z., Serotonin 5-$HT_{1A}$ receptors regulate AMPA receptor channels through inhibiting $Ca^{2+}$/calmodulin-dependent kinase II in prefrontal cortical pyramidal neurons, *J Biol Chem,* 277, 36553, 2002.
22. Calalb, M.B., Kincaid, R.L., and Soderling, T.R., Phosphorylation of calcineurin, effect on calmodulin binding, *Biochem Biophys Res Commun,* 172, 551, 1990.
23. Chen, H., Li, H., and Chuang, D.M., Role of second messengers in agonist up-regulation of 5-$HT_{2A}$ receptor binding sites in cerebellar granule neurons: involvement of calcium influx and a calmodulin-dependent pathway, *J Pharmacol Exp Ther,* 275, 674, 1995.
24. Cho, W. and Stahelin, R.V., Membrane-protein interactions in cell signaling and membrane trafficking, *Annu Rev Biophys Biomol Struct,* 34, 119, 2005.
25. Corti, C., Leclerc-L'Hostis, E., Quadroni, M., Schmid, H., Durussel, I., Cox, J., Dainese Hatt, P., James, P., and Carafoli, E., Tyrosine phosphorylation modulates the interaction of calmodulin with its target proteins, *Eur J Biochem,* 262, 790, 1999.
26. Couve, A., Kittler, J.T., Uren, J.M., Calver, A.R., Pangalos, M.N., Walsh, F.S., and Moss, S.J., Association of $GABA_B$ receptors and members of the 14-3-3 family of signaling proteins, *Mol Cell Neurosci,*17, 317, 2001.
27. Crivici, A. and Ikura, M., Molecular and structural basis of target recognition by calmodulin, *Annu Rev Biophys Biomol Struct,* 24, 85, 1995.
28. De Frutos, T., Martin-Nieto, J., and Villalobo, A., Phosphorylation of calmodulin by permeabilized fibroblasts overexpressing the human epidermal growth factor receptor, *Biol Chem,* 378, 31, 1997.
29. Della Rocca, G.J., Mukhin, Y.V., Garnovskaya, M.N., Daaka, Y., Clark, G.J., Luttrell, L.M., Lefkowitz, R.J., and Raymond, J.R., Serotonin 5-$HT_{1A}$ receptor-mediated Erk activation requires calcium/calmodulin-dependent receptor endocytosis, *J Biol Chem,* 274, 4749, 1999.
30. Fukami, Y., Nakamura, T., Nakayama, A., and Kanehisa, T., Phosphorylation of tyrosine residues of calmodulin in Rous sarcoma virus-transformed cells, *Proc Natl Acad Sci USA,* 83, 4190, 1986.
31. Garnovskaya, M.N., Mukhin, Y.V., Turner, J.H., Vlasova, T., Ullian, M.E., and Raymond, J.R., Mitogen-induced activation of $Na^+/H^+$ exchange in vascular smooth muscle cells involves janus kinase 2 and $Ca^{2+}$/calmodulin, *Biochemistry,* 42, 7178, 2003.
32. Garnovskaya, M.N., Mukhin, Y.V., Vlasova, T.M., and Raymond, J.R., Hypertonicity activates $Na^+/H^+$ exchange through Janus kinase 2 and calmodulin, *J Biol Chem,* 278, 16908, 2003.
33. Garnovskaya, M.N., Nebigil, C.G., Arthur, J.M., Spurney, R.F., and Raymond, J.R., 5-Hydroxytryptamine$_{2A}$ receptors expressed in rat renal mesangial cells inhibit cyclic AMP accumulation, *Mol Pharmacol,* 48, 230, 1995.
34. Goppelt-Struebe, M., Hahn, A., Stroebel, M., and Reiser, C.O., Independent regulation of cyclo-oxygenase 2 expression by p42/44 mitogen-activated protein kinases and $Ca^{2+}$/calmodulin-dependent kinase, *Biochem J,* 339(Pt. 2), 329, 1999.
35. Graff, J.M., Young, T.M., Johnson, J.D., and Blackshear, P.J., Phosphorylation-regulated calmodulin binding to a prominent cellular substrate for protein kinase C, *J Biol Chem,* 264, 21818, 1989.
36. Guo, D.F., Chenier, I., Lavoie, J.L., Chan, J.S., Hamet, P., Tremblay, J., Chen, X.M., Wang, D.H., and Inagami, T., Development of hypertension and kidney hypertrophy in transgenic mice overexpressing ARAP1 gene in the kidney, *Hypertension,* 48, 453, 2006.

37. Guo, D.F., Chenier, I., Tardif, V., Orlov, S.N., and Inagami, T., Type 1 angiotensin II receptor-associated protein ARAP1 binds and recycles the receptor to the plasma membrane, *Biochem Biophys Res Commun,* 310, 1254, 2003.
38. Heppel, L.A., Newton, D.L., Klee, C.B., and Draetta, G.F., The phosphorylation of calmodulin and calmodulin fragments by kinase fractions from bovine brain, *Biochim Biophys Acta,* 972, 69, 1988.
39. Holohean, A.M. and Hackman, J.C., Mechanisms intrinsic to 5-$HT_{2B}$ receptor-induced potentiation of NMDA receptor responses in frog motoneurones, *Br J Pharmacol,* 143, 351, 2004.
40. Ishida, A., Kameshita, I., Okuno, S., Kitani, T., and Fujisawa, H., Phosphorylation of calmodulin by $Ca^{2+}$/calmodulin-dependent protein kinase IV, *Arch Biochem Biophys,* 407, 72, 2002.
41. Joubert, L., Hanson, B., Barthet, G., Sebben, M., Claeysen, S., Hong, W., Marin, P., Dumuis, A., and Bockaert, J., New sorting nexin (SNX27) and NHERF specifically interact with the 5-$HT_{4a}$ receptor splice variant: roles in receptor targeting, *J Cell Sci,* 117, 5367, 2004.
42. Joyal, J.L., Crimmins, D.L., Thoma, R.S., and Sacks, D.B., Identification of insulin-stimulated phosphorylation sites on calmodulin, *Biochemistry,* 35, 6267, 1996.
43. Kagaya, A., Mikuni, M., Muraoka, S., Saitoh, K., Ogawa, T., Shinno, H., Yamawaki, S., and Takahashi, K., Homologous desensitization of serotonin 5-HT2 receptor-stimulated intracellular calcium mobilization in C6BU-1 glioma cells via a mechanism involving a calmodulin pathway, *J Neurochem,* 61, 1050, 1993.
44. Krasel, C., Dammeier, S., Winstel, R., Brockmann, J., Mischak, H., and Lohse, M.J., Phosphorylation of GRK2 by protein kinase C abolishes its inhibition by calmodulin, *J Biol Chem,* 276, 1911, 2001.
45. Kushwaha, N. and Albert, P.R., Coupling of 5-$HT_{1A}$ autoreceptors to inhibition of mitogen-activated protein kinase activation via G beta gamma subunit signaling, *Eur J Neurosci,* 21, 721, 2005.
46. Kushwaha, N., Harwood, S.C., Wilson, A.M., Berger, M., Tecott, L.H., Roth, B.L., and Albert, P.R., Molecular determinants in the second intracellular loop of the 5-hydroxytryptamine-1A receptor for G-protein coupling, *Mol Pharmacol,* 69, 1518–1526, 2006.
47. Lee, S.P., Xie, Z., Varghese, G., Nguyen, T., O'Dowd, B.F., and George, S.R., Oligomerization of dopamine and serotonin receptors, *Neuropsychopharmacology,* 23, S32, 2000.
48. Lefkowitz, R.J., Pitcher, J., Krueger, K., and Daaka, Y., Mechanisms of beta-adrenergic receptor desensitization and resensitization, *Adv Pharmacol,* 42, 416, 1998.
49. Lefler, D., Mukhin, Y.V., Pettus, T., Leeb-Lundberg, L.M.F., Garnovskaya, M.N., and Raymond, J.R., Jak2 and $Ca^{2+}$/calmodulin are key intermediates for bradykinin $B_2$ receptor-mediated activation of $Na^+/H^+$ exchange in KNRK and CHO cells, *Assay Drug Dev Technol,* 1, 281, 2003.
50. Lembo, P.M. and Albert, P.R., Multiple phosphorylation sites are required for pathway-selective uncoupling of the 5-$hydroxytryptamine_{1A}$ receptor by protein kinase C, *Mol Pharmacol,* 48, 1024, 1995.
51. Lembo, P.M., Ghahremani, M.H., Morris, S.J., and Albert, P.R., A conserved threonine residue in the second intracellular loop of the 5-hydroxytryptamine 1A receptor directs signaling specificity, *Mol Pharmacol,* 52, 164, 1997.
52. McIlroy, B.K., Walters, J.D., Blackshear, P.J., and Johnson, J.D., Phosphorylation-dependent binding of a synthetic MARCKS peptide to calmodulin, *J Biol Chem,* 266, 4959, 1991.

53. McLatchie, L.M., Fraser, N.J., Main, M.J., Wise, A., Brown, J., Thompson, N., Solari, R., Lee, M.G., and Foord, S.M., RAMPs regulate the transport and ligand specificity of the calcitonin-receptor-like receptor, *Nature,* 393, 333, 1998.
54. Medeiros, Y.S. and Calixto, J.B., Influence of calcium entry blockers and calmodulin inhibitors on 5-hydroxytryptamine-, potassium- and calcium-induced contractions in human umbilical artery *in-vitro, J Pharm Pharmacol,* 43, 411, 1991.
55. Meller, R., Babity, J.M., and Grahame-Smith, D.G., 5-$HT_{2A}$ receptor activation leads to increased BDNF mRNA expression in C6 glioma cells, *Neuromol Med,* 1, 197, 2002.
56. Miller, C., Biophysics: Lonely voltage sensor seeks protons for permeation, *Science,* 312, 2006, p. 534.
57. Moyano, S., Del Rio, J., and Frechilla, D., Role of hippocampal CaMKII in serotonin 5-$HT_{1A}$ receptor-mediated learning deficit in rats, *Neuropsychopharmacology,* 29, 2216, 2004.
58. Muff, R., Buhlmann, N., Fischer, J.A., and Born, W., An amylin receptor is revealed following co-transfection of a calcitonin receptor with receptor activity modifying proteins-1 or -3, *Endocrinology,* 140, 2924, 1999.
59. Mukhin, Y.V., Garnovskaya, M.N., Collinsworth, G., Bell, J.L., Tholanikunnel, B.G., Pettus, T., Jaffa, A.A., Fitzgibbon, W., Ploth, D.W., Raymond, J.R., and Garnovskaya, M.N., Bradykinin $B_2$ receptors activate $Na^+/H^+$ exchange in mIMCD-3 cells via Janus kinase 2 and $Ca^{2+}$/calmodulin, *J Biol Chem,* 276, 17339, 2001.
60. Nishikawa, Y., Doi, M., Koji, T., Watanabe, M., Kimura, S., Kawasaki, S., Ogawa, A., and Sasaki, K., The role of rho and rho-dependent kinase in serotonin-induced contraction observed in bovine middle cerebral artery, *Tohoku J Exp Med,* 201, 239, 2003.
61. Okeley, N.M. and Gelb, M.H., A designed probe for acidic phospholipids reveals the unique enriched anionic character of the cytosolic face of the mammalian plasma membrane, *J Biol Chem,* 279, 21833, 2004.
62. Olayioye, M.A., Guthridge, M.A., Stomski, F.C., Lopez, A.F., Visvader, J.E., and Lindeman, G.J., Threonine 391 phosphorylation of the human prolactin receptor mediates a novel interaction with 14-3-3 proteins, *J Biol Chem,* 278, 32929, 2003.
63. Pitcher, J.A., Freedman, N.J., and Lefkowitz, R.J., G protein-coupled receptor kinases, *Annu Rev Biochem,* 67, 653, 1998.
64. Premont, R.T., Inglese, J., and Lefkowitz, R.J., Protein kinases that phosphorylate activated G protein-coupled receptors, *FASAB J,* 9, 175, 1995.
65. Prezeau, L., Richman, J.G., Edwards, S.W., and Limbird, L.E., The zeta isoform of 14-3-3 proteins interacts with the third intracellular loop of different $a_2$-adrenergic receptor subtypes, *J Biol Chem,* 274, 13462, 1999.
66. Pronin, A.N., Satpaev, D.K., Slepak, V.Z., and Benovic, J.L., Regulation of G protein-coupled receptor kinases by calmodulin and localization of the calmodulin binding domain, *J Biol Chem,* 272, 18273, 1997.
67. Putkey, J.A., Kleerekoper, Q., Gaertner, T.A., and Waxham, M.N., A new role for IQ motif proteins in regulating calmodulin function, *J Biol Chem,* 278, 49667, 2003.
68. Quadroni, M., James, P., and Carafoli, E., Isolation of phosphorylated calmodulin from rat liver and identification of the *in vivo* phosphorylation sites, *J Biol Chem,* 269, 16116, 1994.
69. Quinn, J.C., Johnson-Farley, N.N., Yoon, J.Y., and Cowen, D.S., Activation of extracellular-regulated kinase by 5-hydroxytryptamine$_{2A}$ receptors in PC12 cells is protein kinase C-independent and requires calmodulin and tyrosine kinases, *J Pharmacol Exp Ther,* 303, 746, 2002.

70. Qvigstad, E., Sjaastad, I., Brattelid, T., Nunn, C., Swift, F., Birkeland, J.A., Krobert, K.A., Andersen, G.O., Sejersted, O.M., Osnes, J.B., Levy, F.O., and Skomedal, T., Dual serotonergic regulation of ventricular contractile force through 5-HT$_{2A}$ and 5-HT$_4$ receptors induced in the acute failing heart, *Circ Res,* 97, 268, 2005.
71. Rahman, S. and Neuman, R.S., Multiple mechanisms of serotonin 5-HT$_2$ receptor desensitization, *Eur J Pharmacol,* 238, 173, 1993.
72. Raymond, J.R., Protein kinase C induces phosphorylation and desensitization of the human 5-HT$_{1A}$ receptor, *J Biol Chem,* 266, 14747, 1991.
73. Raymond, J.R., Mukhin, Y.V., Gelasco, A., Turner, J., Collinsworth, G., Gettys, T.W., Grewal, J.S., and Garnovskaya, M.N., Multiplicity of mechanisms of serotonin receptor signal transduction, *Pharmacol Ther,* 92, 179, 2001.
74. Rhoads, A.R. and Friedberg, F., Sequence motifs for calmodulin recognition, *FASEB J,* 11, 331, 1997.
75. Richman, J.G., Brady, A.E., Wang, Q., Hensel, J.L., Colbran, R.J., and Limbird, L.E., Agonist-regulated Interaction between a$_2$-adrenergic receptors and spinophilin, *J Biol Chem,* 276, 15003, 2001.
76. Rios, C.D., Jordan, B.A., Gomes, I., and Devi, L.A., G-protein-coupled receptor dimerization: modulation of receptor function, *Pharmacol Ther,* 92, 71, 2001.
77. Sacks, D.B., Davis, H.W., Crimmins, D.L., and McDonald, J.M., Insulin-stimulated phosphorylation of calmodulin, *Biochem J,* 286(Pt. 1), 211, 1992.
78. Sacks, D.B., Davis, H.W., Williams, J.P., Sheehan, E.L., Garcia, J.G., and McDonald, J.M., Phosphorylation by casein kinase II alters the biological activity of calmodulin, *Biochem J,* 283(Pt. 1), 21, 1992.
79. Sacks, D.B., Fujita-Yamaguchi, Y., Gale, R.D., and McDonald, J.M., Tyrosine-specific phosphorylation of calmodulin by the insulin receptor kinase purified from human placenta, *Biochem J,* 263, 803, 1989.
80. Sacks, D.B., Mazus, B., and Joyal, J.L., The activity of calmodulin is altered by phosphorylation: modulation of calmodulin function by the site of phosphate incorporation, *Biochem J,* 312(Pt. 1), 197, 1995.
81. Satoh, A., Nakanishi, H., Obaishi, H., Wada, M., Takahashi, K., Satoh, K., Hirao, K., Nishioka, H., Hata, Y., Mizoguchi, A., and Takai, Y., Neurabin-II/spinophilin. An actin filament-binding protein with one PDZ domain localized at cadherin-based cell-cell adhesion sites, *J Biol Chem,* 273, 3470, 1998.
82. Saunders, C. and Limbird, L.E., Disruption of microtubules reveals two independent apical targeting mechanisms for G-protein-coupled receptors in polarized renal epithelial cells, *J Biol Chem,* 272, 19035, 1997.
83. Saunders, C. and Limbird, L.E., Microtubule-dependent regulation of a$_{2B}$ adrenergic receptors in polarized MDCKII cells requires the third intracellular loop but not G protein coupling, *Mol Pharmacol,* 57, 44, 2000.
84. Schiapparelli, L., Del Rio, J., and Frechilla, D., Serotonin 5-HT receptor blockade enhances $Ca^{2+}$/calmodulin-dependent protein kinase II function and membrane expression of AMPA receptor subunits in the rat hippocampus: implications for memory formation, *J Neurochem,* 94, 884, 2005.
85. Schiapparelli, L., Simon, A.M., Del Rio, J., and Frechilla, D., Opposing effects of AMPA and 5-HT$_{1A}$ receptor blockade on passive avoidance and object recognition performance: correlation with AMPA receptor subunit expression in rat hippocampus, *Neuropharmacology,* 50, 897, 2006.
86. Smith, F.D., Oxford, G.S., and Milgram, S.L., Association of the D$_2$ dopamine receptor third cytoplasmic loop with spinophilin, a protein phosphatase-1-interacting protein, *J Biol Chem,* 274, 19894, 1999.

87. Streb, J.W. and Miano, J.M., Cross-species sequence analysis reveals multiple charged residue-rich domains that regulate nuclear/cytoplasmic partitioning and membrane localization of a kinase anchoring protein 12 (SSeCKS/Gravin), *J Biol Chem,* 280, 28007, 2005.
88. Stroebel, M. and Goppelt-Struebe, M., Signal transduction pathways responsible for serotonin-mediated prostaglandin G/H synthase expression in rat mesangial cells, *J Biol Chem,* 269, 22952, 1994.
89. Suzuki, K., The mechanism of enhanced platelet intracellular calcium mobilization stimulated by serotonin — in the pathophysiology of mood disorders, *Hokkaido Igaku Zasshi,* 76, 277, 2001.
90. Tazawa, H., Takahashi, S., and Zilliacus, J., Interaction of the parathyroid hormone receptor with the 14-3-3 protein, *Biochim Biophys Acta,* 1620, 32, 2003.
91. Turner, J.H., Gelasco, A.K., and Raymond, J.R., Calmodulin interacts with the third intracellular loop of the serotonin 5-hydroxytryptamine$_{1A}$ receptor at two distinct sites: putative role in receptor phosphorylation by protein kinase C, *J Biol Chem,* 279, 17027, 2004.
92. Turner, J.H. and Raymond, J.R., Interaction of calmodulin with the serotonin 5-hydroxytryptamine$_{2A}$ receptor. A putative regulator of G protein coupling and receptor phosphorylation by protein kinase C, *J Biol Chem,* 280, 30741, 2005.
93. Uneyama, H., Ueno, S., and Akaike, N., Serotonin-operated potassium current in CA1 neurons dissociated from rat hippocampus, *J Neurophysiol,* 69, 1044, 1993.
94. Wang, D., Sadee, W., and Quillan, J.M., Calmodulin binding to G protein-coupling domain of opioid receptors, *J Biol Chem,* 274, 22081, 1999.
95. Xia, Z., Gray, J.A., Compton-Toth, B.A., and Roth, B.L., A direct interaction of PSD-95 with 5-HT$_{2A}$ serotonin receptors regulates receptor trafficking and signal transduction, *J Biol Chem,* 278, 21901, 2003.
96. Xia, Z., Hufeisen, S.J., Gray, J.A., and Roth, B.L., The PDZ-binding domain is essential for the dendritic targeting of 5-HT$_{2A}$ serotonin receptors in cortical pyramidal neurons *in vitro*, *Neuroscience,* 122, 907, 2003.
97. Yap, K.L., Ames, J.B., Swindells, M.B., and Ikura, M., Diversity of conformational states and changes within the EF-hand protein superfamily, *Proteins,* 37, 499, 1999.
98. Yap, K.L., Kim, J., Truong, K., Sherman, M., Yuan, T., and Ikura, M., Calmodulin target database, *J Struct Funct Genomics,* 1, 8, 2000.
99. Yuen, E.Y., Jiang, Q., Chen, P., Gu, Z., Feng, J., and Yan, Z., Serotonin 5-HT$_{1A}$ receptors regulate NMDA receptor channels through a microtubule-dependent mechanism, *J Neurosci,* 25, 5488, 2005.

# 5 Identification of Novel Transcriptional Regulators in the Nervous System

*Federico Remes-Lenicov, Kirsten X. Jacobsen, Anastasia Rogaeva, Margaret Czesak, Mahmoud Hadjighasem, Mireille Daigle, and Paul R. Albert*

**CONTENTS**

Introduction ... 82
  Transcriptional Regulators and Behavior ... 82
  The Serotonin System ... 83
Identification of Transcription Start Site ... 83
  General Strategies ... 83
  5′-RACE ... 85
  RNase Protection Assay ... 86
Promoter Characterization ... 86
  Characteristics of Promoters ... 86
  Bioinformatic Analysis of Promoters ... 87
  Gene Reporter Assays ... 87
  Transfection of Cell Lines ... 88
  Transfection of Primary Neurons ... 89
  Interpretation of Reporter Assays ... 89
Identification of DNA Elements ... 90
  EMSA (Gel Retardation Assay) ... 90
    Probe Design ... 90
    Recombinant Protein ... 91
    Binding Reaction ... 91
  Supershift EMSA ... 92
Identification of DNA Binding Proteins ... 92
  Yeast One-Hybrid Analysis ... 93
    One-Hybrid Screening Protocol ... 94
  Chromatin Immunoprecipitation (CHIP) Assay ... 94

Transcription Factor Activity........ 95
GAL4-DBD Hybrid System........ 95
siRNA or Antisense Probes........ 97
Summary: Transcriptional Approach to Discovery........ 98
References........ 98

## INTRODUCTION

Increasingly, it is becoming recognized that transcriptional modulation of genes by sequence variations, DNA methylation, and alterations in transcription factor expression can play a major role in the behavioral phenotype throughout its development. Hence, the understanding of the transcriptional mechanisms that regulate key candidate genes for mental illness has taken on new significance. In a step-by-step approach we describe the crucial experiments to identify and characterize mechanisms of transcription regulation of the 5-HT1A receptor gene, which has been implicated in major depression, anxiety disorders, and suicide. These steps include: identification of transcription start sites by RT-PCR, primer extension, 5′-RACE, and RNase protection assay; characterization of upstream promoter, enhancer, and repressor regions using 5′-deletion luciferase reporter constructs and transfection assays in cell lines and primary cultures; identification of protein-DNA interactions using EMSA, supershift and CHIP assays; cloning of novel transcription factors using yeast one-hybrid screening and functional characterization using Gal4-DBD system and by siRNA approaches. Using these approaches in combination with bioinformatics searches it is possible to identify and characterize the basic transcriptional regulatory mechanisms of virtually any candidate gene of interest and to identify specific proximal DNA elements and transcriptional factors that mediate its regulation.

### Transcriptional Regulators and Behavior

The roles of transcriptional regulation in determining behavioral phenotype are beginning to be defined by exploiting new knowledge of the DNA elements and transcription factors that control brain-specific gene expression. For example, the brain-derived neurotrophin (BDNF) promoter has long been known as a target for the transcription factor CREB, which mediates calcium- and cAMP-dependent induction.[1] More detailed characterization has identified at least seven different promoters in the BDNF gene and differential regulation for these promoters,[2,3] and a novel calcium-regulated transcription factor (CaRF).[4] In addition, novel roles for DNA methylation and methyl binding proteins at specific BDNF promoter sites are associated with distinct behavioral (social defeat stress) or pharmacological (depolarization, cocaine) stimuli leading to BDNF-induced alterations in synaptic plasticity.[5–9] Similarly, enhanced maternal care of rat pups increases serotonergic activation of the NGFIA transcription factor, which binds to its DNA element in exon $1_7$ GR promoter, leading to altered acetylation and methylation, increased glucocorticoid receptor expression, and correlating with a reduced stress response in adult rats.[10] Thus, early lifetime changes in transcription factor activity due to environment can lead to lifelong changes in promoter methylation status, gene expression, and consequent behavioral alterations. The challenge is to understand how these changes take place, and the first step is the identification of promoter/enhancer regions of behavior-modifying genes.

### The Serotonin System

The serotonin system of the brain is a major determinant of mood and emotional status,[11,12] and altered regulation of this system has been strongly implicated in mental illness. Serotonin is synthesized in a small population of neurons from the raphe nuclei that express the enzyme tryptophan hydroxylase (TPH).[13] Another important site of regulation of the 5-HT system is the 5-HT reuptake transporter (5-HTT), which eliminates 5-HT from the synapse and is the target of serotonin-specific reuptake inhibitor antidepressants.[14] The release of 5-HT triggers activation of a wide variety of postsynaptic receptors. The 5-HT1A receptor is one of the most abundant 5-HT receptors in the brain[12] and functions as the predominant somatodendritic autoreceptor to inhibit the firing activity of serotonin neurons.[15] Desensitization of presynaptic 5-HT1A receptors is thought to accelerate antidepressant action,[16] whereas recent studies in animal models indicate a predominant role for postsynaptic 5-HT1A receptors in anxiety and antidepressant action.[17,18]

Because of the potential importance of 5-HT1A receptors in major depression and anxiety, we have focused on identifying the transcriptional regulators of the 5′-HT1A receptor gene.[19] The 5-HT1A gene has the important advantage of being an intronless gene[20,21]; hence, the contribution of intron sequences to its regulation is not a complication. The protocols described focus mainly on the 5′-regulatory region of the 5-HT1A gene but can be applied to regulatory regions of other genes, including those identified in introns or 3′-flanking sequence.

## IDENTIFICATION OF TRANSCRIPTION START SITE

The first step in characterizing a promoter of interest is to identify the transcription start site or sites (TSS). The use of different TSS is an important source of genetic diversity for differential tissue expression, with different promoters driving expression of RNA variants. Thus, it is important to identify the number and location of the TSS of a gene in order to identify promoters and transcriptional regulatory mechanisms of interest in a particular tissue.

### General Strategies

Initially, a DNA sequence spanning the candidate promoter site can be examined through one of several Web-based programs that detect eukaryotic promoters.[22,23] As programs are algorithms based on previously described promoter features, a negative result (i.e., the program does not recognize the sequence as a promoter) does not mean absence of a promoter, whereas a positive result does not guarantee its presence.

In order to experimentally verify a putative TSS identified by bioinformatics, several methods can be used including primer extension, 5′-RACE, and RNase protection assays. These methods require some knowledge of the candidate region; in particular, a stretch of the 5′ end of the coding sequence long enough to design a primer. One way to verify or narrow down the TSS region is by RT-PCR of isolated RNA using an antisense primer within the coding region and amplification using progressively 5′ upstream primers.[20] As shown in Figure 5.1, primers that are upstream of the TSS will fail to amplify, hence localizing the approximate TSS region. Once the TSS region

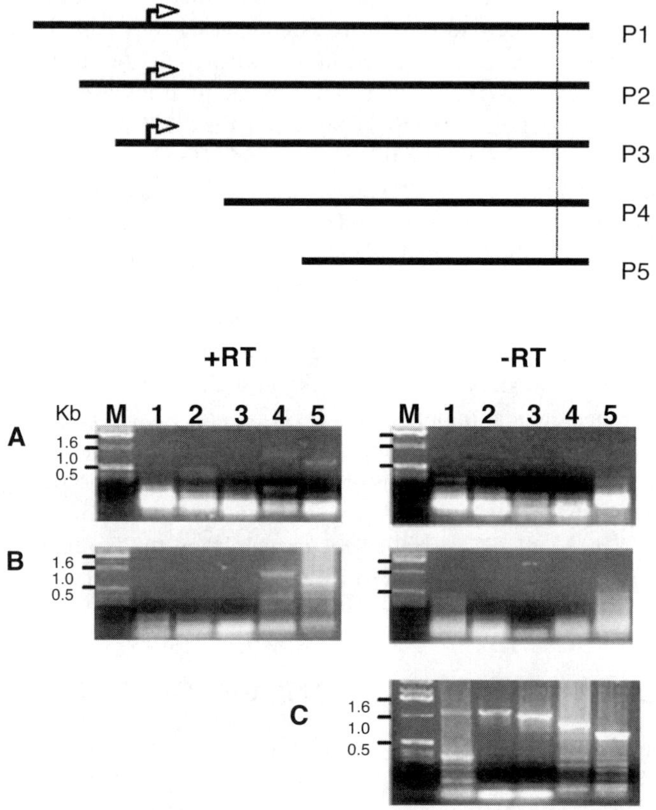


**FIGURE 5.1 Progressive 5′ RT-PCR strategy to identify the rat 5-HT1A receptor transcriptional initiation region.** Shown above is the location on the rat 5-HT1A receptor 5 region of the predicted PCR products from the common 3′-primer adjacent to the initiator ATG codon (dashed lines) and progressively 5′-primers P1-P5; the transcription start site (arrow) was later identified by primer extension and RNase protection analysis. Poly-A+ RNA (0.25 μg/lane) was isolated from hippocampal tissue (A) or RN46A cells (B) and was reverse-transcribed (+RT) or not (RT) and subjected to PCR using the common 3AS primer adjacent to the initiator ATG codon and primers P1-P5 (lanes 1–5). The presence of RT-PCR products of the correct size in both hippocampus and RN46A cDNA samples using P4 and P5, but not P1 to P3, indicates the location of the start site between P3 and P4. No specific PCR products were observed in RNA samples that were not reverse-transcribed. (C) PCR amplification of the correct products from rat genomic DNA (1 μg) by P1-P5 is a positive control to demonstrate that the PCR conditions were correct for all primers. (Reproduced with permission from Storring, J.M., Charest, A., Cheng, P., and Albert, P.R., *J Neurochem,* 72[6], 2238–2247, 1999; see Figure 2.)

has been localized to a short sequence, a primer extension experiment can be performed to pinpoint the exact start site.[20] In this protocol, a gene-specific labeled oligonucleotide primes reverse transcription of the 5′ end of the target mRNA; the length of the cDNA is measured by urea-polyacrylamide gel electrophoresis. The reverse transcription will terminate right at the TSS. The labeled product is run on a polyacrylamide gel to determine its precise size and therefore the TSS location.

## 5′-RACE

The method of choice for identification of the 5′-TSS is rapid amplification of cDNA 5′-ends (5′-RACE).[24,25] Following ligation of a linker primer to the 5′-end of the cDNA, PCR amplification of the 5′ end of a cDNA molecule is initiated between two nested gene-specific antisense primers and nested linker primers, followed by purification and DNA sequencing of the resulting products. The advantage of this procedure over primer extension is that the precise DNA sequences of the 5′ end of the RNA transcripts are determined, which can verify the predicted TSS or identify novel TSS that would not be predicted. A major caveat with 5′-RACE (and primer extension) is that the cDNA may not be full-length, particularly in CG-rich regions due to secondary structure that stalls the reverse transcriptase. Although commercial additives (e.g., DMSO, etc.) can reduce this problem, the best way to confirm TSS identification is by using RNase protection, which does not depend on reverse transcription (see the following).

We have performed 5′-RACE making use of BD Biosciences Marathon-Ready cDNA library[26] (Remes-Lenicov et al., unpublished). Alternately, tagged cDNA libraries can be homemade, depending on the characteristics of the target RNA. For example, to investigate a transcript that is expressed only at low levels in a subset of neurons, it might be better to isolate RNA from enriched tissue rather than use a commercial whole-brain cDNA library. In any event, when purchasing or constructing a library, it should be remembered that active TSS might change, depending on tissue or stage of development.

When designing the specific primer for the target gene, the user-designed primer may be located in a region that is spliced out in a particular transcript (false negative PCR). On the other hand, choosing a primer too far from the 5′ end (or with an unexpectedly long 5′-UTR) may also cause a false negative result. The latter problem may be resolved by longer reverse transcription times, by using high processivity DNA polymerases, or setting a longer extension period in the PCR program. Another problem is that the cDNA may not be full-length due to stalling of reverse transcriptase, particularly in CG-rich sequences. As the forward primer is the common tag and the user-designed antisense primer is specific for the target gene, the specificity of amplification is dependent on the specificity of the antisense primers. By using two sets of primers, specificity is enhanced but even if specific, it should be considered that a single band can represent multiple distinct products of similar size coming from distinct RNA variants. It is also possible that multiple bands arise from the same TSS as by-products of RNA degradation before the 5′-tag was added. In any 5′-RACE protocol, the final answer should come after purification and sequencing of each product. If the PCR results in too many products (a smear), a Southern blot using the relevant labeled genomic fragment can be done to identify products from the gene of interest, which can be verified by DNA sequencing. The identification of the TSS is an essential component in the study of the control of gene expression and opens the way to further investigate the promoter region(s).

### RNase Protection Assay

An important and complementary method to identify the TSS is the RNase protection assay.[27,28] In this assay, a synthetic RNA probe antisense to the target DNA and extending beyond the putative TSS is synthesized and labeled with [$^{32}$P]-UTP by *in vitro* transcription from a plasmid- containing genomic sequence including the target TSS and 5-flanking sequence.[20] Typically, plasmids such as KS+ (Stratagene) or pcDNA3 (Invitrogen) can be used as they contain initiation sites for T7, T3, or SP6 RNA polymerase. The plasmid is linearized to terminate the probe, and the reaction is terminated using RNase-free DNase to remove plasmid DNA and the probe purified on urea-polyacrylamide gel. The labeled RNA probe is hybridized overnight to denatured poly-A+ RNA or total RNA from tissue that expresses the target gene and is digested with RNase at conditions that are optimized. RNase preferentially removes the single-stranded nonhybridized portion of the RNA–RNA hybrid. The remaining portion is then electophoresed on urea-polyacrylamide gel to determine the molecular size. For accurate size determination, the hybridized portion should be less that 150 bp. Because this method relies on RNase digestion rather than reverse transcription from an antisense primer as is the case for primer extension or 5′-RACE, it is not subject to the problem of incomplete cDNA synthesis. Hence, the RNase protection assay provides a complementary approach to identify the TSS.

## PROMOTER CHARACTERIZATION

### Characteristics of Promoters

Once the transcriptional start site of a gene has been identified, the promoter region upstream of that start site can be characterized to elucidate its transcriptional regulation. There are two general types of promoters: TATA box and TATA-less promoters. TATA-containing promoters typically initiate transcription 25–35 nucleotides downstream of the TATA box and often have a CCAAT box sequence located with 50–100 nucleotides upstream.[29,30] TATA-less promoters are often CG-rich with multiple GC boxes (sites for Sp1/MAZ) and initiate at multiple sites or at an initiator (Inr) motif.[31–34] Examination of potential promoter regions of over 1000 genes by Suzuki and collaborators (2001) revealed that almost all putative promoter regions contained GC boxes (97%) and initiator motifs (85%), whereas substantially fewer contained CAAT or TATA boxes (64% and 34%, respectively), and only a fraction of promoters are composite promoters (TATA and Initiator dependent). However, the presence or absence of a TATA box is not necessary for correct expression. The human and mouse 5-HT1A receptor promoters, for example, have a TATA-less promoter that instead responds to MAZ and Sp1, and have multiple initiation sites.[21] By contrast, the rat 5-HT1A receptor promoter has CCAAT and TATA boxes, in addition to MAZ/Sp1 sites, and initiates at a single major TSS located 58 bp downstream of the TATA box.[20] Nevertheless, both mouse and rat 5-HT1A receptors are expressed with a similar tissue and brain regional distribution.[17,35] Beyond these well-characterized essential promoter elements, there are multiple enhancer and repressor elements that coordinately regulate developmental and tissue-specific gene expression.[36] It is also important to note that, whereas promoters are typically located immediately upstream of the transcription start site, there are many examples of distally located, intron or downstream enhancer or repressor elements.[37–39]

## Bioinformatic Analysis of Promoters

For determining where known or user-defined transcriptional regulatory elements might reside in putative promoters, programs like the TESS (Transcription Element Search System, www.cbil.upenn.edu/cgi-bin/tess/tess) can be indispensable. Many other programs exist to facilitate identification of transcriptional regulatory elements, including TRANSFAC and JASPAR,[22,23] and the methodology for determining consensus-binding elements has been well described.[23] However, identification of a consensus element in a gene does not mean that it is a truly functional element, and this must be tested by functional assays as described below. Oppositely, many DNA elements have degenerate or poorly defined consensus sequences that may not be identified by search programs or may appear numerous times in the sequence as false hits.

## Gene Reporter Assays

Once a putative promoter or regulatory sequence has been identified, gene reporter assays are widely used to functionally characterize putative promoter, enhancer, and repressor regions. Reporter assays provide a sensitive readout of the transcriptional activity of a DNA segment and can be used to test the effect of specific treatments on the transcriptional activity of a promoter.

Reporter constructs are generated by subcloning putative promoter sequences upstream of the transcriptional start site into vectors carrying an assayable nonmammalian reporter gene such as luciferase, chloramphenicol acetyltransferase (CAT), enhanced green fluorescent protein (EGFP), or β-galactosidase. The rationale for using a nonmammalian reporter is that it removes background due to endogenous mammalian transcripts and protein that are present in transfected mammalian cells. Also, the reporter construct removes mRNA sequences of the gene that could be regulated by posttranscriptional mechanisms, allowing for a discrete and sensitive assay of transcriptional activity. Typically, a series of 5′-deletion reporter constructs are generated that terminate at a common transcription start site so that the activity of equivalently initiated transcripts can be assessed.[20]

Reporter assays involve transfection of the reporter plasmid into cells and preparation of cell extracts for assay of the reporter protein activity. The luciferase enzyme, for example, catalyses a reaction that emits light upon addition of its substrate luciferin (D-luciferin for firefly luciferase or coelenterazine for Renilla luciferase), which can be monitored by a luminometer such as the Lmax II (Molecular Devices) that permits rapid measurement of reporter activity in 96-well plates.

Such sensitive monitoring systems also permits the use of smaller numbers of cells, less transfection reagent and less DNA for reporter assays. Assuming that all constructs have the same start sites and posttranscriptional processing, the activity of the reporter protein should be directly proportional to transcriptional activity of the transfected plasmid. To compare different plasmids and transfections, a second reporter plasmid, such as β-galactosidase or Renilla luciferase under control of a strong viral promoter (e.g., SV40), is cotransfected in order to control for transfection efficiency. The activity of this second reporter can also be measured and used to normalize the respective luciferase values. The fast decay kinetics of luminescence can be a problem when reading a large number of samples. As reactions are read in serial order, the reaction times of the first and last samples can differ considerably. Using the luciferase encoded in Promega plasmids pGL3 or pGL4, we suggest setting the integration time such that all reactions are read within 2 min. Finally, although buffer, substrate, and cell lysates are best kept on ice, the reaction should proceed at room temperature. Typical controls include transfection of the vector (without insert), and a positive control using a strong promoter to drive reporter expression. To determine whether the insert is a promoter or enhancer, the insert is cloned in the antisense orientation relative to the reporter as promoters lack activity in the antisense orientation, whereas enhancers or repressors display orientation-independent activity. In addition to measuring basal activity of the constructs, induced activity can be measured upon treatment with pharmacological regulators (second messenger activators such as forskolin, phorbol esters, growth factors, etc.) or cotransfected regulators to identify regulatory elements in the promoter.

Once a region of interest has been narrowed down by the deletion analysis described above, the specific element can be defined by excising the region of interest and placing it adjacent to minimal promoter–reporter construct.[40] If possible, the element of interest can be further narrowed down using bioinformatics to identify potential consensus sequences or palindrome (inverted repeat) sites. Alternately, DNase protection analysis of the region of interest can be used to experimentally identify sites of protein-DNA interaction.[40] Once the element of interest is identified, its function can be verified by deletion or site-directed mutagenesis of conserved nucleotides to disrupt the element, and reporter assays done. In addition, the activity of the element subcloned upstream of a minimal promoter should be examined. Upon identification of a known DNA element, the role of cognate transcription factors can be examined by cotransfection of their expression plasmids with reporter constructs. Binding of these factors to the DNA element can be verified by EMSA or supershift (see the following). If the element is novel, can be used yeast one-hybrid screening to identify transcription factors that bind to the element (see the next section).

## Transfection of Cell Lines

Transfection of cells in culture is a basic technique of molecular research; however, optimal methods for different systems can vary widely. One of the more commonly used and efficient transfection methods is liposome-mediated transfections,[41] which package DNA into lipophilic vesicles that can fuse to the cell membrane and release the DNA they carry. Many companies market transfection reagents for different purposes, with optimal working protocols inevitably being determined by the end user. It is important to optimize the cell density, lowest effective concentration and ratio of DNA-lipid reagent, the time of transfection, and the length of time the cells continue to grow after the transfection has been stopped. Common lipid transfection protocols from Invitrogen (i.e., Lipofectamine Plus) and Qiagen (i.e., TransMessenger) call for 3–6 h transfections, followed by 24–72 h recovery posttransfection. One of the least expensive transfection protocols is the calcium phosphate[42] method, which was one of the first transfection methods used but, has the downfall of using substantially more DNA and at times being unreliable. Other methods for DNA delivery include DEAE-dextran, dendrimers, "gene guns," viral infection, electroporation, and direct microinjection.

### Transfection of Primary Neurons

The mechanism for transfecting primary neurons in a culture with any lipid-based reagent differs from that acting on other cells. The lipid and the plasmid form a complex that allows the latter to be taken up by the cell. Once inside, in actively dividing cells, the DNA becomes trapped in the nucleus, where it can be expressed.[43,44] On the other hand, the DNA enters the neuronal nucleus by other means that remain to be determined. In practical terms, this means that neurons and nonneuronal cells should be plated at different densities. Although proliferating cells should be given space to divide following transfection, it is better to maximize the number of plated neurons. Confluence around 85 to 90% is optimal, as overcrowding can negatively affect transfection efficiency.[45]

### Interpretation of Reporter Assays

Although reporter assays represent an essential tool for examining regulatory elements, it is important to bear in mind that reporter constructs transfected in cell lines do not replicate the full regulation of genes that occurs *in vivo*. For example, the action of distal enhancers or regulators is lost, together with the effect of elements present in introns, the coding sequence or the 3′-UTR. Secondly, evidence is growing for the effects of a specific nuclear location or genomic environment on transcription, factors that do not apply to transfected plasmids.[46] In addition, the epigenetic history that might control a promoter *in vivo* cannot be reproduced unless purposefully replicated for the experiment. The activities of promoters driving effector plasmids will likely also lead to levels of these proteins that do not normally occur *in vivo*. Other features are lacking in transfected plasmids, such as histone packaging and chromatin condensation, which also may not reflect the *in vivo* situation. Thus, interpretation of reporter assays must be tentative until the role of specific DNA elements or transcription factors can be demonstrated in the endogenous gene as described below.

Transformed cell lines are often used for transfection experiments as their transfection efficiency is high and very reproducible; however, cell lines are transformed and may not accurately reflect regulation of genes in normal tissues, hence, the regulation should be verified in primary cultures or tissues if possible. However, some cell lines do retain characteristics of normal cells. For example, rat RN46A cells provide a model of serotonergic raphe cells,[47] whereas human SK-N-AS neuroblastoma cells model presynaptic dopamine neurons.[48,49] Nonetheless, even in cell lines that retain differentiated properties, the levels of expression may be different than that observed in normal cells, and hence the transcriptional activity of the promoter may be weak or subject to strong repression. For example, in RN46A cells the level of 5-HT1A receptors is much lower than in raphe neurons,[50] in part due to strong repression by the novel repressor Freud-1.[40,51] To obtain optimal responses in reporter assays it is important to try to match the species of transfected promoter DNA with the cell species. It is also important to compare reporter activity in cells that express the gene of interest to activity in cell lines that do not express this gene (i.e., the use of the nonneuronal, serotonin-deficient rat L6 myoblasts for studying 5-HT1A promoter elements).

## IDENTIFICATION OF DNA ELEMENTS

### EMSA (Gel Retardation Assay)

The interaction of proteins with DNA is important to control DNA repair, recombination, and transcription processes. Electrophoresis mobility shift assay (EMSA) is useful in studying the gene regulation and determining the DNA-protein interaction. The assay is based on the observation that complexes of protein and DNA migrate through a nondenaturing polyacrylamide gel more slowly than the free DNA fragments. The gel shift assay is performed by incubating a purified recombinant protein, nuclear or cell extract with [$^{32}$P]-end-labeled DNA fragment containing the putative protein binding site. The reaction products are then analyzed on a nondenaturing polyacrylamide gel. The specificity of this interaction is confirmed by competition experiment using unlabeled DNA fragment of interest.

### Probe Design

The first step in EMSA is probe design. Double-stranded linear DNA fragments of the binding site of interest are used in the gel retardation assay and are typically 10–30 nucleotides in length. Synthesized target sequence oligonucleotides (either sense or antisense), including a surrounding sequence with 5′ single-stranded overhangs (for labeling with Klenow) are purified by polyacrylamide gel electrophoresis to remove incomplete products or HPLC- (high performance liquid chromatography) purified primers are used.[40] The primers are annealed and radiolabeled by incorporating [$\alpha$-$^{32}$P] dNTP during a 3′ fill-in reaction using Klenow fragment of *E. coli* DNA polymerase.[52] Alternately, 5′-end labeling is done using [$\gamma$-$^{32}$P] ATP and T4 polynucleotide kinase, although we find that Klenow results in better and more stable labeling. For nonradioactive DNA labeling, a biotinylated or hapten-labeled dNTP is incorporated, then probed and detected by the sensitive or chemiluminescent substrate. Free radioactivity probe is removed using Sephadex G-50 column chromatography. For novel DNA elements it is important to identify the minimal sequence that is required for binding of a transcription factor to minimize nonspecific or irrelevant protein complexes. This can be done by DNase protection analysis[40] or by designing a series of unlabeled probes of shorter sequences and examining their ability to either compete for specific binding, or by using the shorter sequence as a probe to bind to the protein of interest.

## Recombinant Protein

Secondly, a source of transcription factor is required: either cell extract or recombinant protein can be used. For cell extracts, a cell line or tissue that preferentially expresses the protein of interest is chosen: the presence of the transcription factor can be identified by EMSA or Western blot. For generation of recombinant proteins, several *in vitro* transcription/translation protein expression systems are presently available such as EcoPro (Novagen), TNT® Coupled Reticulocyte Lysate (Promega), and TNT® Coupled Wheat Germ (Promega). The major limitation of these systems is their low levels of protein expression, making it difficult to perform more than one experiment with the same preparation of protein as well as limiting detection abilities in EMSA. In addition, proteins in the lysate sometimes produce nonspecific background. Alternately, the most common method for recombinant protein expression is the use of bacterial expression systems. Multiple vectors (pTriEX4 and pET30, Novagen; and pGEX, GE Healthcare) allow for expression of large amounts of protein. The major limitation to this expression system is potentially inadequate posttranslational modification, and decreased solubility and stability of proteins in the lysis and elution buffers, sometimes limiting the quantity and quality of the resultant protein. In some cases these problems can be fixed with the use of denaturing conditions and dialysis for renaturing, but not all proteins refold properly and degradation/aggregation may occur following denaturation. Finally, another method for protein expression uses eukaryotic expression systems with transfection reagents and mammalian expressing vectors such as pcDNA3 (Invitrogen) and pTriEX4 (Novagen). A mammalian cell expression system is optimal because it provides protein cofactors and modifications that might be essential for protein–DNA interaction. However, the presence of endogenous proteins can often cause serious background problems or interfere with detection of the recombinant protein.

## Binding Reaction

For EMSA, the labeled primer, protein extract, and competition primers are combined at room temperature. As a negative control, the protein extract is omitted. To initiate DNA binding, purified protein is premixed with Poly [d[I-C]] and incubated 20 min, then a 100- to 200-fold excess of unlabeled competitor sequence is added and incubated for 20 min. Finally, 50,000 cpm of labeled probe is added to the mixture. Any specific band should be eliminated by the presence of excess unlabeled specific competitor. The addition of a mutant or unrelated sequence will not compete with the labeled target, and the band will be preserved. The competition assay is a particularly important control for cell extracts, which may contain multiple nonspecific or other protein–DNA complexes. At high excess (>50-fold), nonspecific competitor primers with weak sequence similarity to the labeled primer may partially compete for the band of interest, and competitors with no sequence similarity to probe and lacking repetitive sequences should be chosen. It is also essential to use Poly [d[I-C]] (Roche) or herring sperm DNA to reduce nonspecific binding. Poly [d[I-C]] is thought to be better at blocking nonspecific binding but may eliminate specific interactions, in which case use of herring sperm DNA is necessary.[53]

### SUPERSHIFT EMSA

Electrophoretic separation is done using a nondenaturing gel to preserve the protein–DNA complexes and resolves based on the charge and hydrodynamic size, hence the migration of the complex does not necessarily correspond to its absolute molecular weight. In order to identify the proteins in the complex it is necessary to use antibody specific for the protein of interest to supershift the complex. If the protein of interest binds to the target DNA, the antibody will bind to that protein–DNA complex, decreasing furthermore its mobility and resulting in a "supershift." In some cases, the antibody may interfere with the protein–DNA interaction, resulting in a reduction or loss of the specific band, rather than a supershift.[51] An important control is to use preimmune serum to show that the effect of the antibody is specific; in addition, a control with antiserum in the absence of cell extract should be run to rule out nonspecific DNA binding in the antiserum. An alternative identification process would be the "shift Western blot." This involves transferring the resolved protein–DNA complexes to stacked nitrocellulose and anion-exchange membrane. Labeled DNA probe is detected on the anionic membrane whereas protein captured on the nitrocellulose membrane can be probed with a specific antibody (Western blot).[54]

If for some reason EMSA does not result in a clear outcome, it is possible to demonstrate protein–DNA interaction with the use of a CHIP assay or yeast one-hybrid system.[51]

## IDENTIFICATION OF DNA BINDING PROTEINS

Once promoter analysis has revealed a DNA element of interest, if the element identified does not match a known consensus sequence and it is clear from EMSA of relevant nuclear extracts that a specific protein–DNA interaction occurs, then a screen must be performed to identify the interacting proteins. There are several options for screening. We have tried screening lambda-GT11 bacterial expression cDNA library with labeled oligonucleotides for the DNA element but obtained weak and inconsistent signals (Ou and Albert, unpublished data). This may reflect the low DNA binding activity of transcription factors following lysis of bacteria. Another option is to use the DNA element as bait in an affinity purification protocol to isolate DNA binding proteins from relevant nuclear extracts. However, DNA binding proteins are often present in such low abundance that this may require large scale, multistep purification protocols. Finally, we have had success using the yeast one-hybrid approach, which provides an effective method for screening for DNA-binding proteins.

## YEAST ONE-HYBRID ANALYSIS

The yeast one-hybrid system is an *in vivo* genetic assay for isolating novel genes encoding proteins that bind to a target, *cis*-acting regulatory element. Yeast provides a useful system to screen for mammalian transcription factors as the background is low. However, enhancer or promoter elements may produce background problems if the yeast homologues are able to directly bind to target elements and activate transcription. In contrast, repressors reduce the expression of target genes and therefore do not contribute to the background signal.

To screen a library for a gene encoding a DNA binding protein (BP) of interest, a reporter construct is made using oligonucleotides containing at least three tandem repeats of the specific target DNA element placed upstream of the reporter (Figure 5.2). Multiple copies of the DNA element enhance sensitivity and specificity of the screen. We use a dual His3/LacZ reporter system to reduce background and maximize stringency. Once the target-reporter strain is isolated and selection conditions established, a cDNA-library fused to the Gal4 activation domain (GAL4-AD) is transformed into the strain for screening. Clones that encode a GAL4-AD fusion protein that are able to bind to the target DNA element recruit GAL4 to activate transcription of the selectable marker, allowing growth of the yeast clone (Figure 5.2). This clone is then subjected to β-galactosidase assay to finally examine the activity of the isolated clone.

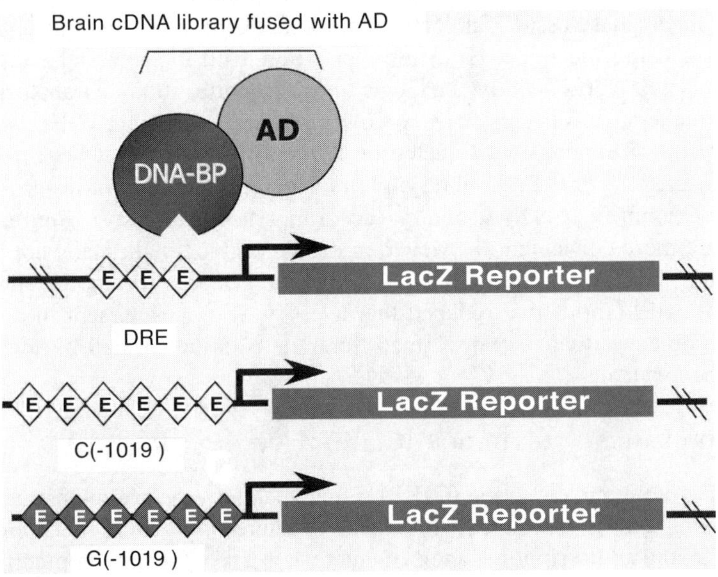


**FIGURE 5.2 (Color figure follows p. 110.) Yeast one-hybrid strategy for cloning DNA binding proteins.** Yeast one-hybrid cloning was done using a human brain cDNA library fused to the GAL4 activation domain (AD). The target sequence was composed of either 3 copies of the DRE (for cloning of Freud-1[51]) or 6 copies of the 26-bp palindrome C(-1019) element to clone NUDR/Deaf-1 and Hes5,[76] and these were fused to the LacZ reporter or other selectable markers. To show specificity for the C allele, we used 6 copies of the 26-bp G(-1019) element. (Reproduced with permission from Albert, P.R. and Lemonde, S., *Neuroscientist,* 10[6], 575–593, 2004; see Figure 5.)

Here, it is important to isolate DNA and retransform naïve reporter yeast strain to verify that the activity is due to the GAL4-AD clone. The positive clones are sequenced to confirm the identity of the cDNA insert and that it is in-frame with the GAL4-AD. The positive clone must then be tested for its DNA binding activity by EMSA, and for its function using promoter analysis in mammalian cells as described above.

### One-Hybrid Screening Protocol

The plasmids provided with the Matchmaker One-Hybrid System (www.clontech.com/clontech/techinfo/manuals/PDF/PT1031-1.pdf) are: pHISi, pHISi-1, and pLacZi, which contain for selection the yeast HIS3 or URA3 genes, respectively. The target-reporter constructs are linearized, transformed into the yeast strain YM4271 (Clontech) (MAT *a*, strain for integration of the element) using PEG4000/Li acetate followed by DMSO shock, and plated on the appropriate selection plates (SD-his or SD-ura) at 30°C. For pLacZi, we use a qualitative β-galactosidase assay[55] directly on the plate and for pHISi and pHISi-1 to determine the optimum concentration of 3-AT (3-amino-1,2,4-triazole; a competitive inhibitor of the HIS3 gene product) and identify which of the two HIS reporter strains has the lower background of HIS3 expression. If the β-galactosidase assay indicates that the background *lacZ* expression is low, you can prepare a dual reporter strain for library screening by integrating the target-HIS3 with the lower background HIS3 strain into the target lacZi strain. If the β-galactosidase assay[55] indicates that the background is high, the library can be screened using the target-HIS3 reporter strain with the lowest background of HIS3 expression. Subsequently, the single or dual reporter strain is transformed with the Matchmaker cDNA library fused to GAL4-AD and containing a LEU2 selectable marker for transformants, and selected on SD-leu-his+3-AT (at the optimized concentration) plates. For the AD-library, pretransformed cDNA libraries are available or you can amplify a cDNA library (according to availability). Pretransformed libraries are more convenient. However, in our hands we obtain many copies of the same clones and fewer different clones. This could be due to amplification of the pretransformed libraries that reduced the diversity of clones present in the library. Another problem is that pretransformed libraries of tissues or cell lines of interest may not be available.

## Chromatin Immunoprecipitation (CHIP) Assay

Chromatin immunoprecipitation (CHIP) is an excellent way to demonstrate that the transcription factor is associated with a gene of interest *in vivo*. In this approach, an antibody to the transcription factor of interest is used to immunoprecipitate the protein–DNA complex that has been cross-linked in whole cells. The immunoprecipitate is then subjected to PCR using primers that flank the DNA element of interest. The products are quantified using electrophoresis or Q-PCR. In order to visualize these complexes it is essential to have a tissue or cell line that expresses the protein of interest, as the transcription factors bound to a gene may vary depending on cell type and level of gene expression. Another limitation for this technique could be the sequence surrounding the DNA element, as it might be challenging at times to design an oligonucleotide sequence that will yield an adequate PCR product.

A critical step is to identify an antibody that is able to immunoprecipitate the protein (IP) of interest once it has been fixed to its cofactors and DNA target. Alternately, if no such antibody is available, a transfected cell line that stably expresses the transcription factor (preferably containing an epitope tag to allow the use of anti-epitope antibody) can be generated.[56] Many commercially available antibodies will now indicate if they are suitable for CHIP assays. The initial step to this technique is fixation of cells that should be kept at the same confluency from experiment to experiment to achieve reproducibility. Many fixation protocols have now been developed. Some use formaldehyde (FA) only, whereas others combine *N*-hydroxysuccinimide (NHS) ether chemistry with FA to crosslink protein complexes.[57] Optimizing the fixation time is important, as overfixation can lead to protein–protein or protein–DNA aggregation while insufficient fixation will fail to stabilize protein-DNA complexes.[58] Following fixation, it might be advantageous for some proteins to perform nuclear extraction in order to eliminate background from cytosolic proteins.[57] The most widely used kit and protocol is from Upstate with many modified versions, one of which includes addition of 212- to 300-$\mu$m diameter glass beads during the critical sonication step (Sigma).[59] The most important modification to the Upstate protocol is an increase in the number of washes to reduce high background in the elution fractions.[5] It is also critical to use silanized tubes with low DNA and protein binding capacity, as well as to change tubes during the washes. The analysis of bound DNA can be done either by regular PCR or Q-PCR for quantitative results. If regular PCR is used, it is essential to identify the number of cycles required for amplification, so as not to reach a plateau level as this will prevent accurate data analyses. On the other hand, the data analysis with Q-PCR is complicated by a lack of internal standard. It is then essential to find a control that will allow for normalization between experiments, such as a no antibody control. For reproducible results it is important to use consistent amounts of tissue, sonication, wash, and solution conditions.

Although use of CHIP assay may present several challenges, particularly for characterizing novel transcription factors, it is a powerful approach to demonstrate protein-DNA interactions *in vivo*, and complements the DNase I protection analysis and EMSA approaches.

## TRANSCRIPTION FACTOR ACTIVITY

### GAL4-DBD Hybrid System

Once a putative transcription factor has been identified, it is important to understand whether transcriptional regulation is an intrinsic property or whether it is indirect occurring through competition with other DNA binding proteins. Repressor-operator systems, such as the Gal4-DNA Binding Domain (DBD) heterologous hybrid system,[60] provide useful methods to address the intrinsic transcriptional regulatory activity of a protein. The *S. Cerevisiae* transcription activator Gal4 is a protein of

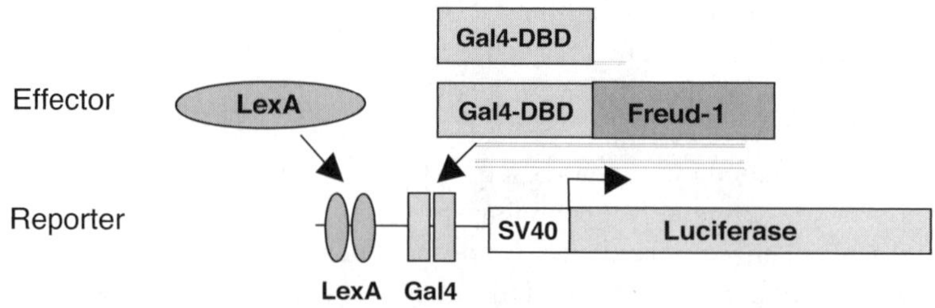


**FIGURE 5.3 Gal4-DBD-hybrid to assess intrinsic repressor activity of Freud-1.** To assess the intrinsic repressor activity of Freud-1, the indicated plasmids were constructed including Gal4-DBD (Gal vector), Gal4-Freud-1, or LexA (positive control) and were cotransfected with the X2G2P reporter construct containing two LexA and Gal4 sites upstream of SV40 promoter-luciferase in HEK 293 cells and luciferase activity measured in triplicate. Luciferase activity was corrected for transfection efficiency by cotransfection of pCMVβgal and normalized to the Gal vector. (Reproduced with permission from Lemonde, S., Rogaeva, A., and Albert, P.R., *J Neurochem,* 88[4], 857–868, 2004; see Figure 9.)

881 amino acids[61] and is required for the proper expression of genes that encode galactose-metabolizing enzymes. The first 147 amino acids of Gal4 contains a cysteine-rich DBD, which binds to specific DNA sequences ($UAS_G$), whereas the C-terminal domain of Gal4 contains the acidic transactivation domain (AD) which recruits RNA polymerase.[60] The Gal4 DBD and AD can function independently as "cassettes" to confer their activities on fusion proteins (Figure 5.3). By fusion of the transcription factor to be examined with the Gal4 DBD, the activity of this transcription factor (enhancer or repressor) is conferred upon the fusion protein. Although mammalian cells lack endogenous Gal4, yeast Gal4 protein can be a functional transcription factor in these cells by using the firefly enzyme luciferase as a typical reporter gene. When transfected into mammalian cells, the Gal4-DBD fusion protein can bind to the $UAS(_G)$ sequence. By cotransfection of a luciferase reporter plasmid that has incorporated the $UAS(_G)$, the activity of the fusion protein is assayed.

For example, a Gal4-DBD fusion construct was generated by subcloning the protein of interest (Deaf-1, Freud-1) into a vector (EcoR1 site of the pBXG-1 vector, obtained from Reference 62), ensuring ORF is in frame with the Gal4-DBD[52,63] (Figure 5.3). A reporter construct (X2G2P) containing two LexA and two Gal4 UAS sites upstream of the SV40 promoter-luciferase was cotransfected with the fusion construct to determine whether the protein of interest confers repressor or enhancer activity when it is recruited to the heterologous element and promoter. As the transfected reporter construct is at a 10- to 100-fold molar excess compared to endogenous promoters, this provides a reliable measure of the intrinsic activity of the fusion protein at the reporter construct. Using this assay we found that although Freud-1 displays repressor activity in all cell types examined, Deaf-1 displayed intrinsic repressor or enhancer activities, depending on the cell type.[52,63] A bacterial repressor, LexA, has been shown to reduce gene expression in eukaryotes[64] and is used as a positive control for repressor activity. A protein that lacks intrinsic activity would not affect the activity of this construct. However, a negative result must be interpreted with caution as the structure and function of the protein may be compromised in the fusion. As a positive control, the fusion should be tested for activity at its cognate DNA element.

The mammalian two-hybrid system is another version of this technique that is useful for detecting protein–protein interactions in mammalian cells. In this system, both Gal4-DBD and Gal4-AD fusion proteins are cotransfected with the reporter construct containing Gal4-UAS and minimal promoter. If the two fusion proteins interact, this leads to recruitment to the reporter of the Gal4-AD fusion by the Gal4-DBD fusion protein. It is important to note that transcription activation domains other than Gal4, such as VP16 (viral protein 16) or B42, can also be used, each with its own associated advantages and disadvantages.

## siRNA or Antisense Probes

Many gene targets are proposed for transcription factors but *in vivo* analysis of their activity is necessary to conclusively state their involvement in gene regulation. One useful approach is to use RNAi (RNA interference)[65] to knock down the transcription factor of interest and measure the effect on the gene of interest using QPCR to quantify RNA levels, Western blot to measure protein, or by cotransfecting reporter constructs to assay transcriptional activity. siRNA (small interfering RNA) has proven a useful approach for specific knockdown of gene expression. The RNAi-target RNA hybrid recruits an RNA-inducing silencing complex (RISC) which aids in cleavage of targeted RNA with the use of endo- and exonucleases.[66] The use of siRNA is advantageous due to fast production and lower interferon response compared to other methods such as shRNA (short hairpin RNA).[67,68] One disadvantage to the use of siRNA is the inability to make stably expressing cells lacking the protein of interest, which is possible with the use of antisense or shRNA technology. The knockdown time is also limited due to the lability of RNAi. This method is also quite costly, especially if the target protein is not knocked down with the first couple of siRNAs, in which case multiple siRNAs must be tested. If there is a need to target the same gene in multiple species, it can be challenging to find a common sequence that would specifically target the gene of interest in all species.

The most important step in siRNA technology is the design of siRNA molecules that specifically target the sequence of the gene of interest only, using BLAST search for validation. The targeting sequence is usually 21 ribonucleic acids in length. It is essential to make a negative control, ideally containing the same nucleotide content but lacking recognition, with similar %GC content. Several free Web-based design software programs (e.g., Invitrogen, https://rnaidesigner.invitrogen.com/rnaiexpress/setOption.do?designOption=stealth) as well as many predesigned siRNAs for known genes are available which guarantee "specific" knockdown. Importantly, at least one additional siRNA that targets your protein should be designed to verify the specificity of the effect. siRNA can be readily transfected into cultured cells using available reagents such as Lipofectamine 2000 (Invitrogen) and HiPerFect (Qiagen). A positive control such as BLOCK-iT™ Fluorescent Oligo (Invitrogen) for transfection efficiency is useful to optimize delivery of siRNA. The time period for efficient protein knockdown will vary, depending on the RNA and protein half-life of turnover, and it is critical to optimize this time for each target. In our hands 72-h exposure to siRNA leads to ~90% down-regulation of protein expression (unpublished data), but this will vary, depending on the protein. The siRNA approach can provide evidence that the transcription factor is functionally important for the regulation of the gene of interest. However, it cannot be concluded whether the effect is direct (via a specific DNA element on the gene) or indirect (via modulation of other genes such as transcription factor genes).

## SUMMARY: TRANSCRIPTIONAL APPROACH TO DISCOVERY

The methods described above provide a step-by-step approach to characterize the transcriptional regulation and regulators of a gene of interest. This approach provides mechanistic insight into the fundamental regulation of the gene, but other sites of regulation including enhancer or repressor elements located in introns or 3-flanking sequence. The identification of specific DNA elements and transcription factors provides a starting point for further studies of epigenetic changes such as alterations in histone modification (lysine acetylation, methylation, sumoylation, etc.) using CHIP assay or of DNA methylation (at CG dinucleotides) that occur at these regulatory sites.[6,8,69] Recently, modification of histone acetylation and DNA methylation have been used to produce alterations in learning and behavior.[6,7,70,71]

The finding of novel transcription regulators and DNA elements provides an important tool for understanding global gene regulation in the nervous system. As transcription factors generally regulate multiple gene targets that have the appropriate DNA element, these can be identified by bioinformatics searches for consensus DNA elements, or by gene array studies.[72,73] Gene knockout studies of novel transcription factors have also revealed specific functions in systems regulation for specific transcription factors (Pet-1/DREAM).[74,75] Because these factors provide a global regulation of multiple genes within a system, they may become useful therapeutic targets for global modulation of gene expression.

The identification of transcriptional mechanisms provides a novel method for addressing the function of promoter polymorphisms. For example, the 5-HT1A receptor promoter polymorphism C(-1019)G was initially identified in the repressor region of the 5-HT1A gene. We have been able to show that the polymorphism is the target of at least two repressor proteins (Hes-5 and Deaf-1) and that the G-allele binds weakly or not at all to these factors, leading to derepression of the 5-HT1A autoreceptor.[76] We are currently using the novel DNA elements that we have identified to search for conserved elements and nearby functional polymorphisms in other candidate genes involved in mental illness.

## REFERENCES

1. West, A.E., Chen, W.G., Dalva, M.B., Dolmetsch, R.E., Kornhauser, J.M., Shaywitz, A.J., Takasu, M.A., Tao, X., and Greenberg, M.E., Calcium regulation of neuronal gene expression, *Proc Natl Acad Sci USA,* 98(20), 11024–11031, 2001.
2. Liu, Q.R., Lu, L., Zhu, X.G., Gong, J.P., Shaham, Y., and Uhl, G.R., Rodent BDNF genes, novel promoters, novel splice variants, and regulation by cocaine, *Brain Res,* 1067(1), 1–12, 2006.

3. Hansson, A.C., Sommer, W.H., Metsis, M., Stromberg, I., Agnati, L.F., and Fuxe, K., Corticosterone actions on the hippocampal brain-derived neurotrophic factor expression are mediated by exon IV promoter, *J Neuroendocrinol,* 18(2), 104–14, 2006.
4. Tao, X., West, A.E., Chen, W.G., Corfas, G., and Greenberg, M.E., A calcium-responsive transcription factor, CaRF, that regulates neuronal activity-dependent expression of BDNF, *Neuron,* 33(3), 383–95, 2002.
5. Chen, W.G., Chang, Q., Lin, Y., Meissner, A., West, A.E., Griffith, E.C., Jaenisch, R., and Greenberg, M.E., Derepression of BDNF Transcription Involves Calcium-Dependent Phosphorylation of MeCP2, *Science,* 302(5646), 885–889, 2003.
6. Tsankova, N.M., Berton, O., Renthal, W., Kumar, A., Neve, R.L., and Nestler, E.J., Sustained hippocampal chromatin regulation in a mouse model of depression and antidepressant action, *Nat Neurosci,* 9(4), 519–525, 2006.
7. Levenson, J.M., Roth, T.L., Lubin, F.D., Miller, C.A., Huang, I.C., Desai, P., Malone, L.M., and Sweatt, J.D., Evidence that DNA (cytosine-5) methyltransferase regulates synaptic plasticity in the hippocampus, *J Biol Chem,* 281(23), 15763–15773, 2006.
8. Kumar, A., Choi, K.H., Renthal, W., Tsankova, N.M., Theobald, D.E., Truong, H.T., Russo, S.J., Laplant, Q., Sasaki, T.S., Whistler, K.N., Neve, R.L., Self, D.W., and Nestler, E.J., Chromatin remodeling is a key mechanism underlying cocaine-induced plasticity in striatum, *Neuron,* 48(2), 303–314, 2005.
9. Martinowich, K., Hattori, D., Wu, H., Fouse, S., He, F., Hu, Y., Fan, G., and Sun, Y.E., DNA Methylation-Related Chromatin Remodeling in Activity-Dependent BDNF Gene Regulation, *Science,* 302(5646), 890–893, 2003.
10. Weaver, I.C., Cervoni, N., Champagne, F.A., D'Alessio, A.C., Sharma, S., Seckl, J.R., Dymov, S., Szyf, M., and Meaney, M.J., Epigenetic programming by maternal behavior, *Nat Neurosci*, 2004.
11. Jacobs, B.L. and Azmitia, E.C., Structure and function of the brain serotonin system, *Physiol Rev,* 72(1), 165–229, 1992.
12. Barnes, N.M. and Sharp, T., A review of central 5-HT receptors and their function, *Neuropharmacology,* 38(8), 1083–152, 1999.
13. Hornung, J.P., The human raphe nuclei and the serotonergic system, *J Chem Neuroanat,* 26(4), 331–343, 2003.
14. Adell, A., Celada, P., Abellan, M.T., and Artigas, F., Origin and functional role of the extracellular serotonin in the midbrain raphe nuclei, *Brain Res Brain Res Rev,* 39(2–3), 154–180, 2002.
15. Haddjeri, N., Lavoie, N., and Blier, P., Electrophysiological evidence for the tonic activation of 5-HT(1A) autoreceptors in the rat dorsal raphe nucleus, *Neuropsychopharmacology,* 29, 1800–1806, 2004.
16. Celada, P., Puig, M., Amargos-Bosch, M., Adell, A., and Artigas, F., The therapeutic role of 5-HT(1A) and 5-HT(2A) receptors in depression, *J Psychiatry Neurosci,* 29(4), 252–265, 2004.
17. Gross, C., Zhuang, X., Stark, K., Ramboz, S., Oosting, R., Kirby, L., Santarelli, L., Beck, S., and Hen, R., Serotonin1A receptor acts during development to establish normal anxiety-like behavior in the adult, *Nature,* 416(6879), 396–400, 2002.
18. Santarelli, L., Saxe, M., Gross, C., Surget, A., Battaglia, F., Dulawa, S., Weisstaub, N., Lee, J., Duman, R., Arancio, O., Belzung, C., and Hen, R., Requirement of Hippocampal Neurogenesis for the Behavioral Effects of Antidepressants, *Science,* 301(5634), 805–809, 2003.
19. Albert, P.R. and Lemonde, S., 5-HT1A Receptors, gene repression, and depression: guilt by association, *Neuroscientist,* 10(6), 575–593, 2004.

20. Storring, J.M., Charest, A., Cheng, P., and Albert, P.R., TATA-driven transcriptional initiation and regulation of the rat serotonin 5-HT1A receptor gene, *J Neurochem,* 72(6), 2238–2247, 1999.
21. Parks, C.L. and Shenk, T., The serotonin 1a receptor gene contains a TATA-less promoter that responds to MAZ and Sp1, *J Biol Chem,* 271(8), 4417–4430, 1996.
22. Vlieghe, D., Sandelin, A., De Bleser, P.J., Vleminckx, K., Wasserman, W.W., van Roy, F., and Lenhard, B., A new generation of JASPAR, the open-access repository for transcription factor binding site profiles, *Nucl Acids Res,* 34(Database issue), D95–D97, 2006.
23. Wasserman, W.W. and Sandelin, A., Applied bioinformatics for the identification of regulatory elements, *Nat Rev Genet,* 5(4), 276–287, 2004.
24. Frohman, M.A., Rapid amplification of complementary DNA ends for generation of full-length complementary DNAs: thermal RACE, *Methods Enzymol,* 218, 340–356, 1993.
25. Zhang, Y. and Frohman, M.A., Using rapid amplification of cDNA ends (RACE) to obtain full-length cDNAs, *Methods Mol Biol,* 69, 61–87, 1997.
26. Mao, H., Zhao, Q., Daigle, M., Ghahremani, M.H., Chidiac, P., and Albert, P.R., RGS17/RGSZ2, a novel regulator of Gi/o, Gz, and Gq signaling, *J Biol Chem,* 279(25), 26314–26322, 2004.
27. Prediger, E.A., Detection and quantitation of mRNAs using ribonuclease protection assays, *Methods Mol Biol* 160, 495–505, 2001.
28. Krieg, P.A. and Melton, D.A., *In vitro* RNA synthesis with SP6 RNA polymerase, *Methods Enzymol,* 155, 397–415, 1987.
29. Naar, A.M., Lemon, B.D., and Tjian, R., Transcriptional coactivator complexes, *Annu Rev Biochem,* 70, 475–501, 2001.
30. Stargell, L.A. and Struhl, K., Mechanisms of transcriptional activation *in vivo*: two steps forward, *Trends Genet,* 12(8), 311–315, 1996.
31. Suzuki, Y., Tsunoda, T., Sese, J., Taira, H., Mizushima-Sugano, J., Hata, H., Ota, T., Isogai, T., Tanaka, T., Nakamura, Y., Suyama, A., Sakaki, Y., Morishita, S., Okubo, K., and Sugano, S., Identification and characterization of the potential promoter regions of 1031 kinds of human genes, *Genome Res,* 11(5), 677–684, 2001.
32. Benoist, C. and Chambon, P., *In vivo* sequence requirements of the SV40 early promotor region, *Nature,* 290(5804), 304–310, 1981.
33. Mathis, D.J. and Chambon, P., The SV40 early region TATA box is required for accurate *in vitro* initiation of transcription, *Nature,* 290(5804), 310–315, 1981.
34. Ambrosini, G., Praz, V., Jagannathan, V., and Bucher, P., Signal search analysis server, *Nucl Acids Res,* 31(13), 3618–3620, 2003.
35. Pompeiano, M., Palacios, J.M., and Mengod, G., Distribution and cellular localization of mRNA coding for 5-HT1A receptor in the rat brain: correlation with receptor binding, *J Neurosci,* 12(2), 440–453, 1992.
36. Hochheimer, A. and Tjian, R., Diversified transcription initiation complexes expand promoter selectivity and tissue-specific gene expression, *Genes Dev,* 17(11), 1309–1320, 2003.
37. Ogbourne, S. and Antalis, T.M., Transcriptional control and the role of silencers in transcriptional regulation in eukaryotes, *Biochem J,* 331(Pt. 1), 1–14, 1998.
38. Suen, T.C. and Goss, P.E., Characterization of a repressor element and a juxtaposed tissue-restricted activator element located on the distal neu gene promoter, *J Biol Chem,* 275(9), 6600–6607, 2000.
39. Ainscough, J.F., John, R.M., Barton, S.C., and Surani, M.A., A Skeletal Muscle-Specific Mouse Igf2 Repressor Lies 40 kb Downstream of the Gene, *Development,* 127(18), 3923–3930, 2000.

40. Ou, X.M., Jafar-Nejad, H., Storring, J.M., Meng, J.H., Lemonde, S., and Albert, P.R., Novel dual repressor elements for neuronal cell-specific transcription of the rat 5-HT1A receptor gene, *J Biol Chem,* 275(11), 8161–8168, 2000.
41. Fraley, R., Subramani, S., Berg, P., and Papahadjopoulos, D., Introduction of liposome-encapsulated SV40 DNA into cells, *J Biol Chem,* 255(21), 10431–10435, 1980.
42. Graham, F.L. and van der Eb, A.J., A new technique for the assay of infectivity of human adenovirus 5 DNA, *Virology,* 52(2), 456–467, 1973.
43. Escriou, V., Carriere, M., Bussone, F., Wils, P., and Scherman, D., Critical assessment of the nuclear import of plasmid during cationic lipid-mediated gene transfer, *J Gene Med,* 3(2), 179–187, 2001.
44. Zuhorn, I.S. and Hoekstra, D., On the mechanism of cationic amphiphile-mediated transfection. To fuse or not to fuse: is that the question?, *J Membr Biol,* 189(3), 167–179, 2002.
45. Dalby, B., Cates, S., Harris, A., Ohki, E.C., Tilkins, M.L., Price, P.J., and Ciccarone, V.C., Advanced transfection with Lipofectamine 2000 reagent: primary neurons, siRNA, and high-throughput applications, *Methods,* 33(2), 95–103, 2004.
46. Dillon, N., Gene regulation and large-scale chromatin organization in the nucleus, *Chromosome Res,* 14(1), 117–126, 2006.
47. White, L.A., Eaton, M.J., Castro, M.C., Klose, K.J., Globus, M.Y., Shaw, G., and Whittemore, S.R., Distinct regulatory pathways control neurofilament expression and neurotransmitter synthesis in immortalized serotonergic neurons, *J Neurosci,* 14(11 Pt. 1), 6744–6753, 1994.
48. Sugimoto, T., Tatsumi, E., Kemshead, J.T., Helson, L., Green, A.A., and Minowada, J., Determination of cell surface membrane antigens common to both human neuroblastoma and leukemia-lymphoma cell lines by a panel of 38 monoclonal antibodies, *J Natl Cancer Inst,* 73(1), 51–57, 1984.
49. Wang, J. and Bannon, M.J., Sp1 and Sp3 activate transcription of the human dopamine transporter gene, *J Neurochem,* 93(2), 474–482, 2005.
50. Kushwaha, N. and Albert, P.R., Coupling of 5-HT1A autoreceptors to inhibition of mitogen-activated protein kinase activation via Gbetagamma subunit signaling, *Eur J Neurosci,* 21(3), 721–732, 2005.
51. Ou, X.M., Lemonde, S., Jafar-Nejad, H., Bown, C.D., Goto, A., Rogaeva, A., and Albert, P.R., Freud-1: a neuronal calcium-regulated repressor of the 5-HT1A receptor gene, *J Neurosci,* 23(19), 7415–7425, 2003.
52. Lemonde, S., Rogaeva, A., and Albert, P.R., Cell type-dependent recruitment of trichostatin A-sensitive repression of the human 5-HT1A receptor gene, *J Neurochem,* 88(4), 857–868, 2004.
53. Ohshiro, T. and Saigo, K., Transcriptional regulation of breathless FGF receptor gene by binding of TRACHEALESS/dARNT heterodimers to three central midline elements in Drosophila developing trachea, *Development,* 124(20), 3975–3986, 1997.
54. Demczuk, S., Harbers, M., and Vennstrom, B., Identification and analysis of all components of a gel retardation assay by combination with immunoblotting, *Proc Natl Acad Sci USA,* 90(7), 2574–2578, 1993.
55. Duttweiler, H.M., A highly sensitive and non-lethal beta-galactosidase plate assay for yeast, *Trends Genet,* 12(9), 340–341, 1996.
56. Shi, Y., Sawada, J., Sui, G., Affar el, B., Whetstine, J.R., Lan, F., Ogawa, H., Luke, M.P., Nakatani, Y., and Shi, Y., Coordinated histone modifications mediated by a CtBP co-repressor complex, *Nature,* 422(6933), 735–738, 2003.
57. Nowak, D.E., Tian, B., and Brasier, A.R., Two-step cross-linking method for identification of NF-kappaB gene network by chromatin immunoprecipitation, *Biotechniques,* 39(5), 715–725, 2005.

58. Spencer, V.A., Sun, J.M., Li, L., and Davie, J.R., Chromatin immunoprecipitation: a tool for studying histone acetylation and transcription factor binding, *Methods,* 31(1), 67–75, 2003.
59. Weinmann, A.S., Bartley, S.M., Zhang, T., Zhang, M.Q., and Farnham, P.J., Use of chromatin immunoprecipitation to clone novel E2F target promoters, *Mol Cell Biol,* 21(20), 6820–6832, 2001.
60. Fields, S. and Song, O., A novel genetic system to detect protein-protein interactions, *Nature,* 340(6230), 245–246, 1989.
61. Laughon, A. and Gesteland, R.F., Primary structure of the Saccharomyces cerevisiae GAL4 gene, *Mol Cell Biol,* 4(2), 260–267, 1984.
62. Bremner, R., Cohen, B.L., Sopta, M., Hamel, P.A., Ingles, C.J., Gallie, B.L., and Phillips, R.A., Direct transcriptional repression by pRB and its reversal by specific cyclins, *Mol Cell Biol,* 15(6), 3256–3265, 1995.
63. Czesak, M., Lemonde, S., Peterson, E.A., Rogaeva, A., and Albert, P.R., Cell-specific repressor or enhancer activities of Deaf-1 at a serotonin 1A receptor gene polymorphism, *J Neurosci,* 26(6), 1864–1871, 2006.
64. Brent, R. and Ptashne, M., A bacterial repressor protein or a yeast transcriptional terminator can block upstream activation of a yeast gene, *Nature,* 312(5995), 612–615, 1984.
65. Fire, A., Xu, S., Montgomery, M.K., Kostas, S.A., Driver, S.E., and Mello, C.C., Potent and specific genetic interference by double-stranded RNA in Caenorhabditis elegans, *Nature,* 391(6669), 806–811, 1998.
66. Lee, S.H. and Sinko, P.J., siRNA-getting the message out, *Eur J Pharm Sci,* 27(5), 401–410, 2006.
67. Heidel, J.D., Hu, S., Liu, X.F., Triche, T.J., and Davis, M.E., Lack of interferon response in animals to naked siRNAs, *Nat Biotechnol,* 22(12), 1579–1582, 2004.
68. Bridge, A.J., Pebernard, S., Ducraux, A., Nicoulaz, A.L., and Iggo, R., Induction of an interferon response by RNAi vectors in mammalian cells, *Nat Genet,* 34(3), 263–264, 2003.
69. Colvis, C.M., Pollock, J.D., Goodman, R.H., Impey, S., Dunn, J., Mandel, G., Champagne, F.A., Mayford, M., Korzus, E., Kumar, A., Renthal, W., Theobald, D.E., and Nestler, E.J., Epigenetic mechanisms and gene networks in the nervous system, *J Neurosci,* 25(45), 10379–10389, 2005.
70. Korzus, E., Rosenfeld, M.G., and Mayford, M., CBP histone acetyltransferase activity is a critical component of memory consolidation, *Neuron,* 42(6), 961–972, 2004.
71. Weaver, I.C., Champagne, F.A., Brown, S.E., Dymov, S., Sharma, S., Meaney, M.J., and Szyf, M., Reversal of maternal programming of stress responses in adult offspring through methyl supplementation: altering epigenetic marking later in life, *J Neurosci,* 25(47), 11045–11054, 2005.
72. Bruce, A.W., Donaldson, I.J., Wood, I.C., Yerbury, S.A., Sadowski, M.I., Chapman, M., Gottgens, B., and Buckley, N.J., Genome-wide analysis of repressor element 1 silencing transcription factor/neuron-restrictive silencing factor (REST/NRSF) target genes, *Proc Natl Acad Sci USA,* 101(28), 10458–10463, 2004.
73. Impey, S., McCorkle, S.R., Cha-Molstad, H., Dwyer, J.M., Yochum, G.S., Boss, J.M., McWeeney, S., Dunn, J.J., Mandel, G., and Goodman, R.H., Defining the CREB regulon: a genome-wide analysis of transcription factor regulatory regions, *Cell,* 119(7), 1041–1054, 2004.
74. Hendricks, T.J., Fyodorov, D.V., Wegman, L.J., Lelutiu, N.B., Pehek, E.A., Yamamoto, B., Silver, J., Weeber, E.J., Sweatt, J.D., and Deneris, E.S., Pet-1 ETS gene plays a critical role in 5-HT neuron development and is required for normal anxiety-like and aggressive behavior, *Neuron,* 37(2), 233–247, 2003.

75. Cheng, H.Y., Pitcher, G.M., Laviolette, S.R., Whishaw, I.Q., Tong, K.I., Kockeritz, L.K., Wada, T., Joza, N.A., Crackower, M., Goncalves, J., Sarosi, I., Woodgett, J.R., Oliveira-dos-Santos, A.J., Ikura, M., van der Kooy, D., Salter, M.W., and Penninger, J.M., DREAM is a critical transcriptional repressor for pain modulation, *Cell,* 108(1), 31–43, 2002.
76. Lemonde, S., Turecki, G., Bakish, D., Du, L., Hrdina, P.D., Bown, C.D., Sequeira, A., Kushwaha, N., Morris, S.J., Basak, A., Ou, X.M., and Albert, P.R., Impaired repression at a 5-hydroxytryptamine 1A receptor gene polymorphism associated with major depression and suicide, *J Neuroscience,* 23(25), 8788–8799, 2003.

# 6 Serotonin 2A (5-HT$_{2A}$) Receptor Function: Ligand-Dependent Mechanisms and Pathways

*Ishier Raote,* Aditi Bhattacharya,* and Mitradas M. Panicker*

## CONTENTS

Localization of 5-HT$_{2A}$ mRNA and Receptor Protein in the Brain ..................... 106
Implication in Diseases ..................... 108
Intracellular Signaling Cascades Activated by the 5-HT$_{2A}$ Receptor ..................... 109
Activation Studies ..................... 109
Calcium Release Experiments ..................... 110
IP$_3$ Activation ..................... 111
5-HT$_{2A}$ Receptor Functional Selectivity ..................... 111
Intracellular Localization of the 5-HT$_{2A}$ Receptor ..................... 115
Intracellular Trafficking of the 5-HT$_{2A}$ Receptor ..................... 116
Long-Term Regulation of 5-HT$_{2A}$ Receptor ..................... 122
Conclusion ..................... 124
Acknowledgments ..................... 124
References ..................... 124

G-protein-coupled receptors or GPCRs have been a major area of study in the past decades. This is not surprising as this group, comprising a thousand-odd proteins, constitutes the largest group of signal transducers across the cell membrane and is also implicated in a vast number of diseases [1]. All major neuromodulators and some fast neurotransmitters act through either a single GPCR or a family of GPCRs, collectively termed a neuromodulator receptor family (such as the serotonin receptor

* Authors have contributed equally.

family). This dependence of proper brain function on GPCR signaling is reflected in the fact that most psychiatric diseases implicate one or many malfunctioning GPCRs in their pathophysiology. Consequently, drugs aimed at treating these disorders are primarily targeted to GPCRs.

The serotonin 2A receptor (5-$HT_{2A}$) has been implicated in mental disorders with complex etiologies that are still not clearly understood, in processes such as learning and memory, and also in neurogenesis. There are a large number of drugs targeted to this receptor. Though the receptor has been studied largely in relation to its multiple functions in the CNS, high levels of receptor expression in other areas such as the intestine, platelets and endothelial cells suggest that it could play crucial roles in other aspects of physiology. Research shows that some GPCRs, including the 5-$HT_{2A}$ receptor, exhibit critical differences in aspects of functional regulation from those seen in conventionally studied model GPCRs such as the β-2-adrenergic receptor. The receptor also couples to a number of intracellular signaling cascades, making it an important receptor to study. The 5-$HT_{2A}$ receptor could well serve as an important alternate paradigm in the study of GPCR function.

The 5-$HT_{2A}$ receptor was initially identified by hybridization using conserved elements of the cloned 5-$HT_{2C}$ receptor, followed by functional expression [2,3]. It bears strong similarity in primary sequence to the two other members of its subfamily, i.e., 5-$HT_{2B}$ and 5-$HT_{2C}$. All members of the 5-$HT_2$ receptor subfamily primarily couple to PLC on activation. As with most GPCRs, 5-$HT_{2A}$ functional regulation also involves desensitization and resensitization—regulatory processes that help prevent overstimulation and allow recuperation of signaling competence, respectively. Internalization and recycling of the receptor represent two processes that regulate this desensitization and resensitization. The receptor contains numerous recognition motifs for interacting partners to dock and facilitate receptor signaling, desensitization or trafficking. In some GPCRs, these processes are mediated by posttranslational modifications such as phosphorylation of one or more amino acid residues of the receptor. What makes these processes an area of consistent interest in 5-$HT_{2A}$ research is that drugs, psychotropic or therapeutic, modify many aspects of receptor functionality. Conversely, it has been suggested that the pathophysiology of psychiatric disorders is based on malfunctions in one or more aspects of GPCR function.

The importance of the role played by the 5-$HT_{2A}$ receptor in mediating CNS function can be seen by its considerable expression in various regions of the CNS and its wide-ranging effects.

## LOCALIZATION OF 5-$HT_{2A}$ MRNA AND RECEPTOR PROTEIN IN THE BRAIN

Initial localization studies of the 5-$HT_2$ class of receptors were based on radioactive ligand binding in tissues. These studies did not fully distinguish between various subtypes of 5-$HT_2$ receptors because sequence information was not available to design subtype-specific probes or generate subtype-specific antibodies. With the cloning and identification of 5-$HT_2$ receptor subtypes, initially from the rat and later

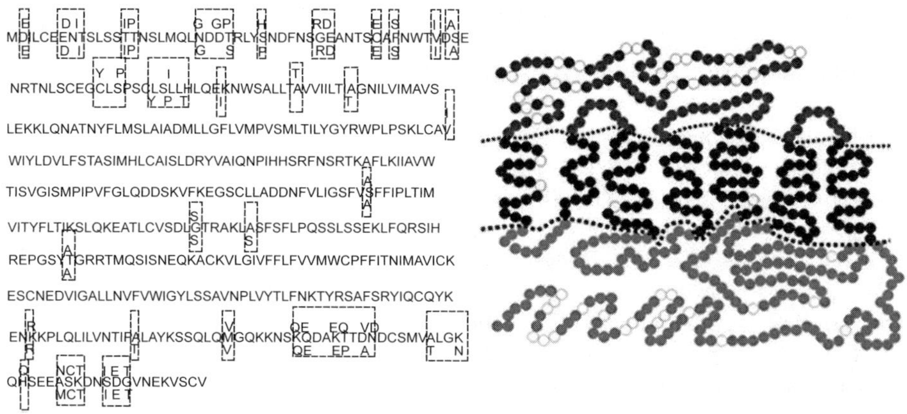


**FIGURE 6.1** Comparison of amino acid composition of the human, rat and mouse 5-HT$_{2A}$ receptors. The amino acid sequence of the human 5-HT$_{2A}$ receptor is shown with differences in amino acid residues of the rat and mouse receptors shown above and below the human sequence, respectively. The right panel is a topological depiction of the 5-HT$_{2A}$ receptor. Extracellular and intracellular domains are delineated by the dotted lines depicting the plasma membrane. Amino acid positions where there are differences among various isoforms have been depicted by unfilled circles.

other species (Figure 6.1), it became possible to determine subtype-specific localization of transcripts within the CNS. This could then be correlated with radioactive ligand binding in the same location. Development of subtype-specific antibodies has provided a detailed description of receptor localization as well as changes in levels in various experimental conditions.

The initial report of 5-HT$_{2A}$ receptor mRNA expression in the rat CNS was based on northern blots of RNA extracted from various regions of the CNS [2]. Further studies using *in situ* hybridization indicated high expression levels in layers 1, 4, and 5a of the cerebral cortex, the entorhinal and the piriform cortex. The olfactory bulb and some brainstem areas like the hypoglossal, pontine, motor trigeminal, and facial nuclei, also showed expression. Intermediate expressing areas were the limbic system and basal ganglia. These studies did not detect transcripts in the cerebellum, thalamic nuclei and found low expression in the hippocampus. In humans, no 5-HT$_{2A}$ receptor mRNA was detected in the striatum and cerebellum. In addition, 5-HT$_{2A}$ mRNA expression has been found in the dorsal horn of the spinal cord [4,5].

5-HT$_{2A}$ protein expression has been ascertained by a variety of immuno-histochemical techniques, coupled with light, fluorescence, and electron microscopy. In keeping with mRNA expression data, 5-HT$_{2A}$ protein is highly expressed in all layers of the cortex with layer 5 having the highest concentration. Pyramidal neurons of the frontal, insular, orbital, parietal, entorhinal, cingulate, perirhinal, piriform, insular, and deep layer of the cingulated cortex express 5HT$_{2A}$ receptors. In the cortex, the 5-HT$_{2A}$ is also found on certain astrocytes—an observation that has implicated glia in playing a role in schizophrenia [6,7]. In the basal ganglia and forebrain, the 5-HT$_{2A}$ has been immunologically localized to medium- and large-sized neurons in the lateral septal nuclei. In the hippocampus, the receptor is found in pyramidal cells in CA1-3 regions and granular cells of the dentate gyrus. 5-$HT_{2A}$ receptor immunoreactivity has also been shown in the developing cerebellum [8].

## Implication in Diseases

Given the extensive localization of this receptor to brain areas that mediate cognitive functions and social interaction, it suggests that the 5-$HT_{2A}$ receptor might be involved in diseases in which these functions are impaired. Disorders in which the 5-$HT_{2A}$ receptor seems to be involved range from schizophrenia, depression, obsessive compulsive disorder (OCD), and attention deficit–hyperactivity disorder (ADHD), to eating disorders such as anorexia nervosa, to autism spectrum disorders. Evidence to support such connections varies from genetic screens to binding and protein expression data and some molecular data.

*Schizophrenia*: 5-$HT_{2A}$ is thought to play a significant role in this disorder. This is partly due to the fact that most typical and almost all atypical antipsychotic drugs used in therapy, bind to 5-$HT_{2A}$ receptors, as do many hallucinogens, which mimic schizophrenia-like symptoms, bind to this receptor [9–11]. Another line of investigation has led to the hypothesis that 5-$HT_{2A}$, by modulating dopamine release in the striatum and cortex causes the motor and cognitive defects seen in schizophrenia [12]. Adding to these observations are postmortem studies that show altered 5-$HT_{2A}$ receptor expression in areas like the Brodmann area 9 of the cortex in schizophrenic patients [13] and increased expression in suicide cases diagnosed with schizoid symptoms. Genetic studies, however, have not yielded conclusive results; the promoter polymorphism 1438 bp upstream of the coding region involving a G/A change has been the focus of much debate, with some studies claiming a correlation to schizophrenia, increased aggression, etc., and others refuting the claim [14–16]. To date, no polymorphism in the 5-$HT_{2A}$ gene has been consistently associated with the disease or any of its symptoms across numerous studies and populations [17].

*Depression/anxiety*: It is known that long-term administration of tri-cyclic antidepressants causes a decrease in the cortical density of 5-$HT_{2A}$ receptors [18]. PET binding studies have reported a decrease in 5-$HT_{2A}$ levels in the hippocampus and platelets in patients with major depression [19]. It is also unclear whether there is a correlation between 5-$HT_{2A}$ receptor polymorphisms and the propensity or onset of the disease [20–22]. Recently, studies using a transgenic mouse with little or no expression of the 5-$HT_{2A}$ receptor in the brain indicate that it is required for modulation of anxiety-related behavior in mice [23]. 5-$HT_{2A}$ receptor agonists have been reported to result in significant but differential regulation of BDNF mRNA levels in the hippocampus and neocortex. BDNF mRNA expression has been reported to affect neurogenesis in the dentate gyrus and has bearing on the depressive phenotype [24].

*OCD/ADHD*: Patients with obsessive compulsive disorders show increased binding to 5-$HT_{2A}$ receptors in brain areas like the caudate nuclei [25].

Studies monitoring correlations between polymorphisms in the receptor as well as its promoter and disease onset have not yielded results that can be generalized across different ethnic groups [26]. 5-$HT_{2A}$ receptor levels seem unaffected in patients suffering from attention deficit hyperactivity disorder (ADHD) [27], yet the receptor polymorphisms may be a modulating factor in ADHD.

In addition, the 5-$HT_{2A}$ receptor has also been implicated in other diseases with less established etiologies. PET scans of patients suffering from *Anorexia nervosa* reveal decreased 5-$HT_{2A}$ receptor binding in the hippocampus, amygdala, and the cingulate cortex [28,29]. Autism and Asperger's syndrome have been associated with increased platelet serotonin levels and decreased 5-$HT_{2A}$ binding in cortical regions [30]. The receptor has also been implicated in the pathophysiology of diseases like bipolar disorders, Alzheimer's disease, progressive multifocal leukoencephalopathy, obstructive sleep apnea syndrome, and thermal or inflammatory pain [31–37].

In recent years, a handful of studies have attempted to integrate increased understanding about signaling pathways modulated by the receptor as well as its interacting partners and its possible roles in cellular and physiologic processes. Such studies will no doubt help elucidate the role 5-$HT_{2A}$ receptors play in the etiology of mental and peripheral disorders. A recent report identified the role of platelet-derived serotonin in mediating liver regeneration via the 5-$HT_{2A}$ receptor [38]. A signaling cascade linking the 5-$HT_{2A}$ receptor to cellular proliferation has also been recently delineated [39]. The involvement in platelet aggregation of the 5-$HT_{2A}$ receptor expressed in platelets has been widely investigated. In addition, there appears to be a role that the drosophila 5-$HT_2$ homolog plays in mediating adhesion in drosophila [40].

## INTRACELLULAR SIGNALING CASCADES ACTIVATED BY THE 5-$HT_{2A}$ RECEPTOR

Signaling pathways are fundamental mechanisms that form the basis of cell physiology and the relation of a cell to its environment. These pathways interact and are interconnected in networks that regulate most cellular functions. Signaling pathways are the focus of current research in both experimental and theoretical studies.

### ACTIVATION STUDIES

The 5-$HT_{2A}$ receptor activates PLC through $G_q$ and leads to an accumulation of $IP_3$, di-acylglycerol (DAG) and activation of protein kinase C (PKC) [41]. Increase in cytoplasmic $IP_3$ causes a release of calcium from intracellular endoplasmic reticulum stores—a characteristic activation signature of many GPCRs. This cascade has been the most studied and is perhaps the most important signal transduction pathway regulated by this receptor. Two assays associated with characterization of this pathway are discussed below. It has also been found that the 5-$HT_{2A}$ receptor activates other signal transduction cascades in a ligand-dependent manner. This ligand-specific functional activation of intracellular signaling cascades is discussed later.

## Calcium Release Experiments

Calcium release has been monitored by loading cells with $Ca^{2+}$ sensitive fluorescent dyes like Fura-2AM, fluo-3AM and rhod-2AM, to name a few. Descriptions of the large number and types of dyes available, labeling substrate molecules and calculating quanta of $Ca^{2+}$ release, are usually available with the manufacturers. Briefly, a $Ca^{2+}$ release experiment involves loading live cells with a $Ca^{2+}$ sensor dye in the form of a cell-permeable neutral AM (acetoxy-methoxy) ester. The dye is usually in solution in DMSO and is added along with a detergent to facilitate its entry into cells. Cellular esterases cleave the dye ester to a charged form, thus sequestering the dye in the cytoplasm. The ligand under study is then applied to cells, and fluctuations in $Ca^{2+}$ levels in the cells are measured. The dye alters its fluorescence intensity or causes a spectral shift in emission or excitation wavelength when bound to the released $Ca^{2+}$. Experimental considerations are:

- The choice of dye, i.e., its affinity to calcium and the changes expected.
- The nonuniformity of dye loading (dyes may be charged species that get sequestered selectively in some organelles exhibiting a potential difference across their membrane like Rhod-2AM in mitochondria).
- The type of the measurement desired, i.e., qualitative or quantitative. Ratiometric dyes are preferred for quantitative measurements as measurements with such dyes can be made independent of bleaching, uneven dye uptake, or leakage or variation in cell thickness.

Imaging in such experiments is also a crucial factor. As one has to contend with dye bleaching, accurate background corrections have to be obtained in images to ascertain precise amounts of fluorescence enhancement. In summary, once standardized, this technique has the potential to answer questions on signaling pathways activated by receptors, the time course of activation, ligand-specific variation in efficacy of receptor activation, and intracellular $Ca^{2+}$ fluctuations that might occur due to constitutive activity of the receptor [42–44]. It is also important to note that experimental observations are very sensitive to temperature. Other than temperature affecting the kinetics of release, heat stress also attenuates $Ca^{2+}$ mobilization. This temperature-sensitive attenuation of $Ca^{2+}$ release can be rescued by the expression of Hsp 70, possibly via a protease mechanism [45,46].

Studies have reported elevated platelet $Ca^{2+}$ levels upon activation of 5-$HT_{2A}$ receptors in patients with bipolar disorders. Such studies should help in understanding receptor dysfunction in depression and could be incorporated in designing patient-specific drug regimens [47–53].

New methods of visualizing intracellular calcium dynamics are also being introduced and should lead to increased sensitivity in studying spatiotemporal regulation of $Ca^{2+}$ in cells in response to receptor function [54].

5-$HT_{2A}$ activation has also been shown to effect $Ca^{2+}$ signaling in cells by regulating voltage-gated $Ca^{2+}$ channels. In acutely isolated prefrontal pyramidal cortical neurons, 5-$HT_{2A}$ receptors, via their classical pathway, activate calcineurin and inhibit activation of Cav1.2 L-type $Ca^{2+}$ currents. This modulation and its

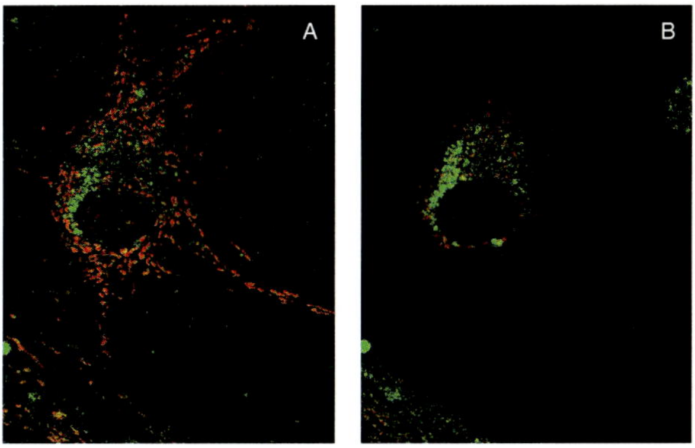


FIGURE 1.9

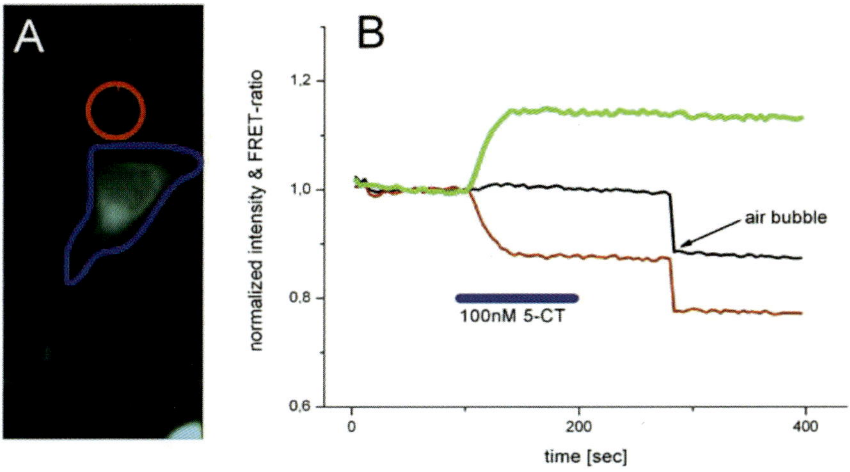


FIGURE 2.6

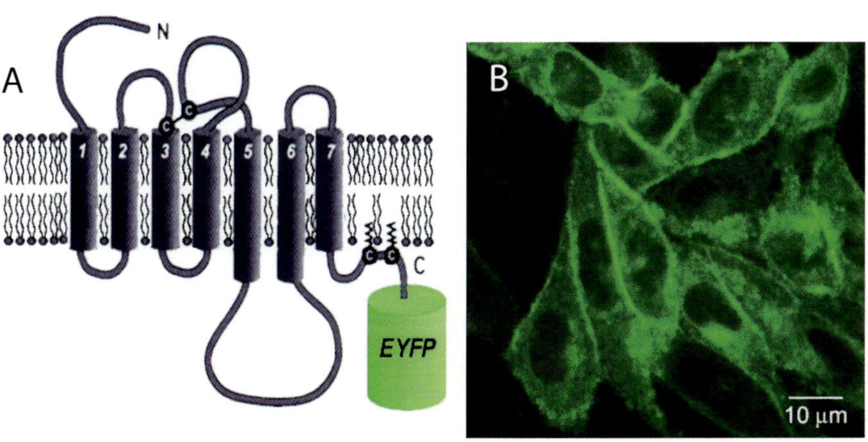


FIGURE 3.1

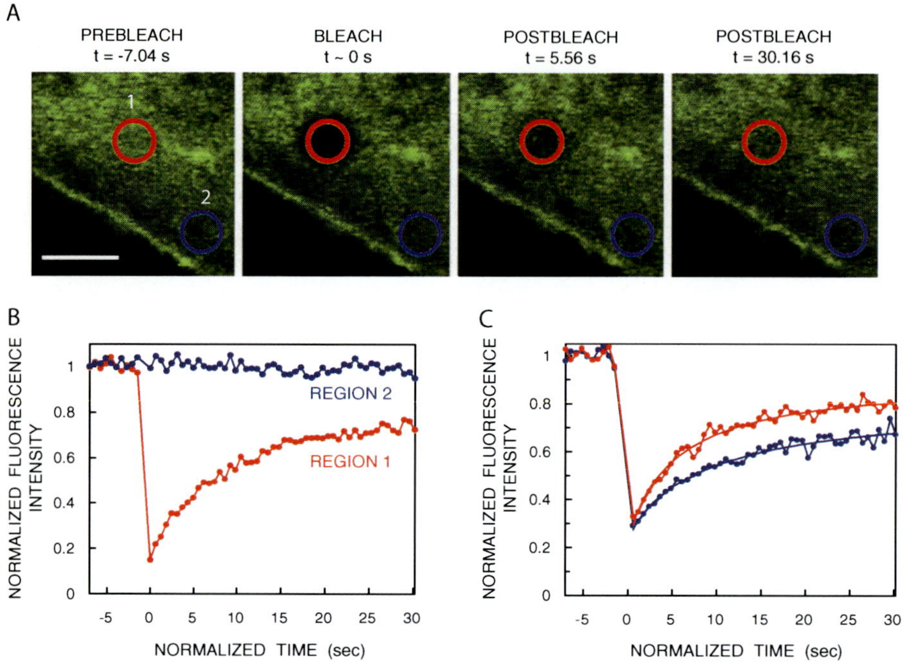


**FIGURE 3.3**

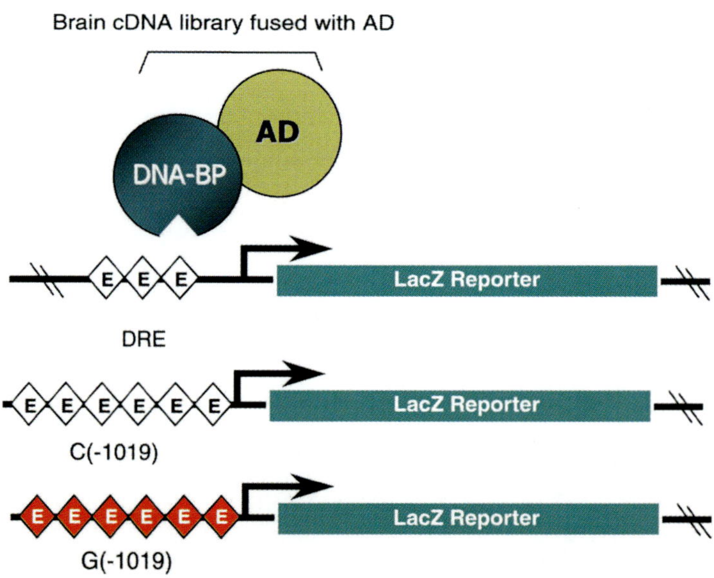


**FIGURE 5.2**

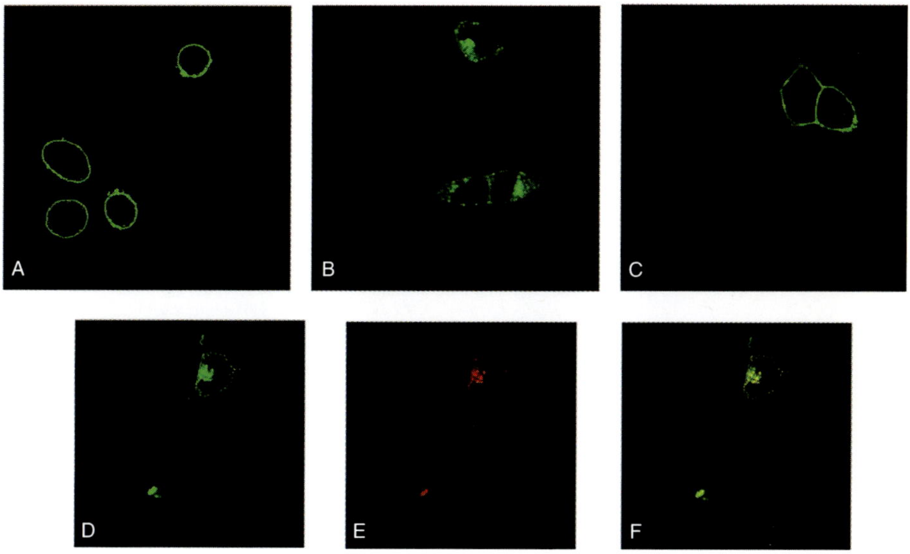


FIGURE 6.2

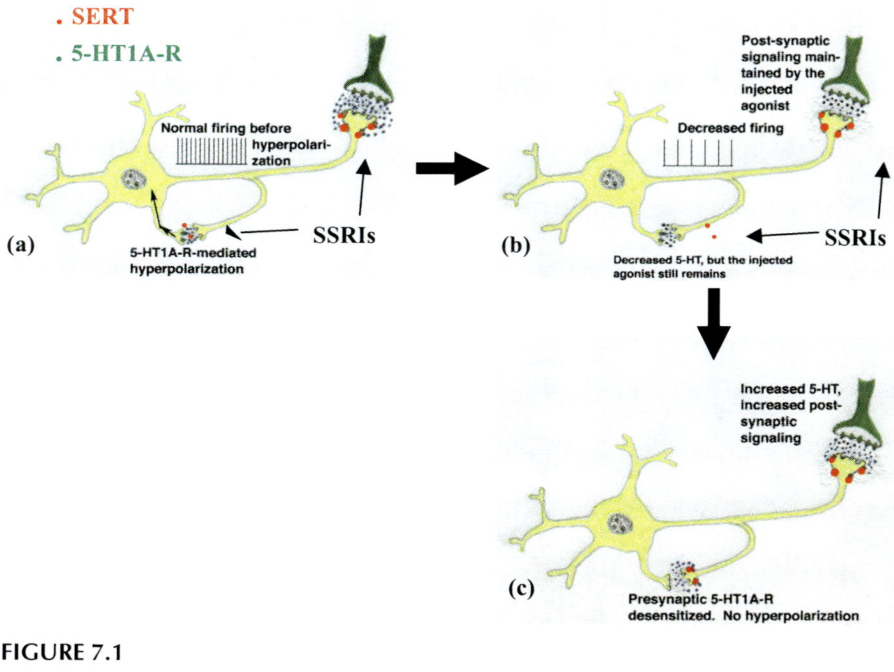


FIGURE 7.1

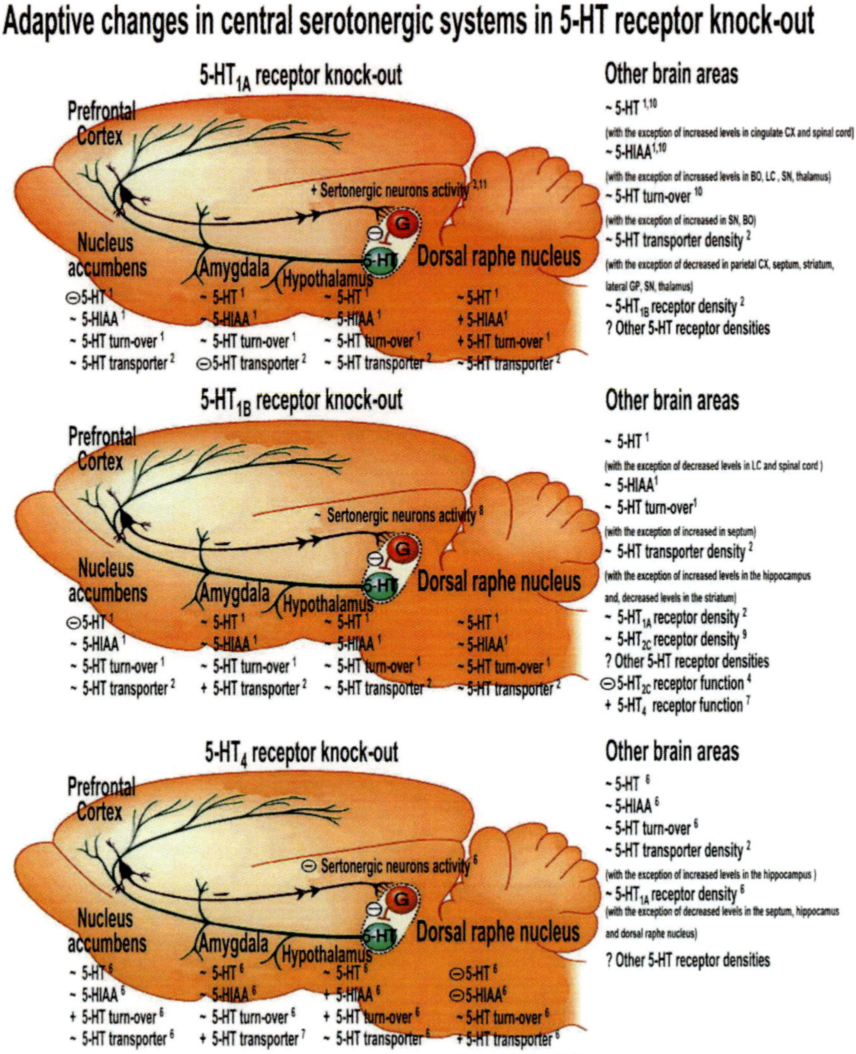

Adaptive changes in central serotonergic systems in 5-HT receptor knock-out
5-HT1A receptor knock-out
Prefrontal Cortex
+ Sertonergic neurons activity 3,11
Nucleus accumbens
Amygdala
Hypothalamus
5-HT
G
Dorsal raphe nucleus
⊖5-HT 1
~ 5-HIAA 1
~ 5-HT turn-over 1
~ 5-HT transporter 2
~ 5-HT 1
~ 5-HIAA 1
~ 5-HT turn-over 1
⊖5-HT transporter 2
~ 5-HT 1
~ 5-HIAA 1
~ 5-HT turn-over 1
~ 5-HT transporter 2
~ 5-HT 1
+ 5-HIAA 1
+ 5-HT turn-over 1
~ 5-HT transporter 2
Other brain areas
~ 5-HT 1,10
(with the exception of increased levels in cingulate CX and spinal cord)
~ 5-HIAA 1,10
(with the exception of increased levels in BO, LC, SN, thalamus)
~ 5-HT turn-over 10
(with the exception of increased in SN, BO)
~ 5-HT transporter density 2
(with the exception of decreased in parietal CX, septum, striatum, lateral GP, SN, thalamus)
~ 5-HT1B receptor density 2
? Other 5-HT receptor densities
5-HT1B receptor knock-out
Prefrontal Cortex
~ Sertonergic neurons activity 8
Nucleus accumbens
Amygdala
Hypothalamus
5-HT
G
Dorsal raphe nucleus
⊖5-HT 1
~ 5-HIAA 1
~ 5-HT turn-over 1
~ 5-HT transporter 2
~ 5-HT 1
~ 5-HIAA 1
~ 5-HT turn-over 1
+ 5-HT transporter 2
~ 5-HT 1
~ 5-HIAA 1
~ 5-HT turn-over 1
~ 5-HT transporter 2
~ 5-HT 1
~ 5-HIAA 1
~ 5-HT turn-over 1
~ 5-HT transporter 2
Other brain areas
~ 5-HT 1
(with the exception of decreased levels in LC and spinal cord)
~ 5-HIAA 1
~ 5-HT turn-over 1
(with the exception of increased in septum)
~ 5-HT transporter density 2
(with the exception of increased levels in the hippocampus and, decreased levels in the striatum)
~ 5-HT1A receptor density 2
~ 5-HT2C receptor density 9
? Other 5-HT receptor densities
⊖5-HT2C receptor function 4
+ 5-HT4 receptor function 7
5-HT4 receptor knock-out
Prefrontal Cortex
⊖ Sertonergic neurons activity 6
Nucleus accumbens
Amygdala
Hypothalamus
5-HT
G
Dorsal raphe nucleus
~ 5-HT 6
~ 5-HIAA 6
+ 5-HT turn-over 6
~ 5-HT transporter 6
~ 5-HT 6
~ 5-HIAA 6
~ 5-HT turn-over 6
+ 5-HT transporter 7
~ 5-HT 6
+ 5-HIAA 6
+ 5-HT turn-over 6
~ 5-HT transporter 6
⊖5-HT 6
⊖5-HIAA 6
~ 5-HT turn-over 6
+ 5-HT transporter 6
(with increased levels in mRNA 5-HT transporter)
Other brain areas
~ 5-HT 6
~ 5-HIAA 6
~ 5-HT turn-over 6
~ 5-HT transporter density 2
(with the exception of increased levels in the hippocampus)
~ 5-HT1A receptor density 6
(with the exception of decreased levels in the septum, hippocamus and dorsal raphe nucleus)
? Other 5-HT receptor densities


FIGURE 8.1

blockade by atypical neuroleptics are postulated to have wide-ranging effects on synaptic integration and long-term gene expression in the deep-layer prefrontal pyramidal regions of the brain [55]. In contrast, in astrocytes, 5-$HT_{2A}$ stimulation opens voltage-independent $Ca^{2+}$ channels, resembling depletion-operated calcium channels (DOCCs) [56].

### $IP_3$ Activation

Inositol 1,4,5-tris-phosphate accumulation by 5-$HT_{2A}$ receptor activation has also been used, not only to monitor and quantify receptor activation, but also to measure desensitization. $IP_3$ assays routinely measure phosphoinositol hydrolysis using cells preincubated with radioactive myo-inositol or inositol, which acts as the source of labeled $IP_3$. Following agonist or antagonist application, accumulation of total radiolabeled IP (inositol mono-phosphate, inositol bis-phosphate, and inositol tris-phosphate) is determined using ion exchange chromatography and scintillation counting. This assay not only provides quantitative estimates but also temporal data regarding receptor activation. The assay has yielded many insights into the mechanism of 5-$HT_{2A}$ desensitization and activation, as well as allowed identification of specific ligands against the 5-$HT_{2A}$ receptor [57–64]. In addition, many antagonists act as inverse agonists, inhibiting ligand-independent basal receptor activation. Decrease in basal levels of $IP_3$ thereby provides a measure of antagonist efficacy.

## 5-$HT_{2A}$ Receptor Functional Selectivity

Many GPCRs display a phenomenon whereby different ligands can differentially activate signaling pathways *via* the same receptor. This differential activation arises independently of differences in the affinity of the ligand to the receptor. Ligands often display differences in efficacy and/or potency at one signaling pathway vs. another. A number of terms have been coined to describe this phenomenon, of which *functional selectivity* is gaining acceptance [65].

The 5-$HT_{2A}$ receptor is one of the first receptors to have been characterized as displaying the phenomenon of functional selectivity [66,67]. This was first hypothesized on the basis of the observation that hallucinogenic effects of drugs such as LSD do not correlate with their activation of the $IP_3$/diacylglycerol pathway. The receptor has since been extensively studied and several conditions identified under which functional selectivity may be observed, and ligands that can bring about this selectivity have been characterized.

Stimulation of the 5-$HT_{2A}$ receptor leads to the production of at least three distinct biochemical signals, $IP_3$/diacylglycerol, arachidonic acid (AA), 2-arachidonylglycerol (2-AG), and the relative activation of these pathways varies with the ligand used [68]. By measuring simultaneously at least two of these pathways using radioactively labeled $IP_3$ and arachidonic acid (the signal transduction outputs of each of these pathways from the same cells) it has been seen that ligands vary in the efficacy with which they stimulate each of the different pathways.

More recently, identification of molecular players involved in the signaling cascade resulting in 5-$HT_{2A}$ receptor-induced arachidonic acid release, revealed that more than one signaling pathway is involved [69]. It raises the possibility that using a single readout such as arachidonic acid to characterize a biochemical pathway may not be sufficient. Different ligands may activate different pathways that finally impinge on the same molecule.

Numerous studies in different cell lines using antibodies to look at levels of molecular players have linked regulation of several signaling pathways to the 5-$HT_{2A}$ receptor. In addition to the three major pathways mentioned above, 5-$HT_{2A}$ stimulation can lead to a change in levels and/or activity of several molecular players including $PLA_2$, PLD, ERK1/2, nitric oxide, calmodulin, CREB, Akt, Fos, TGF-β, EGFR, and JAK/STATs [39,70–79], depending on cell line used and the context of stimulation. Some, if not all, of these pathways will undoubtedly be involved in functional selectivity displayed by ligands targeting the receptor. This leads to a situation of ever-increasing complexity wherein accurate characterization of receptor signaling will require simultaneous measurements of a number of key molecular players involved in each of these pathways.

In addition to these signaling pathways, receptor regulation also takes another form, i.e., internalization and recycling of the receptor. Interestingly, this pathway, too, is differentially regulated by different ligands.

Agonists such as 5-HT and those with partial efficacy on the same receptor, such as dopamine, bring about internalization of the rat 5-$HT_{2A}$ receptor, perhaps to different levels. A study performed in HEK293 cells using high resolution epifluorescence microscopy observed GFP-tagged receptors and their trafficking properties. Interestingly, it was observed that PKC activation is necessary for 5-HT-mediated internalization, whereas it is not required in dopamine-mediated internalization. Yet, both dopamine as well as 5-HT internalized receptors take the same amount of time to recycle to the cell surface. A further level of complexity is introduced when the receptor is primed toward dopamine-mediated internalization by low levels of 5-HT. Thus, two agonists (5-HT and dopamine) interact at the 5-$HT_{2A}$ receptor in a defined temporal sequence to bring about receptor internalization at concentrations lower than either of the two ligands could achieve individually [43,44].

Some antagonists (antipsychotic drugs) have also been reported to bring about an internalization of the receptor as well. The biochemical pathways stimulated by these ligands and the mechanisms involved in receptor internalization remain to be characterized [10,80,81].

Ligands differ in the extent to which they can stimulate signaling pathways as well as the biochemical pathways of receptor trafficking. It is conceivable that signaling pathways affected by the receptor will vary depending on its intracellular localization. This raises yet another issue in 5-$HT_{2A}$ signaling—it becomes necessary to consider subcellular localization at a time when the receptor is affecting a signaling pathway. The 5-$HT_{2A}$ receptor may, on internalization, enter a signaling endosome and, depending on the local environment in and around the endosome, affect signaling pathways in the cell. This has particular relevance in a polarized cell such as a neuron wherein the receptor may regulate different pathways in different subcellular locations.

Signaling could then be further affected by a number of factors including how long the receptor is retained in a milieu such as a signaling or recycling endosome.

It is foreseeable that ligands, after bringing about receptor internalization, will generate signals that alter the time taken for receptor recycling. This could change the time that the receptor spends in a subcellular environment, adding another dimension to the signaling. Thus ligand-specific modifications to the receptor may introduce yet another factor in receptor regulation of signaling and biochemical pathways.

A detailed study characterizing functional selectivity, not only among signaling pathways but also among trafficking pathways, is essential, and structural features of the receptor that allow for differential stimulation of individual pathways require identification. Interacting partners of the receptor under different conditions when stimulated by agonists and antagonists might yield crucial information. Possible posttranslational modifications such as ligand-specific phosphorylation of the receptor might well have a role to play in functional specificity.

One possible mechanism whereby this functional selectivity is regulated could be by different structural conformations attained by the receptor. Studies have shown different receptor reserves for separate signaling pathways [68]. This further confounds the issue by bringing into focus changes in signaling brought about by altering levels of receptor expression and differences in conformation and/or interacting partners that might bring about alterations in receptor reserve.

In some conformations, receptors may activate effectors in the absence of a stimulating ligand. This receptor activity is decreased by inverse agonists and increased by agonists. This realization led to the development of multistate models of receptor function in which receptors can spontaneously form a variety of "active" conformations that regulate effector mechanisms in the absence of a ligand. Results from these studies have been interpreted as demonstration of multiple populations of receptor, each with a distinct conformation, with each conformation differing in affinity to different ligands and in its ability to stimulate signaling pathways. A naturally occurring single nucleotide polymorphism (SNP) has been identified that causes a histidine to tyrosine (H452Y) change and brings about a destabilization of the signaling conformation of the receptor [82]. The mechanism by which this mutation acts might give insights into the normal functioning of the receptor. Other point mutations in the receptor that change some aspect of receptor signaling and/or trafficking have been identified. For instance, a single amino acid mutation from cysteine to lysine (C322K) in the sixth transmembrane region makes the receptor constitutively active in that it results in continual stimulation of at least one signaling pathway, the $IP_3$/diacylglycerol pathway [83]. In another exhaustive study, all serine and threonine residues in the protein were mutated in groups and the effects on signaling and desensitization of the receptor were characterized [84].

It is important to note that the activation of all these biochemical and signaling pathways is cell-type specific and care should be taken in choosing an appropriate system [57]. It could also be useful to identify commonalities of receptor signaling in different systems. At this point, comprehensive theoretical models of receptor signaling pathways might provide information at a system level that may not be arrived at experimentally.

There is a continuing need for a representation of signaling in dynamic compartments within a cell, with spatio-temporal information obtained from experiment. Such developments are essential for a quantitative understanding of how multiple functions of a cell are coordinated and regulated, and to evaluate the specifics of GPCR signaling.

All studies describing signaling pathways affected and their differential regulation by the 5-$HT_{2A}$ receptor have been carried out in cell culture. These studies are yet to be carried forward to studying signaling in an organism. For such studies perhaps drosophila or zebrafish may be useful initially as systems that allow easy genetic manipulation, and there exist a plethora of well-developed techniques available to study GPCR signaling and trafficking in these systems [85]. While using these organisms as model systems, in order to better understand the roles played by the receptor in mammals, it would also be important to look at both designed and naturally occurring mutants of this receptor and associated molecules.

## Probing 5-$HT_{2A}$ Oligomerization

GPCRs oligomerization has profound implications on GPCR-ligand interactions, receptor functionality and on cross-talk between different signaling systems. GPCR Homodimerization can bring about cooperativity between associating receptors, leading to substantially altered functional interactions between receptor and one or more ligands [44]. Oligomerization of the 5-$HT_{2A}$ receptor has not been demonstrated to date, though it has been shown that 5-$HT_{2C}$ signaling is regulated by homodimerization [133]. The possibility of 5-$HT_{2A}$ multimerization needs to be addressed as characterization of the oligomerization state of the receptor would add to our understanding of 5-$HT_{2A}$ function.

Traditionally, association has been determined by coimmunoprecipitation of two populations of receptor tagged with the two different epitopes. However, coimmunoprecipitation may be unable to distinguish between direct receptor interactions and indirect associations within a complex [134].

Two techniques that are now commonly used that allow real-time identification of interacting partners in live cells are fluorescence resonance energy transfer (FRET) [135,136] and bioluminescence resonance energy transfer or BRET [136–138]. FRET and BRET studies have the added advantage of being able to study complexes in different subcellular locations. Some studies have utilized all three methods FRET, BRET and coimmunoprecipitation to validate their hypothesis [133].

If receptor multimerization is confirmed, it would be of interest to check if receptor multimerization is constitutive, ligand-induced, or even ligand-specific [139,140]. Furthermore, the stoichiometry of a receptor multimer can be determined using cross-linkers to covalently associate receptor assemblies whose molecular size is subsequently determined by gel electrophoresis [141]. TIRF microscopy could also identify stoichiometry of receptor complexes at the cell membrane by looking at bleaching characteristics of fluorescent receptors in these complexes.

Subsequently, residues and regions of the receptor involved in multimerization can be identified by sequential mutation of residues or deletion or portions of the receptor [133,142–146]. This has also been approached either through a variety of bioinformatics techniques [147–154].

## INTRACELLULAR LOCALIZATION OF THE 5-HT$_{2A}$ RECEPTOR

Although existing studies on receptor distribution in tissues using a variety of tissue-based techniques have allowed determination of steady-state levels and localization of the receptor *in vivo*, studies in cell culture have been most successful in addressing mechanistic details of receptor function and intracellular distribution. As the distribution of 5-HT$_{2A}$ mRNA and radioligand binding indicate the 5-HT$_{2A}$ receptor to be present in the mammalian cortex, the rat prefrontal cortex has been the focus of considerable interest in looking at subcellular localization of 5-HT$_{2A}$ receptors. Localization of the receptor to the cortex also correlates with the effects of antagonists and inverse agonists on behavior and cognitive function. Light microscopy using antibodies shows the receptor to be localized to dendritic shafts of pyramidal and local circuit neurons. Electrophysiological studies predict presynaptic localization on glutamatergic cerebellar mossy fiber nerve terminals [86]; while neurochemical, immunoelectron microscopy and immunohistochemical studies suggest that 5-HT$_{2A}$ may be presynaptic on dopaminergic axon terminals in the ventral tegmental area [87–90].

Ultrastructural immuno-electron microscopy also demonstrates that receptors are localized primarily to the dendrites of neurons in the rat prefrontal cortex. Moreover, 5-HT$_{2A}$ receptors are expressed to a greater extent in apical dendrites than basilar dendrites [6].

Information on receptor targeting to various subcellular compartments on activation as well as inhibition of the receptor is available from both *in vivo* studies as well as cell cultures. Although immunohistochemistry has been useful for *in vivo* experiments, modification of the receptor with tags has been useful in cell culture studies. Somato-dendritic targeting of the receptor has been established from both *in vivo* and in vitro studies [91]. To study receptor localization in cell cultures *in vitro*, receptors fluorescently tagged (GFP) at the C-terminus or FLAG-tagged at the *N*-terminus of the receptor have been used [43]. Preliminary observations showed that receptor insertion into the plasma membrane remains unaffected using such tags [91–93]. On the other hand, *N*-terminal GFP tags have not been successful in obtaining receptor useful for experimental studies as plasma membrane targeting seems to be affected (unpublished data). Mutational analyses have also revealed the importance of an aspartate residue (Asp155), critical for targeting the receptor to the membrane [94].

However, it was discovered that GFP, by obscuring a potential PDZ binding domain at the C-terminus of the receptor, results in reduced targeting to the dendritic compartment in neurons. The 5-HT$_{2A}$ receptor is the one of the first GPCRs showing that the PDZ-binding domain may play a critical role in dendritic targeting.

In using tags to study targeting of the receptor, the receptor was targeted correctly to the dendrites if the GFP was shifted 20 amino acids upstream of the C-terminus within the gene to allow the PDZ-binding domain to remain exposed. This provides a caveat that the tag has to be appropriately located such that all motifs are maintained intact for appropriate receptor functioning [91].

Studying the subcellular localization of the receptor and its targeting to different compartments suggests that there exist within the protein, signals for subcellular targeting and trafficking that are yet to be characterized.

Antibodies against the 5-$HT_{2A}$ receptor are also widely used, with most studies using antibodies raised against an *N*-terminal portion of the receptor. Antibodies have been particularly useful for live staining cells expressing the receptor when the epitope is extracellular and the receptor is properly inserted into the cell membrane [93,95,96].

## INTRACELLULAR TRAFFICKING OF THE 5-$HT_{2A}$ RECEPTOR

GPCRs, in addition to initiating intracellular signal transduction cascades, also trigger cellular and molecular mechanisms leading to regulation of receptor signaling. Often, this is achieved by receptor phosphorylation, followed by internalization, wherein the receptor undergoes a series of structural and functional changes before being targeted to their final destination. This regulation of the receptor is brought about by interactions with regulatory molecules and trafficking to various subcellular compartments.

5-$HT_{2A}$ receptors are internalized *in vitro* and *in vivo* by several ligands including agonists as well as some antagonists [10,42–44]. 5-$HT_{2A}$ endocytosis has also been subverted by the human JCV virus, which utilizes 5-$HT_{2A}$ as a cellular receptor. Inhibiting receptor endocytosis reduces the risk of infection, arguing that some receptor antagonists that modulate receptor internalization, could be useful in treatment of progressive multifocal leukoencephalopathy caused by the JCV virus [33]. The biochemical mechanisms responsible for the regulation of 5-$HT_{2A}$ intracellular trafficking are largely unknown.

Internalization in HEK293 cells expressing rat 5-$HT_{2A}$ receptors begins within 2 min after adding 5-HT and is complete within 10 min [43]. Agonist-mediated internalization of the 5-$HT_{2A}$ receptor is seen to occur *via* clathrin-mediated endocytosis [97]. The process of internalization is also dynamin dependent. Dominant-negative mutant forms of clathrin and dynamin both inhibit receptor internalization. The receptor, on internalization is also seen to colocalize with other proteins such as the transferrin receptor (Figure 6.2), known to internalize using clathrin-dependent pathways [43].

Many — but not all — antagonists (including antipsychotic drugs) have been shown to bring about 5-$HT_{2A}$ receptor internalization without receptor activation. Studies both *in vivo* as well as *in vitro* have shown receptor redistribution in response to exposure to antagonists. Antagonist-mediated receptor internalization has also been shown to be *via* clathrin-mediated endocytosis in a dynamin-dependent manner. Incidentally, antagonist-mediated internalization of the rat 5-$HT_{2A}$ receptor, unlike serotonin-mediated internalization, is independent of protein kinase C (PKC) activation [98].

The rat 5-$HT_{2A}$ receptor is unusual among GPCRs, in that it is internalized and desensitized in a $\beta$-arrestin-independent manner. Receptor internalization remains

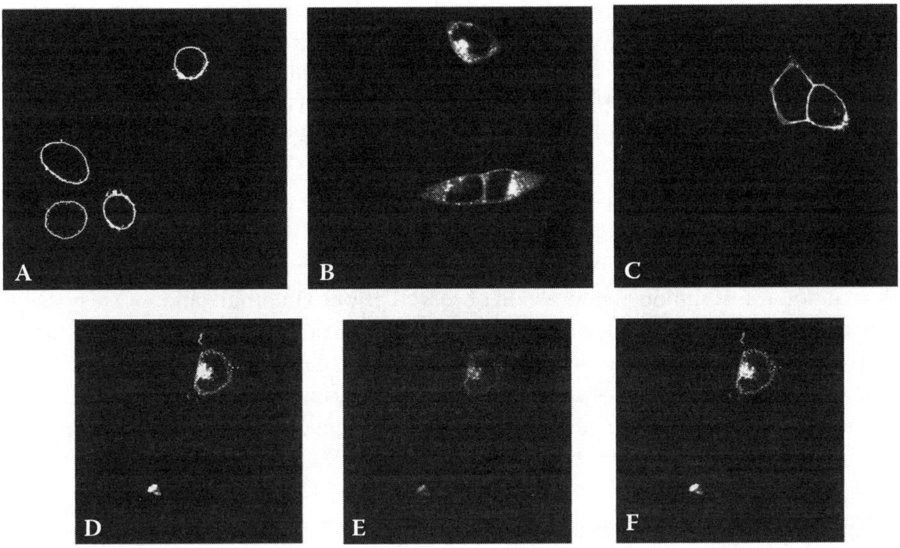


**FIGURE 6.2 (Color figure follows p. 110.)** Expression of 5-$HT_{2A}$ tagged to EGFP in HEK 293 cells. (A) HEK 293 cells expressing EGFP-tagged 5-$HT_{2A}$ receptor (5-$HT_{2A}$-EGFP) show cell surface-localized receptors after cycloheximide treatment. (B) Agonist-induced internalization of EGFP-tagged receptors. (C) Recycled receptors after agonist-mediated internalization. Panels D to F show colocalization of EGFP-tagged receptors with Alexa 568 tagged transferrin after agonist-mediated internalization. (D) 5-$HT_{2A}$-EGFP receptors, (E) Alexa 568-tagged transferrin, and (F) merged image.

unaffected when dominant-negative mutants of different forms of β-arrestin (Arr-2 and Arr-3) are transfected into cells expressing the rat 5-$HT_{2A}$ receptor [57]. Interestingly, though stimulation of the 5-$HT_{2A}$ receptor brings about arrestin-independent internalization, arrestin is sorted into intracellular compartments, distinct from those containing the 5-$HT_{2A}$ receptor. The molecular basis of arrestin-insensitivity of 5-$HT_{2A}$ receptor internalization is not known.

Agonist-induced internalization of 5-$HT_{2A}$ receptors is accompanied by differential sorting of Arr-2, Arr-3, and 5-$HT_{2A}$ receptors into distinct plasma membrane and intracellular compartments. Agonist-induced redistribution of Arr-2 and Arr-3 into intracellular vesicles distinct from those with the 5-$HT_{2A}$ receptor also implies novel roles for Arr-2 and Arr-3 [93]. In another interesting study, it was found that a constitutively active mutant form of arrestin-2 induces 5-$HT_{2A}$ internalization independent of stimulation by a ligand. This effect was not seen if wild-type arrestin is overexpressed in the same cells. Additionally, it was shown that constitutively active arrestin-2 effected a decrease in efficacy of agonist-induced PI hydrolysis with a simultaneous increase in agonist potency. This was explained by postulating that arrestin binds to and stabilizes a receptor conformation that exhibits a higher affinity for agonists, hence allowing for significant changes in properties of receptor trafficking as well as signaling [99]. These novel observations of arrestin function make the 5-$HT_{2A}$ receptor an important and interesting model system for the study of arrestin-GPCR interactions.

As discussed earlier, receptor activation and trafficking can be functionally selective in that different ligands can stimulate different biochemical pathways and signal transduction cascades. It was observed that agonists and antagonists display differential effects on binding to the receptor. Rat 5-$HT_{2A}$ receptor activation by agonists, but not antagonists, induces greater Arr-3 than Arr-2 translocation to the plasma membrane [93].

Arrestin-independent internalization is thought to arise because it is believed that the 5-$HT_{2A}$ receptor is not phosphorylated in response to stimulation by an agonist [100]. This would be unusual in GPCR endocytosis where receptor phosphorylation is a common phenomenon observed prior to internalization. One possible hypothesis is that ligand-specific phosphorylation or other such regulated posttranslational modifications may allow the 5-$HT_{2A}$ receptor to participate in different pathways and traffic differentially in response to various agonists and antagonists.

5-$HT_{2A}$ receptors have also been shown to recycle to the cell surface after activation and internalization by agonists. In HEK 293 cells expressing GFP tagged rat 5-$HT_{2A}$ receptors, activating the receptor with 5-HT or dopamine, brings about internalization followed by receptor recycling, with the entire process taking approximately 2.5 h. These studies provided strong visual evidence that the receptors can recycle to the cell surface after being internalized in the normal course of activation, as has been reported for many other GPCRs. Moreover, the lack of discernible receptors within cells, in the absence of agonist, suggests that receptors are not normally internalized and recycled to any detectable extent unless stimulated. Activation of PKC, in the absence of an agonist also causes at least a portion of the receptors present on the cell surface to internalize and subsequently recycle to the cell surface. The kinetics of internalization and recycling are similar to those seen after activation by 5-HT or dopamine (Figure 6.3). Internalization and recycling could be important processes in desensitization and resensitization of receptor signaling and continue to be areas of intense study [43,44].

A similar study on trafficking of the 5-$HT_{2A}$ receptor on exposure to antagonists is not yet published. It has been speculated that antagonist-internalized receptors are targeted to the lysosomal compartments for degradation, but this has not been proven. Differences in agonist- and antagonist-mediated internalization and recycling/degradation could explain some aspects of functional selectivity in receptor trafficking and signaling.

Studies reported so far have not looked in detail at the process of receptor internalization and the subcellular compartments that receptors localize to during trafficking. Data indicates that agonist-mediated internalization targets the receptor to a recycling endosome (Figure 6.2) [43,44]. The mechanistic details of 5-$HT_{2A}$ recycling and pathways followed during these processes remain to be addressed. The role played by the cytoskeleton at all stages of receptor trafficking and molecular players such as Rab GTPases, which should be involved in these intracellular processes are yet to be identified. How biochemical mechanisms involved in 5-$HT_{2A}$ recycling compare with those observed in other GPCRs also remain to be seen.

A number of studies on 5-$HT_{2A}$ receptor internalization and trafficking have used fluorescently-tagged receptor constructs or fusion proteins wherein the receptor is tagged to an epitope for which an antibody is available. This has allowed for direct

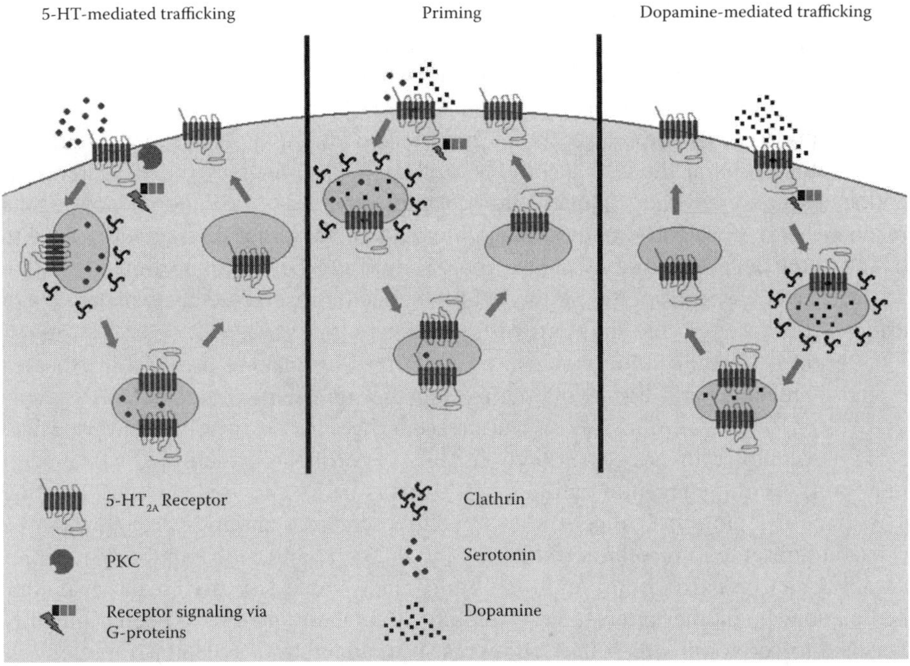


**FIGURE 6.3** Functional selectivity in the 5-$HT_{2A}$ receptor trafficking. Serotonin and dopamine bring about receptor internalization via clathrin and dynamin-dependent pathways. The receptor is more sensitive to dopamine-mediated internalization if serotonin is first added at concentrations that are subthreshold for serotonin-mediated receptor internalization. Biochemical pathways mediating 5-$HT_{2A}$ receptor internalization differ in that PKC activation is *necessary* for serotonin internalization, whereas it is *not involved* in dopamine-mediated internalization.

and occasionally real-time visualization of receptor trafficking in cell cultures and *in vivo* [43,44,93]. Similarly, intracellular transport of molecular players is also being studied for possible roles in functional interactions with the 5-$HT_{2A}$ receptor. In such experiments the proteins studied are often fluorescently-tagged, whereas the receptor is visualized using antibodies.

While studying trafficking using fluorescently tagged receptors, it is often crucial to differentiate between receptors that have been newly synthesized and are being trafficked to the cell membrane and those that are being trafficked in response to, say, a stimulus. In order to do so, cells can be treated with protein synthesis inhibitors such as cycloheximide to ensure that all of the receptors get to the cell surface before the stimulus is provided (Figure 6.2). Though cycloheximide may not affect receptor signaling [101], it is seen that extended treatment of cells with protein synthesis inhibitors proves toxic. If so, alternative methods of reducing protein synthesis can also be employed such as serum starvation. In many cell lines, this also forces cells to enter the $G_0$ phase of the cell cycle. This may serve to reduce cell-cycle dependent variations, if any, in experimental observations.

Alternatively, an inducible system could be used to circumvent the necessity of inhibition of protein synthesis. Photoactivatable GFP (PAGFP) could be ideal in following receptor trafficking over time as PAGFP-tagged receptors, once photoactivated can be chased with no new contribution to the fluorescence from newly synthesized receptors [102]. Other inducible systems activated, for example, by tetracycline or pronesterone may be useful [103].

Biotinylation of surface receptors is a method routinely used in determination of receptor levels at the cell surface as well as in quantitative studies of receptor endocytosis and recycling. This method depends on specific exposed residues of the receptor being covalently modified with biotin. Streptavidin is subsequently used to quantify the levels of biotinylated receptor. It would be useful to remember that artifacts could arise from the biotinylation of specific residues if they change properties of the receptor and interfere with subsequent processes.

Receptor internalization can also be visualized by making use of a fluorescent ligand as receptor and ligand internalize together and remain associated for some time [104,105]. Unfortunately, no fluorescent ligands are available to visualize 5-$HT_{2A}$ receptor trafficking. Antibody feeding experiments which bind to extracellular portions of the receptor partially evade this issue by allowing antibody-labeling of surface receptors in live cells after which the receptor-antibody complex can be chased to observe intracellular trafficking [43,44]. Imaging of the antibody-receptor complex can be carried only until the endosome, where low pH breaks apart this association. In another interesting development, quantum dots have been covalently attached to serotonin which may allow real-time imaging of receptor trafficking as long as receptor and ligand remain associated. As quantum dots are nontoxic and photostable they can be imaged using conventional fluorescence microscopy for an extended period [106,107].

Currently, information on 5-$HT_{2A}$ receptor trafficking has been largely derived from cell cultures. Considering the obvious limitations on extending cell culture studies to *in vivo* situations it would be useful to generate transgenic mice expressing fluorescently tagged or epitope-tagged 5-$HT_{2A}$ receptors. An important aspect of 5-$HT_{2A}$ function that has arisen out of these studies is that a number of observations are cell-type specific. Therefore, choosing an appropriate cell type for study becomes all the more crucial. Observations from such an *in vivo* system could be correlated with data obtained from cell lines. This would also allow simultaneous study over several cell types to carefully document cell-type specific characteristics of receptor trafficking. If receptor expression *in vivo* is also inducible it would add considerably more to our understanding of receptor trafficking lifetimes and region-specific behavior of the receptor. It would be also useful to generate knock-in transgenic mice expressing similar levels of modified receptor to that seen in the wild-type as well as use a model organism wherein the receptor is considerably over-expressed.

In addition, most studies have used the rat 5-$HT_{2A}$ receptor, and it is likely that important differences will arise when the human receptor is characterized. The few amino acid changes between the rat and human receptors may yield a whole new panoply of interacting proteins. Ligands will undoubtedly display completely altered efficacies and potencies at the rat and human receptors as is evident from ligand-binding studies with overexpressed receptors in cell lines [108]. Signaling and trafficking could turn out to be substantially different when using the human receptor as a model.

**TABLE 6.1**
**Proteins That Associate with the 5-$HT_{2A}$ Receptor**

| Interacting Protein | Gene | Region of the Receptor Involved in Interaction | Reference |
|---|---|---|---|
| Amyloid-b precursor protein intracellular domain associated protein-1a | AIDA-1A | i3 | 100 |
| Eukaryotic translation initiation factor 3, subunit 5 e | EIF3S5 | i3 | 100 |
| Neurotrophic tyrosine kinase, receptor, type 3 isoform c precursor | NTRK3 | i3 | 100 |
| Melanoma-associated antigen | MAAT1 | i3 | 100 |
| Paraoxonase 2 | PON2 | i3 | 100 |
| Microtubule associated protein 1A | MAP-1A | i3 | 96,100 |
| Ribosomal protein S6 kinase 2 | RSK2 | i3 | 100 |
| Nucleoside-diphosphate kinase 3 | NME3 | i3 | 100 |
| NADH dehydrogenase (ubiquinone) 1 b subcomplex | NDUFB10 | i3 | 100 |
| Protein phosphatase 5, catalytic subunit | PPP5C | i3 | 100 |
| Glutamine synthetase | GLUL | i3 | 100 |
| ADP ribosylation factor 1 | Arf1 | C-terminus | 128 |
| Post synaptic density protein-95 | PSD-95 | C-terminus (PDZ-binding domain) | 92 |
| Antioxidant protein-2 | AOP-2 | C-terminus (PDZ-binding domain) | 129 |
| Activin receptor-interacting protein 1 | ARIP-1 | C-terminus (PDZ-binding domain) | 129 |
| Synapse-associated protein 97 | SAP97 | C-terminus (PDZ-binding domain) | 129 |
| MAGUK p55 subfamily member-3 | MPP-3 | C-terminus (PDZ-binding domain) | 129 |
| Channel-interacting PDZ domain protein | CIPP | C-terminus (PDZ-binding domain) | 129 |
| Filamin c | | Not known | 130 |
| Receptor for activated protein kinase C | RACK1 | C-terminus | 130 |
| Calmodulin | CAM | i2, C-terminus | 78 |
| Arrestin | ARRB2 | i3 | 131 |
| Caveolin-1 | CAV-1 | Not known | 132 |

## LONG-TERM REGULATION OF 5-$HT_{2A}$ RECEPTOR

Long-term changes in 5-$HT_{2A}$ receptor protein and mRNA expression have been the focus of many studies in the last decade. These time scales are of relevance as drugs that target the receptor are often administered for weeks before their effects are evident. Earlier studies looked at receptor protein levels whereas the advent of more sensitive techniques such as the ribonuclease protection assay (RPA) and reverse transcriptase polymerase chain reaction (RT-PCR) have allowed accurate estimation of receptor mRNA levels.

Long-term regulation of receptor levels in tissues has been monitored by radioligand binding assays. These assays initially involved the use of membrane fractions from tissue samples that express the receptor. Subsequently, cell lines that overexpress the receptor have been used. Receptor levels are determined by measuring binding of a range of concentrations of radioactive ligand to known amounts of the receptor in cell or tissue membranes. The membranes are washed free of excess ligand either by filtration or centrifugation and levels of the radioactive bound ligand determined. The concentration of radio-ligand that produces half-maximal occupancy represents $K_d$, the dissociation constant of the ligand to the receptor [109]. $B_{max}$, a measure of the maximum values of bound ligand, would represent the availability of receptors. Receptor up-regulation and down-regulation cause an increase and reduction in $B_{max}$, respectively, leaving $K_d$ unaltered. The assay is performed under various conditions when looking at regulation of levels of receptor protein.

Designing the assay to have the ligand compete with an already-bound radio-labeled ligand of known affinity would allow for the measurement of relative affinity between various ligands for the receptor. Such competitive binding experiments are helpful in the determination of affinity of drugs to a receptor and also determine variation in affinities among mutant forms of the receptor or its homologues in different species.

Positron emission tomography (PET) is a useful and powerful method to studying receptor levels *in vivo*, particularly in humans in real time. PET studies have also been carried out in rodent models. Radioligands that are often used in these studies are [F18] Altanserin and [C11] MDL 100907 labeled, which have been administered to patients that are either drug-naïve or have undergone treatment. Such studies have shown altered receptor densities in people with enhanced risk of schizophrenia [110], anorexia, and bulimia nervosa [111], and in untreated patients of obsessive compulsive disorders [25]. These methods have also been employed to examine the effectiveness of a certain treatment regime in altering receptor density in a disease [112,113].

mRNA levels of the 5-$HT_{2A}$ receptor have been monitored by *in situ* hybridizations, ribonuclease protection assays, and real time RT-PCR. *In situ* hybridizations have been carried out on brain slices or cell cultures and involve hybridization of a radioactive or nonradioactively labeled antisense probe to the RNA species present in cells in the tissue or in culture. Although levels of mRNA can be estimated using autoradiography with radiolabeled probes, nonradioactive *in situ* hybridizations provide better cellular resolution though changes in transcript levels are not as easily determined [114,115]. Ribonuclease protection assays offer a quantitative and sensitive approach to determine levels of a specific transcript and have been successfully used to determine 5-$HT_{2A}$ transcript levels. In this method, an antisense radiolabeled RNA probe specific to the transcript, is hybridized with RNA extracted from the tissue of interest and then treated with single-stranded specific ribonuclease $T_1$. The presence of the transcript and its hybridization to the probe results in the probe being protected from degradation by RNase $T_1$. Protected probe molecules are visualized and quantified after gel electrophoresis. This method allows quantification of minor variations in RNA levels, at the cost of spatial information. Such a study has yielded information on DOI-mediated long-term changes in 5-$HT_{2A}$ mRNA levels in the prefrontal cortex [116].

PCR based studies have also been used extensively in characterization of 5-$HT_{2A}$ expression and regulation. From identifying expression of receptor transcripts in the dorsal root ganglion in humans [117], understanding the imprinting status of the gene in the human brain [118] to associating receptor polymorphisms in humans with a predisposition toward seasonal affective disorders [119]. Commonly used PCR-based techniques such as reverse-transcriptase polymerase chain reaction (RT-PCR) and quantitative real-time PCR (qRT-PCR) have been used extensively to estimate receptor mRNA levels under various conditions. The two methods are based on mRNA extraction and subsequent generation of cDNA from samples using reverse transcriptase. In RT-PCR studies gene primers designed to amplify a portion of the cDNA containing the receptor sequence are used to determine the presence of mRNA of the gene of interest. RT-PCRs have been employed in examining agonist-induced receptor down-regulation in NIH 3T3 cells and Balb/c-3T3 cells [120,121], identifying neurons that respond to long-term treatment with clozapine by single-cell RT-PCR of rat cortical neurons [122] and mapping the effect of *N*-methyl norsalsolinol on the expression of 5-$HT_{2A}$ receptor transcripts in the caudate nucleus of rats [123]. qRT-PCR is a more accurate method and estimates the number of copies of a gene in the cDNA (and hence mRNA) present in a sample from the kinetics of amplification. qRT-PCR is a far more sensitive method and can be used in determining subtle changes in transcript status.

Microarray analyses have also been utilized by some groups to assess global changes in transcripts upon long-term stimulation or inhibition of the 5-$HT_{2A}$ receptor. An interesting study coupled data from microarray experiments and qRT-PCRs to monitor differential regulation of the cortical transcriptome in response to hallucinogenic vs. nonhallucinogenic drug stimulation of the 5-$HT_{2A}$ receptor using transgenic mice that expressed undetectable levels of the receptor in the cortex [124].

A striking observation that has arisen through studies on 5-$HT_{2A}$ receptor levels is that chronic application of agonists like DOI, 5-HT and LSD, as well as antagonists like mianserin, ketanserin, and pipamperone, cause down-regulation of the receptor. The biochemical mechanisms behind this "paradoxical regulation" of the 5-$HT_{2A}$ receptor are unknown. Agonist-induced long-term regulation has been examined in numerous studies. *In vitro* studies report variable results depending on the cell type used with up-regulation shown in MDCK cells, no difference observed in NIH 3T3 cells, and down-regulation in AtT-20 cells [125]. *In vivo* studies, on the other hand, have consistently reported receptor down-regulation on chronic agonist application [116,126,127].

With the exception of SR 46349B, all currently available 5-HT$_{2A}$ antagonists cause down-regulation of the receptor. It is postulated that antagonist-mediated 5-HT$_{2A}$ receptor down-regulation is brought about by receptor internalization followed by lysosomal degradation of internalized receptors. It has also been suggested that this down-regulation may be the way that antipsychotics wield their therapeutic effect [60,80].

## CONCLUSION

In this chapter we have attempted to describe the serotonin 2A (5-HT$_{2A}$) receptor as a model of GPCR function. We have tried to bring out complexities of signaling paradigms, interacting partners, and characteristics of the receptor in response to different agents/ligands. In keeping with the aim of the book, we have tried to give the reader an understanding of the techniques and methods used in the field to answer the questions being posed. Wherever possible, we have tried to discuss the pros and cons of the method over others that could have been used. In addition, we also suggest some avenues of future research to better understand the role played by the 5-HT$_{2A}$ receptor in mediating CNS functioning, particularly those affected in diseases like schizophrenia, bipolar disorders, depression, and anxiety.

## ACKNOWLEDGMENTS

We would like to acknowledge past and present members of our laboratory, particularly Samarjit Bhattacharyya for valuable discussions. This work was supported by the National Centre for Biological Sciences, Tata Institute of Fundamental Research, Bangalore. A.B. is a recipient of the Kanwal Rekhi Career Development Fellowship from the TIFR endowment fund.

## REFERENCES

1. Marinissen, M. and Gutkind, J., G-protein-coupled receptors and signaling networks: emerging paradigms, *Trends Pharmacol Sci*, 22(7), 368–376, 2001.
2. Julius, D. et al., The 5HT2 receptor defines a family of structurally distinct but functionally conserved serotonin receptors, *Proc Natl Acad Sci USA*, 87(3), 928–932, 1990.
3. Pritchett, D.B. et al., Structure and functional expression of cloned rat serotonin 5HT-2 receptor, *EMBO J*, 7(13), 4135–4140, 1988.
4. Mengod, G. et al., 5-Hydroxytryptamine receptor histochemistry: comparison of receptor mRNA distribution and radioligand autoradiography in the brain, *Serotoninergic Neurons and 5-HT Receptors in the CNS*, Göthert, M. and Baumgarten, H.G., Eds., Springer Verlag, Heidelberg, 1999.
5. Fonseca, M.I. et al., Distribution of serotonin 2A, 2C and 3 receptor RNA in spinal cord and medulla oblongata, *Mol Brain Res*, 89, 11–19, 2001.
6. Miner, L.A.H. et al., Ultrastructural localization of serotonin2a receptors in the middle layers of the rat prelimbic prefrontal cortex, *Neuroscience*, 116, 107–117, 2003.

7. Xu, T. and Pandey, S.C., Cellular localization of serotonin2A (5HT2A) receptors in the rat brain, *Brain Res Bull*, 51(6), 499–505, 2000.
8. Maeshima, T. et al., Serotonin2A receptor-like immunoreactivity in rat cerebellar Purkinje cells, *Neurosci Lett*, 252(1), 72–74, 1998.
9. Kasper, S. et al., Efficacy and safety of aripiprazole vs. haloperidol for long-term maintenance treatment following acute relapse of schizophrenia, *Int J Neuropsychopharmacol*, 6(4), 325–337, 2003.
10. Gray, J.A. and Roth, B.L., Paradoxical trafficking and regulation of 5-HT(2A) receptors by agonists and antagonists, *Brain Res Bull*, 56(5), 441–451, 2001.
11. Meltzer, H.Y. and McGurk, S.R., The effects of clozapine, risperidone, and olanzapine on cognitive function in schizophrenia, *Schizophr Bull*, 25(2), 233–255, 1999.
12. Holmes, A., Lachowicz, J.E., and Sibley, D., Phenotypic analysis of dopamine receptor knockout mice; recent insights into the functional specificity of dopamine receptor subtypes, *Neuropharmacology*, 47, 1117–1134, 2004.
13. Matsumoto, I. et al., 5-HT2A and muscarinic receptors in schizophrenia: a postmortem study, *Neurosci Lett*, 379(3), 164–168, 2005.
14. Li, J. et al., Contribution of 5-HT2A receptor gene -1438A>G polymorphism to outcome of attention-deficit/hyperactivity disorder in adolescents, *Am J Med Genet B Neuropsychiatr Genet*, 141(5), 473–476, 2006.
15. Hamdani, N. et al., Negative symptoms of schizophrenia could explain discrepant data on the association between the 5-HT2A receptor gene and response to antipsychotics, *Neurosci Lett*, 377(1), 69–74, 2005.
16. Ertugrul, A. et al., No association of the T102C polymorphism of the serotonin 2A receptor gene (HTR2A) with suicidality in schizophrenia, *Schizophr Res*, 69(2–3), 301–305, 2004.
17. Basile, V.S. et al., Lack of association between serotonin-2A receptor gene (HTR2A) polymorphisms and tardive dyskinesia in schizophrenia, *Mol Psychiatr*, 6(2), 230–234, 2001.
18. Peroutka, S. and Snyder, S., Long-Term Antidepressant Treatment Decreases Spiroperidol-Labeled Serotonin Receptor Binding, *Science*, 210(4465), 88–90, 1980.
19. Mintun, M.A. et al., Decreased hippocampal 5-HT2A receptor binding in major depressive disorder: *in vivo* measurement with [18F]altanserin positron emission tomography, *Biol Psychiatr*, 55(3), 217–224, 2004.
20. Minov, C. et al., Serotonin-2A-receptor and -transporter polymorphisms: lack of association in patients with major depression, *Neurosci Lett*, 303(2), 119–122, 2001.
21. Khait, V.D. et al., Association of serotonin 5-HT2A receptor binding and the T102C polymorphism in depressed and healthy Caucasian subjects, *Neuropsychopharmacology*, 30(1), 166–172, 2005.
22. Choi, M.J. et al., Serotonin receptor 2A gene polymorphism (-1438A/G) and short-term treatment response to citalopram, *Neuropsychobiology*, 52(3), 155–162, 2005.
23. Weisstaub, N.V. et al., Cortical 5-HT2A Receptor Signaling Modulates Anxiety-Like Behaviors in Mice, *Science*, 313(5786), 536–540, 2006.
24. Vaidya, V.A. et al., 5-HT2A receptor-mediated regulation of brain-derived neurotrophic factor mRNA in the hippocampus and the neocortex, *J Neurosci*, 17(8), 2785–2795, 1997.
25. Adams, K.H. et al., Patients with obsessive-compulsive disorder have increased 5-HT2A receptor binding in the caudate nuclei, *Int J Neuropsychopharmacol*, 8(3), 391–401, 2005.
26. Tot, S. et al., T102C and -1438 G/A polymorphisms of the 5-HT2A receptor gene in Turkish patients with obsessive-compulsive disorder, *Eur Psychiatr*, 18(5), 249–254, 2003.

27. Pornnoppadol, C. et al., No difference between platelet serotonin--5-HT(2A) receptors from children with and without ADHD, *J Child Adolesc Psychopharmacol*, 9(1), 27–33, 1999.
28. Frank, G.K. et al., Reduced 5-HT2A receptor binding after recovery from anorexia nervosa, *Biol Psychiatr*, 52(9), 896–906, 2002.
29. Audenaert, K. et al., Decreased 5-HT2a receptor binding in patients with anorexia nervosa, *J Nucl Med*, 44(2), 163–169, 2003.
30. Murphy, D.G. et al., Cortical serotonin 5-HT2A receptor binding and social communication in adults with Asperger's syndrome: an *in vivo* SPECT study, *Am J Psychiatr*, 163(5), 934–936, 2006.
31. Robertson, E. et al., No association between two polymorphisms at the 5HT2A gene and bipolar affective puerperal psychosis, *Acta Psychiatr Scand*, 108(5), 387–391, 2003.
32. Lam, L.C. et al., 5-HT2A T102C receptor polymorphism and neuropsychiatric symptoms in Alzheimer's disease, *Int J Geriatr Psychiatr*, 19(6), 523–526, 2004.
33. Elphick, G.F. et al., The Human Polyomavirus, JCV, Uses Serotonin Receptors to Infect Cells, *Science*, 306(5700), 2004, pp. 1380–1383.
34. Lai, M.K. et al., Loss of serotonin 5-HT2A receptors in the postmortem temporal cortex correlates with rate of cognitive decline in Alzheimer's disease, *Psychopharmacology (Berl)*, 179(3), 673–677, 2005.
35. Nishiyama, T., Effects of a 5-HT2A receptor antagonist, sarpogrelate on thermal or inflammatory pain, *Eur J Pharmacol*, 516(1), 18–22, 2005.
36. Bayazit, Y.A. et al., Association of the -1438G/A polymorphism of the 5-HT2A receptor gene with obstructive sleep apnea syndrome, *ORL J Otorhinolaryngol Relat Spec*, 68(3), 123–128, 2006.
37. Filippini, N. et al., Influence of serotonin receptor 2A His452Tyr polymorphism on brain temporal structures: a volumetric MR study, *Eur J Hum Genet*, 14(4), 443–449, 2006.
38. Lesurtel, M. et al., Platelet-Derived Serotonin Mediates Liver Regeneration, *Science*, 312(5770), 104–107, 2006.
39. Gooz, M. et al., 5-HT2A receptor induces ERK phosphorylation and proliferation through ADAM-17 tumor necrosis factor-alpha-converting enzyme (TACE) activation and heparin-bound epidermal growth factor-like growth factor (HB-EGF) shedding in mesangial cells, *J Biol Chem*, 281(30), 21004–21012, 2006.
40. Colas, J. et al., Serotonin synchronises convergent extension of ectoderm with morphogenetic gastrulation movements in Drosophila, *Mech Dev*, 87(1–2), 77–91, 1999.
41. Hoyer, D. et al., International union of pharmacology classification of receptors for 5-hydroxytryptamine (Serotonin), *Pharmacol Rev*, 46(2), 157–203, 1994.
42. Ohkura, M., et al., Insulin induces internalization of the 5-HT2A receptor expressed in HEK293 cells, *Eur J Pharmacol*, 518(1), 18–21, 2005.
43. Bhattacharyya, S. et al., Internalization and recycling of 5-HT2A receptors activated by serotonin and protein kinase C-mediated mechanisms, *Proc Natl Acad Sci USA*, 99(22), 14470–14475, 2002.
44. Bhattacharyya, S. et al., Activation, internalization and recycling of the Serotonin 2A receptor by dopamine, *PNAS*, 103(41), 15248–15253, 2006.
45. Jagadeesh, S.R. and Subhash, M.N., Effect of antidepressants on intracellular Ca++ mobilization in human frontal cortex, *Biol Psychiatr*, 44(7), 617–621,1998.
46. Kagaya, A. et al., Effect of heat stress on serotonin-2A receptor-mediated intracellular calcium mobilization in rat C6 glioma cells, *J Neural Transm*, 107(8–9), 919–929, 2000.

47. Yamawaki, S. et al., Enhanced calcium response to serotonin in platelets from patients with affective disorders, *J Psychiatr Neurosci*, 21(5), 321–324, 1996.
48. Fowler, C.J. and Tiger, G., Serotonin-stimulated calcium responses in human platelets: assay buffer dependency, *Cell Signal*, 10(8), 561–568, 1998.
49. Okamoto, Y. et al., The relationship of the platelet 5-HT-induced calcium response to clinical symptoms in eating disorders, *Psychopharmacology (Berl)*, 142(3), 289–294, 1999.
50. Tomiyoshi, R. et al., Serotonin-induced platelet intracellular $Ca^{2+}$ responses in untreated depressed patients and imipramine responders in remission, *Biol Psychiatr*, 45(8), 1042–1048, 1999.
51. Fowler, C.J., Sjoberg, E., and Tiger, G., Serotonin stimulation of calcium mobilisation in human platelets: choice of units of measurement, effects of age and tobacco use, and correlation with serotonin2A receptor density, *Clin Chim Acta*, 287(1–2), 1–18, 1999.
52. Plein, H. and Berk, M., Changes in the platelet intracellular calcium response to serotonin in patients with major depression treated with electroconvulsive therapy: state or trait marker status, *Int Clin Psychopharmacol*, 15(2), 93–98, 2000.
53. Suzuki, K. et al., Serotonin-induced platelet intracellular calcium mobilization in various psychiatric disorders: is it specific to bipolar disorder?, *J Affect Disord*, 64(2–3), 291–29, 62001.
54. George, S.E. et al., A high-throughput glow-type aequorin assay for measuring receptor-mediated changes in intracellular calcium levels, *Anal Biochem*, 286(2), 231–237, 2000.
55. Day, M. et al., Stimulation of 5-HT(2) receptors in prefrontal pyramidal neurons inhibits Ca(v)1.2 L type Ca(2+) currents via a PLCbeta/IP3/calcineurin signaling cascade, *J Neurophysiol*, 87(5), 2490–2504, 2002.
56. Hagberg, G.B. et al., Stimulation of 5-HT2A receptors on astrocytes in primary culture opens voltage-independent $Ca^{2+}$ channels, *Neurochem Int*, 32(2), 153–162, 1998.
57. Gray, J.A. et al., Cell-type specific effects of endocytosis inhibitors on 5-hydroxytryptamine(2A) receptor desensitization and resensitization reveal an arrestin-, GRK2-, and GRK5-independent mode of regulation in human embryonic kidney 293 cells, *Mol Pharmacol*, 60(5), 1020–1030, 2001.
58. Berg, K.A. et al., Differences in rapid desensitization of 5-hydroxytryptamine2A and 5-hydroxytryptamine2C receptor-mediated phospholipase C activation, *J Pharmacol Exp Ther*, 299(2), 593–602, 2001.
59. Pauwels, P.J., Van Gompel, P., and Leysen, J.E., Rapid desensitization and resensitization of 5-HT2 receptor mediated phosphatidyl inositol hydrolysis by serotonin agonists in quiescent calf aortic smooth muscle cells, *Life Sci*, 47(22), 2009–2019, 1990.
60. Van Oekelen, D., Luyten, W.H., and Leysen, J.E., 5-HT2A and 5-HT2C receptors and their atypical regulation properties, *Life Sci*, 72(22), 2429–2449, 2003.
61. Grotewiel, M.S., Chu, H., and Sanders-Bush, E., m-chlorophenylpiperazine and m-trifluoromethylphenylpiperazine are partial agonists at cloned 5-HT2A receptors expressed in fibroblasts, *J Pharmacol Exp Ther*, 271(2), 1122–1126, 1994.
62. Egan, C.T., Herrick-Davis, K., and Teitler, M., Creation of a constitutively activated state of the 5-hydroxytryptamine2A receptor by site-directed mutagenesis: inverse agonist activity of antipsychotic drugs, *J Pharmacol Exp Ther*, 286(1), 85–90, 1998.
63. Arranz, B. et al., Altered platelet serotonin 5-HT2A receptor density but not second messenger inositol trisphosphate levels in drug-free schizophrenic patients, *Psychiatry Res*, 118(2), 165–174, 2003.

64. Rosel, P. et al., Altered 5-HT(2A) binding sites and second messenger inositol tris-phosphate (IP(3)) levels in hippocampus but not in frontal cortex from depressed suicide victims, *Psychiatry Res*, 99(3), 173–181, 2000.
65. Urban, J.D. et al., Functional selectivity and classical concepts of quantitative pharmacology, *J Pharmacol Exp Ther*, 2006.
66. Berg, K.A. et al., Pleiotropic behavior of 5-HT2A and 5-HT2C receptor agonists, *Ann NY Acad Sci*, 861, 104–110, 1998.
67. Berg, K.A. et al., Effector pathway-dependent relative efficacy at serotonin type 2A and 2C receptors: evidence for agonist-directed trafficking of receptor stimulus, *Mol Pharmacol*, 54(1), 94–104, 1998.
68. Kurrasch-Orbaugh, D.M. et al., Serotonin 5-hydroxytryptamine 2A receptor-coupled phospholipase C and phospholipase A2 signaling pathways have different receptor reserves, *J Pharmacol Exp Ther*, 304(1), 229–237, 2003.
69. Kurrasch-Orbaugh, D.M. et al., A complex signaling cascade links the serotonin2A receptor to phospholipase A2 activation: the involvement of MAP kinases, *J Neurochem*, 86(4), 980–991, 2003.
70. Quinn, J.C. et al., Activation of extracellular-regulated kinase by 5-hydroxytryptamine(2A) receptors in PC12 cells is protein kinase C-independent and requires calmodulin and tyrosine kinases, *J Pharmacol Exp Ther*, 303(2), 746–752, 2002.
71. Grewal, J.S., Luttrell, L.M., and Raymond, J.R., G protein-coupled receptors desensitize and down-regulate epidermal growth factor receptors in renal mesangial cells, *J Biol Chem*, 276(29), 27335–27344, 2001.
72. Banes, A.K. et al., Activation of the JAK/STAT pathway in vascular smooth muscle by serotonin, *Am J Physiol Cell Physiol*, 288(4), C805–C812, 2005.
73. Guillet-Deniau, I., Burnol, A.F., and Girard, J., Identification and localization of a skeletal muscle secrotonin 5-HT2A receptor coupled to the Jak/STAT pathway, *J Biol Chem*, 272(23), 14825–14829, 1997.
74. Miller, K.J., Mariano, C.L., and Cruz, W.R., Serotonin 5HT2A receptor activation inhibits inducible nitric oxide synthase activity in C6 glioma cells, *Life Sci*, 61(18), 1819–1827, 1997.
75. Chalecka-Franaszek, E., Chen, H., and Chuang, D.M., 5-Hydroxytryptamine2A receptor stimulation induces activator protein-1 and cyclic AMP-responsive element binding with cyclic AMP-responsive element-binding protein and Jun D as common components in cerebellar neurons, *Neuroscience*, 88(3), 885–898, 1999.
76. Johnson-Farley, N.N. et al., Enhanced activation of Akt and extracellular-regulated kinase pathways by simultaneous occupancy of Gq-coupled 5-HT2A receptors and Gs-coupled 5-HT7A receptors in PC12 cells, *J Neurochem*, 92(1), 72–82, 2005.
77. Gresch, P.J., Strickland, L.V., and Sanders-Bush, E., Lysergic acid diethylamide-induced Fos expression in rat brain: role of serotonin-2A receptors, *Neuroscience*, 114(3), 707–713, 2002.
78. Turner, J.H. and Raymond, J.R., Interaction of calmodulin with the serotonin 5-hydroxytryptamine2A receptor. A putative regulator of G protein coupling and receptor phosphorylation by protein kinase C, *J Biol Chem*, 280(35), 30741–30750, 2005.
79. Grewal, J.S. et al., Serotonin 5-HT2A receptor induces TGF-beta1 expression in mesangial cells via ERK: proliferative and fibrotic signals, *Am J Physiol*, 276(6 Pt. 2), F922–F930, 1999.
80. Willins, D.L. et al., Clozapine and other 5-hydroxytryptamine-2A receptor antagonists alter the subcellular distribution of 5-hydroxytryptamine-2A receptors *in vitro* and *in vivo*, *Neuroscience*, 91(2), 599–606, 1999.

81. Willins, D.L. et al., Serotonergic antagonist effects on trafficking of serotonin 5-HT2A receptors *in vitro* and *in vivo*, *Ann NY Acad Sci*, 861, 121–7, 121998.
82. Hazelwood, L.A. and Sanders-Bush, E., His452Tyr polymorphism in the human 5-HT2A receptor destabilizes the signaling conformation, *Mol Pharmacol*, 66(5), 1293–1300, 2004.
83. Egan, C., Herrick-Davis, K., and Teitler, M., Creation of a constitutively activated state of the 5-HT2A receptor by site-directed mutagenesis: revelation of inverse agonist activity of antagonists, *Ann NY Acad Sci*, 861, 136–139, 1998.
84. Gray, J.A., Compton-Toth, B.A., and Roth, B.L., Identification of two serine residues essential for agonist-induced 5-HT2A receptor desensitization, *Biochemistry*, 42(36), 10853–10862, 2003.
85. Mathuru, A. and Panicker, M.M., *In Vivo* Imaging of Serotonin Receptor Internalization in FENS Forum 2006, Vienna, Austria, 2006.
86. Maura, G. et al., 5-HT2 presynaptic receptors mediate inhibition of glutamate release from cerebellar mossy fibre terminals, *Eur J Pharmacol*, 202(2), 185–190, 1991.
87. Doherty, M.D. and Pickel, V.M., Ultrastructural localization of the serotonin 2A receptor in dopaminergic neurons in the ventral tegmental area, *Brain Res*, 864(2), 176–185, 2000.
88. Nocjar, C., Roth, B.L., and Pehek, E.A., Localization of 5-HT(2A) receptors on dopamine cells in subnuclei of the midbrain A10 cell group, *Neuroscience*, 111(1), 163–176, 2002.
89. Ciccocioppo, R. et al., Autoradiographic mapping of brain 5-HT2A binding sites in P and in AA alcohol-preferring rats, *Brain Res Bull*, 44(1), 33–37, 1997.
90. Ikemoto, K. et al., Human midbrain dopamine neurons express serotonin 2A receptor: an immunohistochemical demonstration, *Brain Res*, 853(2), 377–380, 2000.
91. Xia, Z. et al., The PDZ-binding domain is essential for the dendritic targeting of 5-HT2A serotonin receptors in cortical pyramidal neurons *in vitro*, *Neuroscience*, 122(4), 907–920, 2003.
92. Xia, Z. et al., A direct interaction of PSD-95 with 5-HT2A serotonin receptors regulates receptor trafficking and signal transduction, *J Biol Chem*, 278(24), 21901–21908, 2003.
93. Bhatnagar, A. et al., The dynamin-dependent, arrestin-independent internalization of 5-hydroxytryptamine 2A (5-HT2A) serotonin receptors reveals differential sorting of arrestins and 5-HT2A receptors during endocytosis, *J Biol Chem*, 276(11), 8269–8277, 2001.
94. Kristiansen, K., et al., A highly conserved aspartic acid (Asp-155) anchors the terminal amine moiety of tryptamines and is involved in membrane targeting of the 5-HT(2A) serotonin receptor but does not participate in activation via a "salt-bridge disruption" mechanism, *J Pharmacol Exp Ther*, 293(3), 735–746, 2000.
95. Backstrom, J.R. and Sanders-Bush, E., Generation of anti-peptide antibodies against serotonin 5-HT2A and 5-HT2C receptors, *J Neurosci Methods*, 77(1), 109–117, 1997.
96. Cornea-Hebert, V. et al., Similar ultrastructural distribution of the 5-HT(2A) serotonin receptor and microtubule-associated protein MAP1A in cortical dendrites of adult rat, *Neuroscience*, 113(1), 23–35, 2002.
97. Berry, S.A. et al., Rapid agonist-induced internalization of the 5-hydroxytryptamine2A receptor occurs via the endosome pathway *in vitro*, *Mol Pharmacol*, 50(2), 306–313, 1996.
98. Bhattacharyya, S., Internalization and Recycling of the Serotonin 2A Receptor in Non-Neuronal and Neuronal Cells, in National Centre for Biological Sciences, PhD thesis, TIFR, 2005, Manipal Academy of Higher Education, Bangalore, p. 136.

99. Gray, J.A. et al., The interaction of a constitutively active arrestin with the arrestin-insensitive 5-HT(2A) receptor induces agonist-independent internalization, *Mol Pharmacol*, 63(5), 961–972, 2003.
100. Sheffler, D.J. et al., p90 ribosomal S6 kinase 2 exerts a tonic brake on G protein-coupled receptor signaling, *Proc Natl Acad Sci USA*, 103(12), 4717–4722, 2006.
101. Miyoshi, I. et al., Characterization of 5-HT2A receptor desensitization and the effect of cycloheximide on it in C6 cells, *J Neural Transm*, 108(3), 249–260, 2001.
102. Patterson, G.H. and Lippincott-Schwartz, J., A Photoactivatable GFP for Selective Photolabeling of Proteins and Cells, *Science*, 297(5588), 1873–1877, 2002.
103. No, D., Yao, T.P., and Evans, R.M., Ecdysone-inducible gene expression in mammalian cells and transgenic mice, *Proc Natl Acad Sci USA*, 93(8), 3346–3351, 1996.
104. Roettger, B.F. et al., Antagonist-stimulated internalization of the G protein-coupled cholecystokinin receptor, *Mol Pharmacol*, 51(3), 357–362, 1997.
105. Fabry, M. et al., Monitoring of the internalization of neuropeptide Y on neuroblastoma cell line SK-N-MC, *Eur J Biochem*, 267(17), 5631–5637, 2000.
106. Rosenthal, S.J. et al., Targeting cell surface receptors with ligand-conjugated nanocrystals, *J Am Chem Soc*, 124(17), 4586–4594, 2002.
107. Tomlinson, I.D. et al., Inhibitors of the serotonin transporter protein (SERT): the design and synthesis of biotinylated derivatives of 3-(1,2,3,6-tetrahydro-pyridin-4-yl)-1H-indoles. High-affinity serotonergic ligands for conjugation with quantum dots, *Bioorg Med Chem Lett*, 15(23), 5307–5310, 2005.
108. Johnson, M.P. et al., Species variations in transmembrane region V of the 5-hydroxytryptamine type 2A receptor alter the structure-activity relationship of certain ergolines and tryptamines, *Mol Pharmacol*, 45(2), 277–286, 1994.
109. Bylund, D. B., Deupree, J. D., and Toews, M. L., *Receptor Signal Transduction Protocols (Methods in Molecular Biology)*, Vol. 259, Humana Press, Totowa, NJ, 2004.
110. Hurlemann, R. et al., Analysis of neuroreceptor PET-data based on cytoarchitectonic maximum probability maps: a feasibility study, *Anat Embryol (Berl)*, 210(5–6), 447–453, 2005.
111. Kaye, W.H. et al., Serotonin alterations in anorexia and bulimia nervosa: new insights from imaging studies, *Physiol Behav*, 85(1), 73–81, 2005.
112. Mathews, J. et al., Antidepressant efficacy of olanzapine as monotherapy in major depressive disorder, without psychosis: a pilot study, *Psychiatr Res*, 146(2), 149–155, 2006.
113. Yatham, L.N. et al., A positron emission tomography study of the effects of treatment with valproate on brain 5-HT2A receptors in acute mania, *Bipolar Disord*, 7(Suppl. 5), 53–57, 2005.
114. Lopez-Figueroa, A.L. et al., Serotonin 5-HT1A, 5-HT1B, and 5-HT2A receptor mRNA expression in subjects with major depression, bipolar disorder, and schizophrenia, *Biol Psychiatr*, 55(3), 225–233, 2004.
115. Huang, M. et al., Augmentation by citalopram of risperidone-induced monoamine release in rat prefrontal cortex, *Psychopharmacology (Berl)*, 185(3), 274–281, 2006.
116. Anji, A. et al., Regulation of 5-HT(2A) receptor mRNA levels and binding sites in rat frontal cortex by the agonist DOI and the antagonist mianserin, *Neuropharmacology*, 39(11), 1996–2005, 2000.
117. Pierce, P.A. et al., 5-Hydroxytryptamine receptor subtype messenger RNAs in human dorsal root ganglia: a polymerase chain reaction study, *Neuroscience*, 81(3), 813–819, 1997.
118. Bunzel, R. et al., Polymorphic imprinting of the serotonin-2A (5-HT2A) receptor gene in human adult brain, *Brain Res Mol Brain Res*, 59(1), 90–92 1998.

119. Ozaki, N. et al., Two naturally occurring amino acid substitutions of the 5-HT2A receptor: similar prevalence in patients with seasonal affective disorder and controls, *Biol Psychiatr*, 40(12), 1267–1272, 1996.
120. Saucier, C. and Albert, P.R., Identification of an endogenous 5-hydroxytryptamine2A receptor in NIH-3T3 cells: agonist-induced down-regulation involves decreases in receptor RNA and number, *J Neurochem*, 68(5), 1998–2011, 1997.
121. Saucier, C., Morris, S.J., and Albert, P.R., Endogenous serotonin-2A and -2C receptors in Balb/c-3T3 cells revealed in serotonin-free medium: desensitization and down-regulation by serotonin, *Biochem Pharmacol*, 56(10), 1347–1357, 1998.
122. Vysokanov, A., Flores-Hernandez, J., and Surmeier, D.J., mRNAs for clozapine-sensitive receptors co-localize in rat prefrontal cortex neurons, *Neurosci Lett*, 258(3), 179–182, 1998.
123. Moser, A., Thumen, A., and Qadri, F., Modulation of striatal serotonin and opioid receptor mRNA expression following systemic N-methyl-norsalsolinol administration, *J Neurol Sci*, 216(1), 109–112, 2003.
124. Gonzalez-Maeso, J. et al., Transcriptome fingerprints distinguish hallucinogenic and nonhallucinogenic 5-hydroxytryptamine 2A receptor agonist effects in mouse somatosensory cortex, *J Neurosci*, 23(26), 8836–8843, 2003.
125. Grotewiel, M.S. and Sanders-Bush, E., Regulation of serotonin2A receptors in heterologous expression systems, *J Neurochem*, 63(4), 1255–1260, 1994.
126. Buckholtz, N., Zhou, D., and Freedman, D., Serotonin2 agonist administration down-regulates rat brain serotonin2 receptors, *Life Sci*, 42(24), 2439–2445, 1988.
127. Eison, A. et al., Effects of imipramine and serotonin-2 agonists and antagonists on serotonin-2 and beta-adrenergic receptors following noradrenergic or serotonergic denervation, *Life Sci*, 44(19), 1419–1427, 1989.
128. Robertson, D.N. et al., Selective interaction of ARF1 with the carboxy-terminal tail domain of the 5-HT2A receptor, *Mol Pharmacol*, 64(5), 1239–1250, 2003.
129. Becamel, C. et al., The serotonin 5-HT2A and 5-HT2C receptors interact with specific sets of PDZ proteins, *J Biol Chem*, 279(19), 20257–20266, 2004.
130. Cawston, E.E., Chow, K., Fitches, A.C., Markie D., and Olds, R.J. The cytoplasmic tail of serotonin receptor 2A interacts with filamin-C and RACK1, *FENS Abstr.*, vol. 2 A154.7, 2004.
131. Gelber, E.I. et al., Structure and function of the third intracellular loop of the 5-hydroxytryptamine2A receptor: the third intracellular loop is alpha-helical and binds purified arrestins, *J Neurochem*, 72(5), 2206–2214, 1999.
132. Bhatnagar, A. et al., Caveolin-1 interacts with 5-HT2A serotonin receptors and profoundly modulates the signaling of selected Galphaq-coupled protein receptors, *J Biol Chem*, 279(33), 34614–34623, 2004.
133. Herrick-Davis, K., Grinde, E., and Mazurkiewicz, J., Biochemical and biophysical characterization of serotonin 5-HT2C receptor homodimers on the plasma membrane of living cells, *Biochemistry*, 43(44), 13963–13971, 2004.
134. Bray, D., Signaling complexes: biophysical constraints on intracellular communication, *Annu Rev Biophys Biomol Struct*, 27, 59–75, 1998.
135. Overton, M.C. and Blumer, K.J., Use of fluorescence resonance energy transfer to analyze oligomerization of G-protein-coupled receptors expressed in yeast, *Methods*, 27(4), 324–332, 2002.
136. Xu, Y., Piston, D.W., and Johnson, C.H., A bioluminescence resonance energy transfer (BRET) system: application to interacting circadian clock proteins, *Proc Natl Acad Sci USA*, 96(1), 151–156, 1999.

137. Angers, S., Salahpour, A., and Bouvier, M., Biochemical and biophysical demonstration of GPCR oligomerization in mammalian cells, *Life Sci*, 68(19–20), 2243–2250, 2001.
138. Hanyaloglu, A.C. et al., Homo- and hetero-oligomerization of thyrotropin-releasing hormone (TRH) receptor subtypes. Differential regulation of beta-arrestins 1 and 2, *J Biol Chem*, 277(52), 50422–50430, 2002.
139. Kroeger, K.M. et al., Constitutive and agonist-dependent homo-oligomerization of the thyrotropin-releasing hormone receptor. Detection in living cells using bioluminescence resonance energy transfer, *J Biol Chem*, 276(16), 12736–12743, 2001.
140. Trettel, F. et al., Ligand-independent CXCR2 dimerization, *J Biol Chem*, 278(42), 40980–40988, 2003.
141. Baneres, J. and Parello, J., Structure-based analysis of GPCR function: evidence for a novel pentameric assembly between the dimeric leukotriene B4 receptor BLT1 and the G-protein, *J Mol Biol*, 329(4), 815–829, 2003.
142. Lamey, M. et al., Distinct residues in the carboxyl tail mediate agonist-induced desensitization and internalization of the human dopamine D1 receptor*, *J Biol Chem*, 277(11), 9415–9421, 2002.
143. Jordan, B.A. et al., Oligomerization of opioid receptors with beta 2-adrenergic receptors: a role in trafficking and mitogen-activated protein kinase activation, *Proc Natl Acad Sci USA*, 98(1), 343–348, 2001.
144. Laroche, G. et al., Oligomerization of the alpha and beta isoforms of the thromboxane A2 receptor: relevance to receptor signaling and endocytosis, *Cell Signal*, 17(11), 1373–1383, 2005.
145. Carrillo, J.J., Lopez-Gimenez, J.F., and Milligan, G., Multiple interactions between transmembrane helices generate the oligomeric alpha1b-adrenoceptor, *Mol Pharmacol*, 66(5), 1123–1237, 2004.
146. Guo, W., Shi, L., and Javitch, J.A., The fourth transmembrane segment forms the interface of the dopamine D2 receptor homodimer, *J Biol Chem*, 278(7), 4385–4388, 2003.
147. Fotiadis, D. and Engel, A., High-resolution imaging of bacteriorhodopsin by atomic force microscopy, *Methods Mol Biol*, 242(291–303), 2004.
148. Filizola, M., Olmea, O., and Weinstein, H., Prediction of heterodimerization interfaces of G-protein coupled receptors with a new subtractive correlated mutation method, *Protein Eng*, 15(11), 881–885, 2002.
149. Filizola, M. and Weinstein, H., The study of G-protein coupled receptor oligomerization with computational modeling and bioinformatics, *FEBS J*, 272(12), 2926–2938, 2005.
150. Gouldson, P.R. et al., Dimerization and domain swapping in G-protein-coupled receptors: a computational study, *Neuropsychopharmacology*, 23(4 Suppl.), S60–S77, 2000.
151. Nemoto, W. and Toh, H., Prediction of interfaces for oligomerizations of G-protein coupled receptors, *Proteins*, 58(3), 644–660, 2005.
152. Liang, Y. et al., Organization of the G protein-coupled receptors rhodopsin and opsin in native membranes, *J Biol Chem*, 278(24), 21655–21662, 2003.
153. Fotiadis, D. et al., Atomic-force microscopy: rhodopsin dimers in native disc membranes, *Nature*, 421(6919), 127–128, 2003.
154. Filizola, M. and Weinstein, H., Structural models for dimerization of G-protein coupled receptors: the opioid receptor homodimers, *Biopolymers*, 66(5), 317–325, 2002.

# 7 The 5-$HT_{1A}$ Receptor: A Signaling Hub Linked to Emotional Balance

*Probal Banerjee, Mukti Mehta, and Baishali Kanjilal*

## CONTENTS

Abstract ..... 133
Introduction ..... 134
Distribution and Ontogeny of the 5-$HT_{1A}$-R ..... 135
Presynaptic and Postsynaptic 5-$HT_{1A}$-R and Their Signaling Effects ..... 136
Aberrant 5-$HT_{1A}$-R Expression and Diseases ..... 137
Therapeutic Agents that Function by Regulating 5-$HT_{1A}$-R Signaling ..... 138
Recent Advancement through the Use of the SNRIs ..... 139
Role of 5-$HT_{1A}$-R in the Immune System and Cancer ..... 140
Required Future Studies to Unravel Mechanisms of 5-$HT_{1A}$-R Signaling ..... 140
- SSRIs ..... 141
- SNRIs ..... 141
- 5-HT1A Agonists and 5-$HT_{1A}$-R-Mediated Signaling ..... 141

Examples of Some Studies and Possible Strategies ..... 142
- Early Mouse Brain Development and the 5-$HT_{1A}$-R ..... 143
- Involvement of 5-$HT_{1A}$-R in Clozapine-Evoked Neuronal Activity ..... 144

Concluding Remarks ..... 145
Experimental Procedures ..... 146
Acknowledgments ..... 148
References ..... 149

## ABSTRACT

Serotonin or 5-hydroxytryptamine (5-HT) is an ancient chemical that is synthesized in the brain and also in the peripheral system. It binds to 14 or more receptor proteins, all but one of which are G protein-coupled receptors. Pharmacological, behavioral, and clinical studies have placed one of these receptors, the serotonin 1A (5-$HT_{1A}$) receptor, in the forefront as a protein that binds to 5-HT with high affinity to exert subtle control over emotion and behavior. This review will compare and contrast existing data on expression and signaling activity of the brain 5-HT$_{1A}$ receptor. Our purpose is to critically assess the current understanding of those 5-HT$_{1A}$R-mediated signaling cascades that are physiologically important and also to unravel the poorly understood processes that have yet to be delineated through further experiments.

## INTRODUCTION

Signal transduction through several classes of receptors brings about quick as well as slow effects that regulate neuronal activity, cellular organization, and functional development of the brain. The quick effects are, for example, neuronal excitation and inhibition, which are downstream of the ion channel receptors. In contrast, some of the slow effects are growth cone guidance and cell survival, which are caused by the activation of ephrin/semaphorin receptors and G protein-coupled receptors, respectively. Each of these receptor-mediated effects contributes toward the birth of new neurons as well as formation and maintenance of their connections with targets. Based on their function, the receptors have been classified into two groups: (1) effectors, which produce the final change in neuronal signals, and (2) modulators, which regulate the efficacy of the effectors to produce a timely change that enables the organism to steer itself into an altered state of homeostasis.

The slow-acting receptors regulate phosphorylation of functional proteins, expression of specific genes, and protein–protein interactions. By bringing about growth cone guidance, neurogenesis, and cell survival, these concerted processes play an important role in sculpturing and maintenance of brain compartments. Therefore, it is not surprising that the aberrant function of many G protein-coupled receptors, such as dopamine receptors, noradrenergic receptors, and serotonin receptors, is associated with emotional and cognitive disorders. This article will review the current literature on one serotonin receptor subtype, the serotonin 1A receptor (5-HT$_{1A}$-R), and lay out its known signaling cascades that function independently or through interactions with signaling cascades initiated by other receptors.

A few seminal studies have established the pivotal role of the 5-HT$_{1A}$-R in brain development. First, aberrant expression of this receptor and any disturbance in its signaling activity are concurrent with depression and suicidal tendencies (1–5), which often surface after the infantile and early juvenile stages of brain development. Furthermore, either complete deletion of this receptor or its tissue-specific elimination in the frontal cortex during neonatal development causes elevated anxiety levels in mice (6–9). Finally, many therapeutic agents that are used to ameliorate emotional disorders function through the 5-HT$_{1A}$ receptor (10–13). Despite such observations, there is only limited information available to clearly link the 5-HT$_{1A}$R to early brain development through well-defined biochemical cascades. With this in mind, we will extensively review the existing literature on 5-HT$_{1A}$-R mediated signaling in brain cells.

Although the expression and role of the 5-HT$_{1A}$ receptor in the brain is of prime importance, there is a significant body of literature on the expression of this receptor in T and B cells (14,15). Consequently, these studies link the 5-HT$_{1A}$ receptor also to the functional state of the immune system and, as expected, the schizophrenia drug clozapine, which is a partial agonist at the 5-HT$_{1A}$-R, occasionally causes a dangerous side effect, agranulocytosis (16–18). Therefore, our discussion will also include reports on the functional activity of this receptor in the blood cells and the immune system.

## DISTRIBUTION AND ONTOGENY OF THE 5-HT$_{1A}$-R

Ligand binding, immunohistochemistry, and *in situ* hybridization in rats, mice, cats, tree shrews, and humans have revealed the presence of significant levels of 5-HT$_{1A}$-R in almost all parts of the brain, including the cerebellum (19–26). Positron emission tomography (PET) studies using the 5-HT$_{1A}$-R ligand [$^{18}$F]MPPF in combination with *in vitro* autoradiography with [$^{3}$H]MPPF, [$^{3}$H]8-OH-DPAT, and [$^{3}$H]paroxetine have revealed the highest level of expression in the hippocampus, cingulate, septum, and intralimbic cortex and the lowest in the cerebellum (23).

[$^{3}$H]8-OH-DPAT binding studies have revealed a similar distribution profile for the 5-HT$_{1A}$-R, although the overall level of receptor expression determined from binding assays were lower than those obtained from the more sensitive technique, autoradiography. From PET studies using [$^{11}$C]WAY-100635, Parsey and coworkers have reported regional heterogeneity of 5-HT$_{1A}$R in human cerebellum (25). Lack of 5-HT$_{1A}$-R expression was observed in the cerebellar white matter, whereas the other regions displayed detectable levels of the receptor.

Cellular distribution of this receptor and its mRNA has also been analyzed. Santana and coworkers report that ~60% of glutamatergic cells express the 5-HT$_{1A}$-R transcript and ~25% of GAD-expressing cells contain the 5-HT$_{1A}$-R mRNA (27). Similarly, using immunohistochemistry with multiple antibodies, *in vitro* autoradiography with [$^{3}$H]8-OH-DPAT, and also *in situ* hybridization. Palchauchuri and Függe report 5-HT$_{1A}$-R mRNA expression in pyramidal neurons of layer 2 within the prefrontal, insular, and occipital cortex (21). In contrast, [$^{3}$H]8-OH-DPAT labeling occurred in layers 1 and 2, generating a columnar pattern in the prefrontal and occipital cortex. Pyramidal neurons in the claustrum and anterior olfactory nucleus expressed the receptor. Neurons in the hippocampal CA1 region expressed 5-HT$_{1A}$-R mRNA and [$^{3}$H]8-OH-DPAT labeling was observed in stratum oriens and stratum radiatum. Low receptor expression was observed in CA3 pyramidal neurons, but the granule neurons in the dentate gyrus contained moderate concentrations of the receptor.

Only a few studies describe the ontogeny of this receptor. Using immunohistochemistry, Patel and Zhou have demonstrated that almost all hippocampal neurons begin expressing the 5-HT$_{1A}$-R upon completion of their terminal mitosis (28). At postnatal day 5 (P5) the receptor is expressed mainly on the cell bodies, but at P10 the receptor appears on both cell bodies as well as proximal apical dendrites. Finally, following neuronal maturation (at P21), a relatively sparse distribution is observed on the dendrites of stratum radiatum and stratum oriens of the hippocampus. Curiously, S100 and GFAP-positive glial cells transiently express the 5-HT$_{1A}$-R during early postnatal development of the hippocampus. More than 90% of the S100-positive astrocytes in CA1, CA3, and dentate gyrus also show moderate 5-HT$_{1A}$-R immunoreactivity at P7, which decreases in a dramatic manner by P16. Although tissue-specific distribution of the 5-HT$_{1A}$-R has been studied earlier, this does not ensure that the signaling activity of the receptor will always be proportional to the expression levels of the receptor. Therefore, the regional and temporal profile of $5\text{-HT}_{1A}$-R signaling in the neuronal cells is an important determinant in the functional effect of this receptor. Such signaling effects are regulated by the second messenger coupling of the receptor.

## PRESYNAPTIC AND POSTSYNAPTIC $5\text{-HT}_{1A}$-R AND THEIR SIGNALING EFFECTS

The major electrical effect mediated by the $5\text{-HT}_{1A}$ receptor in neurons involves Go-mediated activation of hyperpolarizing $K^+$ channels (29–32), which in turn causes attenuated firing of action potentials, thus resulting in decreasing firing of neurotransmitters from the synaptic ends of these neurons. Although the hyperpolarizing effect of the $5\text{-HT}_{1A}$-R is observed in both pre- and postsynaptic environments, the profiles of $5\text{-HT}_{1A}$-R desensitization seem to be widely different for the pre- and postsynaptic $5\text{-HT}_{1A}$-R molecules. Sustained administration of a $5\text{-HT}_{1A}$-R agonist or the serotonin reuptake inhibitor (SSRI) causes internalization of the $5\text{-HT}_{1A}$ auto receptors in the raphé neurons but not the postsynaptic $5\text{-HT}_{1A}$ receptors in the hippocampus (33–38). This is also believed to be the basis of action of the SSRIs. Initially, serotonin reuptake inhibition caused by the SSRIs in the presynaptic neurons results in increased serotonin release from these neurons (Figure 7.1). The discharged serotonin molecules bind to the $5\text{-HT}_{1A}$ autoreceptors present on the soma of the raphé neurons, thus causing inhibition of firing from these neurons. Subsequently, the ligand-bound autoreceptors are internalized, thus causing termination of $5\text{-HT}_{1A}$-R signaling in the presnaptic neurons and resumption of serotonin release from the raphé neurons at the synapse with the dendritic terminals of the postsynaptic neurons. In the absence of the $5\text{-HT}_{1A}$ autoreceptors, the released serotonin binds only to the postsynaptic $5\text{-HT}_{1A}$ receptors, thereby eliciting the anxiolytic effect of the SSRIs (Figure 7.1) (36).

Similarly, the antidepressant effects of the $5\text{-HT}_{1A}$ agonists like buspirone or flesinoxan are also caused by the desensitization of only the $5\text{-HT}_{1A}$ autoreceptors (35,37). Thus, after acute treatment, the agonist binds to the autoreceptors on the soma of the raphé neurons. The hyperopolarizing effect of $5\text{-HT}_{1A}$ autoreceptors causes inhibition of serotonin release from the presynaptic terminal (Figure 7.1). Under this condition, the excess agonist acts on the postsynaptic (dendritic) $5\text{-HT}_{1A}$ receptors, thereby causing inhibition of the postsynaptic neurons. Persistent action of the agonist causes internalization and desensitization of the $5\text{-HT}_{1A}$ receptors on the raphé neurons but not the postsynaptic neurons. In the absence of the inhibitory $5\text{-HT}_{1A}$ autoreceptors, the presynaptic raphé neurons elicit uninhibited firing of serotonin, which binds to the $5\text{-HT}_{1A}$-R molecules on the postsynaptic neurons to bring forth the characteristic anxiolytic effect of these agonists (Figure 7.1).

Signaling effects of the pre- and postsynaptic $5\text{-HT}_{1A}$ receptors also appear to be different for at least one biochemical pathway. In hippocampal neuron-derived HN2-5 cells as well as in organotypic cultures of hippocampal slices, agonist activation of the $5\text{-HT}_{1A}$-R causes stimulation of the mitogen-activated protein kinase (MAPK) pathway (39–41). In contrast, Kushwaha and Albert observed that agonist

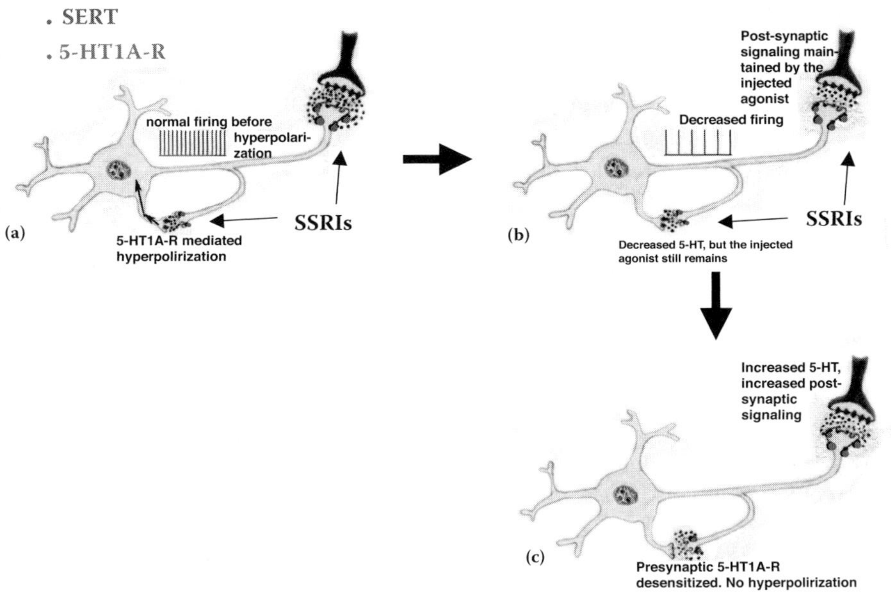


**FIGURE 7.1 (Color figure follows p. 110.)** Increased postsynaptic activity after long-term administration of SSRIs or 5-HT$_{1A}$-R agonists. The SERT molecules are not expressed at the synapses, but they are found on apposed neurons or on the same neurons but away from the synapse (102,103). (a) The SSRIs bind to the SERT molecules, eliciting heightened 5-HT release (blue dots) at the synapses. Binding of these 5-HT molecules to 5-HT$_{1A}$ autoreceptors as well as post-synaptic 5-HT$_{1A}$ hetero-receptors to cause inhibition of action potential in both serotoninergic raphé neurons as well the post-synaptic non-serotoninergic neurons. (b) Inhibition of the serotoninergic raphé neurons causes a short-term decrease in 5-HT release. (c) Prolonged 5-HT binding causes the auto (but not the postsynaptic) 5-HT$_{1A}$-R molecules to desensitize, thus eliminating hyperpolarization, boosting 5-HT release from raphé neurons, which brings about increased postsynaptic 5-HT$_{1A}$-R signaling. Similarly, prolonged treatment with 5-HT$_{1A}$ agonists causes desensitization of the auto 5-HT$_{1A}$-R molecules but not the postsynaptic receptors. This in turn results in increased 5-HT release and elevated postsynaptic 5-HT$_{1A}$-R signaling.

activation of the 5-HT$_{1A}$-R in the raphé-derived cell line RN46A causes a dramatic inhibition of the basal MAPK activity (42). Nonetheless, in both pre- and postsynaptic neurons, agonist activation of the 5-HT$_{1A}$-R causes the customary inhibition of intracellular cAMP (32,42,43). The 5-HT$_{1A}$-R also mediates stimulation of phospholipase C (PLC) in a postsynaptic neuron-derived cell line and in nonneural cells, but this signaling activity has yet to be studied in presynaptic (serotonergic) or raphé-derived neurons (32,39,40).

## ABERRANT 5-HT$_{1A}$-R EXPRESSION AND DISEASES

Expression and signaling activity of the 5-HT$_{1A}$-R play a major role in multiple affective disorders. In schizophrenia (which often appears during adolescence), the majority of postmortem studies have reported increases in 5-HT$_{1A}$-R density in the prefrontal cortex in the approximate range of 15–80% (10). Burnet and coworkers have gone further to show that whereas the 5-$HT_{1A}$-R binding sites are significantly increased (+23%) in the dorsolateral prefrontal cortex, the 5-$HT_{2A}$ binding sites are decreased (27%) in the same region in the postmortem brain of schizophrenics (44). Recent studies in Alzheimer's disease have shown that reduced 5-$HT_{1A}$-R binding in the temporal cortex correlates with aggressive behavior in Alzheimer's disease (45). Thus, altered 5-$HT_{1A}$-R expression is observed in emotional and behavioral disorders.

Sheline and coworkers have reported that subjects with a history of major depression have smaller left and right hippocampal volumes (46). Following this, Pantel and coworkers reported the presence of brain atrophy using 3-D MRI in primary degenerative dementia (47). Drevets and coworkers have shown that the mean gray matter volume in prefrontal cortex, ventral to the genu of the corpus callosum, is reduced by 39% and 48% in bipolar and unipolar samples, respectively (48). A study carried out by Ashtari and coworkers builds a potentially important correlation between hippocampal structure and the expression of major depression in the elderly (49). It is noteworthy that all these morphological changes involve regions that show high expression and activity of the 5-$HT_{1A}$ receptor. In addition to 5-$HT_{1A}$-R expression, its signaling activity is also altered in multiple disorders that affect the brain, such as alcoholism, cocaine abuse, and schizophrenia (10,44,45,50–53).

Genetic analysis performed by Lemonade and coworkers have shown that a C(-1019)G polymorphism in the 5-$HT_{1A}$-R promoter leads to uncontrolled expression of the 5-$HT_{1A}$ receptor. Overexpression of this receptor in the presynaptic raphé neurons results in reduced serotonin firing at the synapse, thus causing decreased postsynaptic 5-$HT_{1A}$-R activity, which in turn is closely associated with major depression and suicide (1). Finally, disruption of the 5-$HT_{1A}$-R gene has been shown to cause elevated anxiety disorder in mice. Furthermore, temporal silencing of this gene up to postnatal day 21 in mouse front brain causes permanent occurrence of elevated anxiety (6–9). Intriguingly, silencing of this gene at a later stage of development does not have any phenotypic consequence. Collectively, such observations establish the profound importance of 5-$HT_{1A}$-R signaling in a number of emotional disorders.

## THERAPEUTIC AGENTS THAT FUNCTION BY REGULATING 5-$HT_{1A}$-R SIGNALING

Over the last decade it has been suggested that 5-$HT_{1A}$ antagonists may have therapeutic utility in such diseases as depression, anxiety, drug- and nicotine-withdrawal. Recently, a compelling rationale has been developed for the therapeutic potential of 5-$HT_{1A}$ receptor antagonists also in Alzheimer's disease and, potentially, other diseases with associated cognitive dysfunction (54). Involvement of the 5-$HT_{1A}$-R in the effects of alcohol intake has been demonstrated by prenatal ethanol treatment of rats, which causes an increase in 5-$HT_{1A}$-R-mediated wet-dog shakes (53). The 5-$HT_{1A}$-R antagonist p-MPPI causes a decrease in ethanol-induced hypothermia and sleep (55). Furthermore, 5-HT$_{1A}$ agonists and antagonists have shown considerable promise in the treatment of depression, cocaine seeking, schizophrenia, and Alzheimer's disease (10,44,54,56,57). Similarly, cocaine treatment of rats has been shown to cause increased serotonergic activity in the hippocampus and nucleus accumbens, which is blocked by a 5-HT$_{1A}$-R antagonist (58). Likewise, the 5-HT$_{1A}$-R antagonist WAY 100635 has been shown to block the locomotor stimulant effect of cocaine (57).

Serotonin 1A receptor-mediated signaling is involved in the therapeutic action of several formulations used for schizophrenia. Thus, the atypical antipsychotics, such as clozapine, olanzapine, risperidone, and perhaps other atypical antipsychotics, function through the 5-HT$_{1A}$ receptor (11,59,60). Pharmacological studies have shown that the atypical antipsychotics function as partial agonists at the 5-HT$_{1A}$-R and as antagonists at 5-HT$_{2A}$ and $D_2$ receptors. The atypical antipsychotics cause increased dopamine release in the brain, which is mediated by simultaneous activation of the 5-HT$_{1A}$-R and blockage of 5-HT$_{2A}$-R and $D_2$ receptors (59,61,62). Additionally, the antipsychotics regulate serotonin release in the brain, and it is believed that their ability to modulate serotonin as well as dopaminergic function is crucial for their therapeutic activity as well as side effects (63). In fact, it is currently believed that the partial 5-HT$_{1A}$ agonist activity of the atypical antipsychotics contribute toward their efficacy to block psychosis without eliciting the extrapyramidal side effects of the typical (or earlier generation) antipsychotics such as haloperidol (60).

Based on such compelling data, it can argued that, in order to understand the etiology of many developmental disorders of the brain and also to make a headway into development of 5-HT$_{1A}$-R-targeted drugs, it is essential to study the profile of signaling activity of this receptor in the developing brain.

## RECENT ADVANCEMENT THROUGH THE USE OF THE SNRIs

Although the SSRIs have been the therapeutic agents of choice for a number of years, increasing lines of evidence suggest that the serotonin and nor-adrenalin reuptake blockers (SNRIs) are sometimes more useful in treating various affective disorders (64). The most widely used SNRI venlafaxine, commercially known as Effexor (Wyeth Chemicals, Inc.), functions as a 5-HT reuptake blocker at lower doses (75 mg/d) and as a 5-HT and nor-adrenalin reuptake inhibitor at higher concentrations (225–375 mg/d) (65). Multiple structural derivatives of venlafaxine and other SNRIs have shown similar efficacy as venlafaxine in treating a spectrum of affective disorders, including obsessive compulsive disorder (OCD) and depression (66,67). Additionally, application of a 5-HT$_{1A}$ receptor blocker WAY100635 would mask the inhibitory 5-HT$_{1A}$ autoreceptors and thereby bolster release of 5-HT from these neurons, thus augmenting the anxiolytic effect of the SNRI (66). However, other studies have also shown that the SNRIs produce some nor-adrenergically mediated side effects, such as increase in blood pressure, dry mouth, and constipation, which can be avoided by using an SSRI such as sertraline (68,69).

## ROLE OF 5-HT$_{1A}$-R IN THE IMMUNE SYSTEM AND CANCER

Expression of the 5-HT$_{1A}$ receptor in the immune system has been the central topic of many articles. Initial studies have shown that 5-HT$_{1A}$-R-mediated signaling is responsible for mitogen-stimulated lymphocyte proliferation in mouse, rat, and fish (70,71). Subsequent experiments demonstrated that treatment of spleen-derived B and T lymphocytes with mitogens, such as lipopolysaccharide or phorbol 12-myristate 13-acetate plus ionomycin, causes induced expression of the 5-HT$_{1A}$-R in these cells. Agonist activation of the newly expressed 5-HT$_{1A}$ receptors triggers proliferation of these cells (14). Finally, 5-HT$_{1A}$-R-mediated promotion of mitogen-activated T and B cell survival and proliferation is associated with activation of NF-kappa B (15). Signaling *via* 5-HT$_{1A}$ signaling also plays an important role in the mononuclear cells, because agonist activation of this receptor causes activation of MAPK in these cells (72).

Based on such reports on the role of 5-HT$_{1A}$-R in the cells of the immune system, it is expected that this receptor would play an important role in cancer, which is often caused by a malfunction of the immune system. Notwithstanding the limited number of publications observed so far on this topic, a few reports have implicated this receptor in the function of cancer cells. Thus, 5-HT$_{1A}$-R mediated signaling exerts mitogenic effect on human small cell lung carcinoma cells (73). Dizeyi and coworkers have shown that high-grade prostate cancer cell lines express the 5-HT$_{1A}$-R, and signaling *via* this receptor causes proliferation of the cancer cells (74). Intriguingly, Rizk and Hesketh report a 5-HT$_{1A}$-R agonist with pronounced antiemetic effect, which could be used to prevent vomiting induced by cancer chemotherapy (75). Collectively, data from multiple laboratories suggest that signaling cascades mediated by the 5-HT$_{1A}$-R cause increased proliferation of cancer cells (76).

Literature reviewed in the preceding sections also indicates that the 5-HT$_{1A}$-R mediates positive regulation toward the immune system, which could be beneficial for health. Thus, the anxiolytic agents described earlier may in fact help stimulate our immune system. However, the effect of such therapeutic agents may also produce adverse effects in cancer patients by causing unwanted proliferation of the transformed cells.

## REQUIRED FUTURE STUDIES TO UNRAVEL MECHANISMS OF 5-HT$_{1A}$-R SIGNALING

This review has so far discussed mechanisms of action of 5-HT$_{1A}$ ligands and 5-HT reuptake inhibitors that eventually cause increased signaling through the 5-HT$_{1A}$-R. But the mechanism of action of most of these therapeutic agents has been defined only up to a limited extent. A thorough analysis of the complete pathway of action of a therapeutic agent, from the action of the drug up to the exact ion channel or set of proteins that are involved in causing anxiolysis, has not been elucidated. Such detailed knowledge is expected to yield improved therapeutic agents and strategies to treat affective disorders with minimum side effects. Therefore, it is imperative to delineate the mechanism of action of each therapeutic agent that is currently used to treat depression, anxiety, OCD, schizophrenia, and other related disorders. This section will summarize some of the mechanisms known so far and also point out the large gap in information that still plagues the field of research on affective disorders.

## SSRIs

These compounds block the 5-HT transporter (SERT) molecules on the soma of the dorsal raphé neurons (DRN). Normally, 5-HT taken up by the serotonergic neurons (DRN) somehow causes an inhibition of tryptophan hydroxylase (TPH), which is the rate limiting enzyme in the conversion of tryptophan to serotonin. This inhibition is eliminated in the presence of the 5-HT uptake blockers, such as the SSRIs. It has been shown that SSRIs, such as fluoxetine (Prozac) and sertraline, cause an activation of TPH (77,78). It is known that the TPH promoter harbors a cAMP response element (79) and a cAMP-protein kinase A (PKA)-mediated process is responsible for the SSRI-evoked transcriptional activation of the TPH gene (78). The inverse relationship between SERT and serotonin synthesis is further corroborated by the observation that 5-HT synthesis is increased 25–79% in the brain of SERT(-/-) mice (80).

Although such findings strongly suggest that intracellular 5-HT somehow inhibits the activity or expression of TPH, molecular mechanisms of this process have not been elucidated yet. The possibility that the end product, 5-HT, would cause a feed-back inhibition of TPH by binding to its catalytic pocket is not applicable here because physiological concentrations of 5-HT (in the micromolar range) do not inhibit TPH (81). Finally, as mentioned earlier, it is not clear how exactly postsynaptic 5-$HT_{1A}$-R signaling functions to keep a check on depression and anxiety.

## SNRIs

In addition to blocking the SERT molecules, the SNRIs also inhibit adrenalin/noradrenalin reuptake into the adrenergic neurons, which are regulated in the same way as the serotoninergic neurons. The monoaminergic hypothesis posits that a general decrease of monoamine neurotransmission occurs in the CNS during depression, whereas excessive monoamine neurotransmission occurs during the manic phase (82). Although it has been claimed in some studies that the SNRIs are more effective in ameliorating symptoms of depression, how the nor-adrenergic neurons regulate depression is far from clear. Furthermore, nonmonoaminergic systems are also involved because only 50% of subjects with depression experience full remission of symptoms with therapeutic agents targeted at the mono-aminergic systems (83,84).

## 5-$HT_{1A}$ Agonists and 5-$HT_{1A}$-R-Mediated Signaling

Antidepressant activity of several 5-$HT_{1A}$-R agonists has been reported (85). The 5-$HT_{1A}$ agonists bind to the cognate receptors at both presynaptic, raphé neuron soma as well as the dendrites of the postsynaptic neurons. As discussed earlier, 5-$HT_{1A}$-R signaling in the presynaptic neurons causes inhibition of 5-HT firing and reduced availability of 5-HT at the dendritic terminals of the synapses (Figure 7.1). In the postsynaptic neurons, the elicited 5-$HT_{1A}$-R signaling causes hyperpolarization and inhibition of action potential, but it also results in other signaling effects that could be crucial for the anxiolytic properties of 5-$HT_{1A}$ signaling. The major observation that the therapeutic effect of 5-$HT_{1A}$ agonists was observed only after prolonged treatment for 2 weeks suggested that the mechanism of the action of antidepressants involved desensitization of the presynaptic 5-$HT_{1A}$ receptors (the postsynaptic 5-$HT_{1A}$-R molecules are not downregulated under this condition; see earlier discussion) and subsequent excitation (lack of hyperpolarization) of the 5-HT-firing raphé neurons (Figure 7.1). However, such prolonged stimulation of the postsynaptic 5-$HT_{1A}$-R molecules could also have other effects, such as increased cell division, which may be crucial for the anxiolytic activity of all agents that function by stimulating postsynaptic 5-$HT_{1A}$-R signaling. Corroborating studies by Mehta and coworkers demonstrate that agonist activation of the hippocampal (postsynaptic) 5-$HT_{1A}$ receptors causes increased neurogenesis through a MAPK-mediated pathway (41). Furthermore, Santarelli and coworkers report that the antidepressant effect of Prozac can be eliminated by blocking Prozac-induced neurogenesis in the hippocampus of mice (13). Thus, the pathway, from the stimulation of postsynaptic 5-$HT_{1A}$-R to the anxiolytic effect of the therapeutic agents, still remains to be elucidated.

Antipsychotics: Conventionally, antipsychotics have been placed into two categories: the typical (or earlier generation) antipsychotics such as haloperidol, and the atypical or newer antipsychotics, for example, clozapine. Most of these antipsychotics have been used to treat schizophrenia and, in some cases, depression. The earlier generation antipsychotics yielded the debilitating side effect of dykinesia or Parkinsonian-type muscle rigidity and movement disorder, which was eliminated when atypical antipsychotics were used. Some mechanistic studies on the atypical antipsychotics have been described earlier. According to the most popular hypothesis, schizophrenia is a result of disturbed dopamine release from the dopamine neurons in the brain (84). Experimental findings in support of this hypothesis show that treatment with antipsychotics causes an increase in dopamine release from the dopaminergic neurons (11,59,60,62). So, is schizophrenia caused by low dopaminergic neurotransmission? This possibility is negated by the observation that schizophrenic patients, when psychotic, show increased dopamine synthesis (86–88) and elevated dopaminergic neurotransmission (89–91). Moreover, other theories also implicate molecules such as calcineurin, neuregulin, and NMDA receptors in the incidence of schizophrenia-associated psychosis (92–95). In fact, the recent understanding is that perturbation of dopamine levels is only an associated effect but not the cause of schizophrenia (84). In view of such observations and also based on 5-$HT_{IA}$ agonist activity of the atypical antipsychotics, it can expected that the 5-$HT_{1A}$-R may play an important role in the etiology of schizophrenia. Thus, further mechanistic studies are required to delineate the role of 5-$HT_{IA}$-R-mediated signaling in schizophrenia.

## EXAMPLES OF SOME STUDIES AND POSSIBLE STRATEGIES

In this section, we will present some mechanistic studies that elucidate the role of the 5-$HT_{1A}$-R in mouse brain development and also shed light on the mechanism of action of the antipsychotic agent clozapine. The main purpose of presenting these results is to add the practical flavor of a multifarious approach, which is required in studying mechanisms underlying brain development and mode of action of antipsychotic drugs.

## Early Mouse Brain Development and the 5-$HT_{1A}$-R

Impairment of 5-$HT_{1A}$-R signaling during this stage leads to emotional disorders, such as heightened anxiety, which indicates that 5-$HT_{1A}$-R signaling is crucial for brain development. Yet, how this receptor is linked to brain development is far from clear. Our studies have shown that 5-$HT_{1A}$-R signaling stimulates division of pre-neuronal cells in neonatal mouse brain (41). Data presented here show that agonist stimulation of the 5-$HT_{1A}$-R causes such widespread activation of MAP kinase in the hippocampus that its averaged effect can be measured by Western blotting (Figure 7.2a,b). Also, this MAPK activation finally results in induced expression of cyclin D1, which is crucial for cell division.

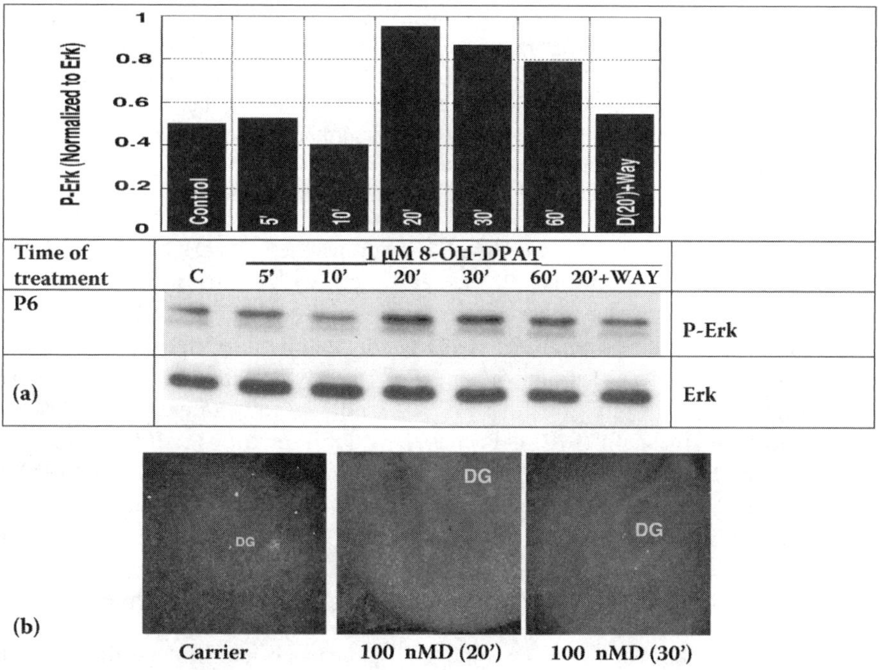


**FIGURE 7.2** Stimulation of 5-$HT_{1A}$-R mediated MAPK activation in mouse hippocampus at P6. Cultured hippocampal slices were placed in serum-free medium and then treated with 8-OHDPAT (D) in the absence and presence of 4 $\mu M$ WAY100635. (a) For Western blotting, the slices were lysed in RIPA buffer, and then aliquots containing 10 μg of protein were analyzed by SDS-PAGE followed by Western blotting for Erk1/2 and then P-Erk. (b) For immunostaining, the slices were fixed and then treated with P-Erk antibody (green) and Nissl stain (red). Western blotting and immunohistochemistry data presented here demonstrate 5-$HT_{1A}$-Rmediated activation of Erk1/2 (which belong to the MAPK family) in the Nissl-positive neuronal cells of the hippocampus.

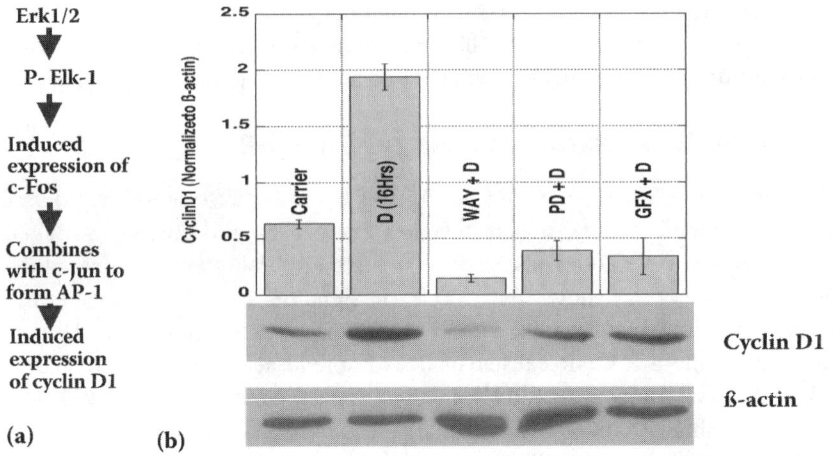


**FIGURE 7.3 Serotonin 1A receptor-mediated stimulation of cyclin D1 in P6 hippocampal slices.** (a) Reported pathway for MAPK-mediated induction of cyclin D1. (b) Cultured, hippocampal slices from P6 mouse brain were placed in serum-free medium and then treated for 16 h with 8OH-DPAT (D) in the absence and presence of a $5\text{-HT}_{1A}$-R antagonist (WAY100635, 4 *μM*), or MEK inhibitor PD98059 (25 *μM*), or a PKC inhibitor (GFX, 2 *μM*). The post-treatment slices were lysed in RIPA buffer and then 10-μg aliquots of protein were analyzed for cyclin D1 and ß-actin levels by SDS-PAGE and Western blotting. The experiment was repeated three times (n = 3) and the data have been presented as mean ± standard deviation.

The mechanism of MAPK-mediated activation of cyclin D1 has been already established (Figure 7.3a). In this pathway, activated Erk1/2 causes phosphorylation-mediated stimulation of the transcription factor Elk-1, which boosts expression of c-Fos. Once induced, c-Fos combines with existing c-Jun molecules to yield elevated levels of the dimeric transcription factor AP-1 (96), which then induces cyclin D1 expression (97,98). Intriguingly, the $5\text{-HT}_{1A}$-R mediated induction of cyclin D1 also requires protein kinase C (PKC), because, in addition to WAY100635 ($5\text{-HT}_{1A}$ antagonist) and PD98059 (blocks phosphorylation-mediated activation of Erk1/2 by MEK), the PKC inhibitor GFX causes reversal of this $5\text{-HT}_{1A}$-R-mediated induction of cyclin D1 (Figure 7.3b).

## Involvement of $5\text{-HT}_{1A}$-R in Clozapine-Evoked Neuronal Activity

Even though it is known that dopaminergic disturbance may not be the cause of schizophrenia, most mechanistic studies of antipsychotic drugs have focused mainly on dopamine release from neurons. Some parts of the brain, like the striatum and ventral tegmental area (VTA), have a high density of dopaminergic neurons, but behavioral abnormalities observed in schizophrenia and other studies on brain activity implicate abnormal function of the prefrontal cortex (PFC) in schizophrenia. Therefore, in our study, we have asked the fundamental question, "Does clozapine cause a change in excitability of neurons in the PFC?" Our studies have revealed the interesting finding that clozapine causes a dramatic increase in activity of the PFC neurons as measured by population spikes (Figure 7.4). Intriguingly, this activity

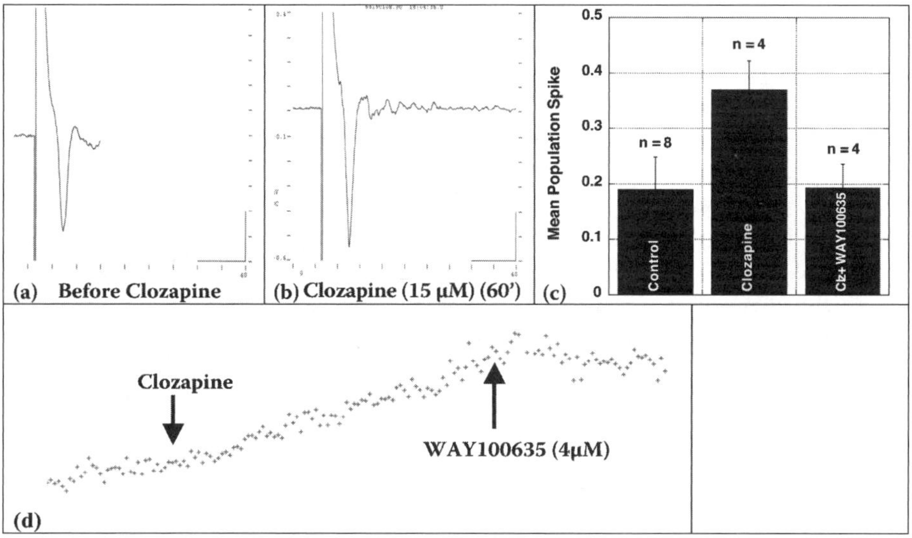


**FIGURE 7.4** Clozapine-evoked increase in population spike is blocked by the 5-$HT_{1A}$-R antagonist WAY100635. Prefrontal cortex (PFC) slices from P20-30 mice were stimulated as described in the methods. After obtaining a stable baseline of population spike (a), clozapine was added to the bath solution to a final concentration of 15 $\mu M$ (b). The maximum increase in population spike (a plateau) was observed after 60 min of clozapine treatment (b and c). This clozapine-evoked increase in activity was completely eliminated when WAY (4 $\mu M$) was included in the bath solution before adding clozapine (c). The increase in neuronal activity could also be truncated before it reached a maximum by adding the 5-$HT_{1A}$ antagonist (WAY100635) 40 min after clozapine treatment (d).

was reversed in the presence of the 5-$HT_{1A}$-R antagonist, WAY100635, thus confirming the involvement of the 5-$HT_{1A}$-R in the clozapine-evoked increase in neuronal activity in the PFC (Figure 2.4). Further analysis of this pathway may help elucidate the mechanism of antipsychotic effects of clozapine.

## CONCLUDING REMARKS

Involvement of the 5-$HT_{1A}$ receptor in crucial physiological processes linked to emotional balance is a well-accepted concept in the current scientific culture. Many scientists still share the notion that the physiologically relevant signaling activity of the 5-$HT_{1A}$-R is limited to the inhibition of adenylyl cyclase. Notwithstanding such dogma, several important studies in cell lines have demonstrated that this receptor is linked to multiple, discreet, yet physiologically important pathways that are not always linked to the inhibition of cAMP (32). Led by such *in vitro* studies, other research teams have conducted similar analysis of signal transduction in mice, rats, and other animals. Due to the complexity of such model systems, results thus obtained have yielded relatively limited mechanistic insight. Nonetheless, they have confirmed the existence of new signaling pathways linked to the 5-$HT_{1A}$-R. Furthermore, importance of this receptor in neonatal brain development has been established by using mutant mice in which receptor expression can be turned on only in the front brain. Despite such findings, it is not clear how this receptor actually helps in brain development.

Similarly, multiple antidepression and antipsychotic drugs are known to function through the 5-$HT_{1A}$ receptor. Yet the mechanisms of action of these therapeutic agents have not been elucidated. Therefore, it is of utmost importance to perform further mechanistic studies to understand how the 5-$HT_{1A}$-R functions in the brain. In addition to the *in vivo* animal models used in some studies, our group has observed that cultured brain slices are more useful in creating a functionally active system that can be placed in a defined medium for appropriate drug treatment (41). By coupling tools like electrophysiology and immunohistochemistry with rigorous biochemical analysis while also using micro-dialysis and genetic manipulation of both animals as well as brain slices, it will now be possible to perform more elaborate studies to unravel the mechanism of involvement of the 5-$HT_{1A}$-R in brain development.

## EXPERIMENTAL PROCEDURES

**Materials:** Antibodies to P-Erk and cyclin D1 were obtained from Cell Signaling (Beverly, MA). The horse radish peroxidase-labeled secondary antibodies were purchased from Santa Cruz Biotechnology (Santa Cruz, CA). The anti-ß-actin antibody, 8-OH-DPAT, WAY100635, wortmannin, and PD98059 were obtained from Sigma Chemicals (St. Louis, MO). Bis-indolylmaleimide or GF109203X (GFX) was purchased from Calbiochem (La Jolla, CA). The Alexafluor-labeled fluorescent secondary antibodies and the fluorescent Nissl stain were obtained from Molecular Probes (Eugene, OR).

**Organotypic culture of hippocampal slices:** The procedure of isolation of mouse brain slices for culture was adapted from publications by Stoppini and coworkers, Xiang and coworkers, and Grimpe and coworkers (99–101). Briefly, mouse pups of specific ages were anaesthetized with ketamine (100 mg/kg) and decapitated. Under sterile conditions, the brains were isolated and then cut at 60° from the longitudinal fissure at the top using a hippocampus dissecting tool to expose the hippocampus. The hemispheres containing the hippocampi were then placed for 30–40 min in modified Gey's balanced salt solution (mGBSS) (prechilled to 4°C) while bubbling a mixture of 95% $O_2$ and 5% $CO_2$. Individual hippocampi were isolated using dissection tools and then 400-$\mu$m thick transverse slices were prepared using a tissue chopper (Stoelting, IL). The slices were placed in ice-cold mGBSS and inspected using a dissection microscope for the presence of uninterrupted bright transparent neuronal layers characteristic of the hippocampal structure. Only such slices were placed on Millicell CM filters (Millipore, Bedford, MA). Up to 3 slices were placed on each filter and the filters were placed in a 6-well dish with 1 ml of medium. The slices were thus kept at the air-medium interphase on high potassium culture medium (25% horse serum, 50% Basal Essential Media-Eagles, 25% Earle's balanced salt solution

(EBSS), 25 m*M* Na-HEPES, 1 m*M* glutamine, 28 m*M* glucose, pH 7.2) for the first 2 d. After incubation at 32°C in a 5% CO2 atmosphere, the culture medium was changed to physiological potassium slice culture medium (20% dialyzed fetal bovine serum, 5% Basal Essential Media-Eagles, and EBSS modified to adjust the potassium concentration to 2.66 m*M*). After 20% dialyzed serum treatment for two days, and before drug treatment, the slices were placed on serum-free medium (the same medium as above, but without serum) for one hour. This was followed by treatment with the inhibitors and antagonists for 30 min, followed by treatment with the agonist for the specified time periods. After drug treatment, the slices were either fixed in 4% paraformaldehyde or lysed as discussed in the following sections. mGBSS composition (in m*M*): $CaCl_2$ (1.5), KCl (4.9), $KH_2PO_4$ (0.2), $MgCl_2$ (11.0), $MgSO_4$ (0.3), NaCl (138), $NaHCO_3$ (2.7), $Na_2HPO_4$ (0.8), NaHEPES (25), glucose 6% (w/v), pH 7.2.

**Drug treatment of slices:** The slices were routinely treated with drugs on the fourth day of culture. The inhibitors were added 30 min before the agonist (100 n*M* 8-OH-DPAT). The concentrations of antagonists and inhibitors were as follows: WAY100635 (4 $\mu M$), PD98059 (25 $\mu M$), and GFX (2 $\mu M$).

**Western blotting:** The drug-treated slices were lysed in 1 ml RIPA buffer (PBS containing 1% NP40, 0.5% sodium deoxycholate, 0.1% SDS, 0.5 m*M* Na3VO4 plus freshly added protease inhibitor cocktail; Boeringer), the lysate (10 µg protein) was resolved using a 716% gradient acrylamide gel, protein bands transferred to a nitrocellulose membrane, in a blocking solution containing 5% solution of defatted milk in 0.1% TWEEN 20 in TBS (20 m*M* Tris-HCl, pH 7.4, 0.8% NaCl) (t-TBS) and then the membrane probed with an ERK1/2 antibody (1: 500) followed by treatment with HRP-linked goat antirabbit IgG (1:50,000). Both antibodies were dissolved in the blocking solution. After probing with the Erk1/2 antibody, the blot was stripped by incubating for 1 h at room temperature in 0.2 *M* glycine (pH 2.5), and then blocked in 5% defatted milk solution and reprobed using a monoclonal, phospho-Erk specific (antiactive) antibody at 1:1000 dilution and then with horseradish peroxidase (HRP)-labeled antimouse IgG (1:5000). The immunoreactive bands were visualized using the Supersignal luminol kit (Pierce). The ERK1/2 bands were used to confirm that the observed increase in P-Erk bands is not due to an increase in the amount of ERK1/2 proteins. As the P-Erk antibody could not be stripped off in successive stripping/reprobing, it was always used in the final probing.

While monitoring cyclin D1 levels, a similar procedure was followed. The cyclin D1 antibody was used at 1:1000 dilution and the ß-actin antibody was diluted to 1:10,000 in 5% defatted milk solution.

**Immunostaining of slices:** The cultured and drug-treated slices were washed quickly with chilled 10 m*M* phosphate buffer (PB) and then fixed overnight at 4°C in 4% paraformaldehyde. The sections were then removed from the membrane with a brush and placed in a 48-well plate in TBS. This was followed by 2–3 washes (15 each) with TBS. For immunofluorescence staining, free-floating sections were first incubated for 30 min in 2N HCl at 37°C, and then rinsed 3X (15 each) with TBS. Sections were then blocked in TBS-X (TBS-0.1% Triton X-1003% serum from the animal used to raise the 2° antibody that was going to be used later) for 1 h at room temperature. This was followed by treatment with primary (1°) antibody in TBS for 48 h at 4°C with gentle rocking. Antibody concentration used: P-Erk (1:400). The sections were next washed 3 × 15′ at room temperature with TBS and then treated with fluorescent, 2° antibodies covalently linked to AlexaFluor488 (green) (1:200). After 48 h of 2° antibody treatment at 4°C, the sections were washed in TBS and then treated with Nissl stain. Slices were incubated in NeuroTrace red fluorescent Nissl stain (Molecular Probes) (1:200) in blocking solution without Triton-X-100 at room temperature for 20–30′. Following this, the slices were mounted on slides with ProLong Gold antifade reagent (Molecular Probes, Eugene, CA) for visualization and photography using a Laser confocal microscope.

**Confocal microscopy of the immunstained slices, cell counting, and statistical analysis:** Using a Nikon C1-LU3 laser scanning confocal system and 488 nm exciting wavelength for P-Erk (green) and 568 nm for Nissl-labeled (red) cells, the slices were viewed at 4× and 20×. The EZC1-system software was used to determine the total thickness of each slice after adjusting channels to obtain pictures from each exciting wavelength separately while blocking the laser beam of the other, exciting wavelength.

**Recording neuronal activity from acutely isolated prefrontal cortex slices:** Prefrontal cortex slices from postnatal day 20 to day 30 Swiss–Webster mice. Thick coronal sections (300-$\mu$m thick) from Prefrontal cortex were used for the electrical recording the stimulating electrode was placed on Layer VI and population spike was measured from Layer IV. Low frequency (0.03 Hz) repeated stimulation was applied at every 30-sec time interval. Once a steady basal level of population spike was obtained for about 10 min, clozapine (15 $\mu M$) was added to the bath (Ringer buffer, containing, in m$M$: NaCl 124, KCl 3.1, $KH_2PO_4$ 1.3, $MgSO_4$ 1.3, $CaCl_2$ 3.1, $NaHCO_3$ 25.5, glucose 10.0) and the recording was continued until a plateau was reached (60 min). The same experiment was repeated four times (n = 4). In all experiments clozapine resulted in a significant increase in the population spike. The average value of the clozapine-boosted population spike obtained from 4 sets of experiments was compared with the average of population spike observed before drug treatment, and also in the presence of WAY100635 (4 $\mu M$) plus clozapine (Figure 7.4).

## ACKNOWLEDGMENTS

We thank Dr. G. Merz (Digital Microscopy Lab, IBR, NY) for expert advice and help in performing confocal microscopy. Fellowship supports were provided by the NYNYS–OMRDD (Mukti Mehta) and Macromolecular Assembly Institute (CUNY) (Baishali Kanjilal). Part support obtained from NIH grant MH071376.

## REFERENCES


1. Lemonde, S., Turecki, G., Bakish, D., Du, L., Hrdina, P.D., Bown, C.D., Sequeira, A., Kushwaha, N., Morris, S.J., Basak, A., Ou, X.M., and Albert, P.R., Impaired repression at a 5-Hydroxytryptamine $_{1A}$ receptor gene polymorphism associated with major depression and suicide, *J Neurosci,* 23, 8788–8799, 2003.
2. Hansenne, M., Pitchot, W., Pinto, E., Reggers, J., Scantamburto, G., Fuchs, S., Pirard, S., and Ansseau, M., 5-$HT_{1A}$ dysfunction in borderline personality disorder, *Psychol Med,* 32, 935–941, 2002.
3. Lowther, S., De Paermentier, F., Cheetham, S.C., Crompton, M.R., Katona, C.L., and Horton, R.W., 5-$HT_{1A}$ receptor binding sites in post-mortem brain samples from depressed suicides and controls, *J Affect Disord,* 42, 199–207, 1997.
4. Hsiung, S.C., Adlersberg, M., Arango, V., Mann, J.J., Tamir, H., and Liu, K.P., Attenuated 5-$HT_{1A}$ receptor signaling in brains of suicide victims: involvement of adenylyl cyclase, phosphatidylinositol 3-kinase Akt and mitogen-activated protein kinase, *J Neurochem,* 87, 182–194, 2003.
5. Mann, J.J., Neurobiology of suicidal behaviour, *Nat Rev Neurosc,* 4, 819–828, 2003.
6. Ramboz, S., Oosting, R., Amara, D.A., Kung, H.F., Blier, P., Mendelsohn, M., Mann, J.J., Brunner, D., and Hen, R., Serotonin $receptor_{1A}$ knockout: an animal model of anxiety-related disorder, *Proc Natl Acad Sci USA,* 95, 14476–14481, 1998.
7. Parks, C.L., Robinson, P.S., Sibille, E., Shenk, T., and Toth, M., Increased anxiety of mice lacking the $serotonin_{1A}$ receptor, Proc *Natl Acad Sci USA,* 95, 10734–10739, 1998.
8. Heisler, L.K., Chu, H-M., Brennan, T.J., Danao, J.A., Bajwa, P., Parsons, L.H., and Tecott, L.H., Elevated anxiety and antidepressant-like responses in serotonin 5-$HT_{1A}$ receptor mutant mice, *Proc Natl Acad Sci USA,* 95, 15049–15054, 1998.
9. Gross, C., Zhuang, X., Stark, K., Ramboz, S., Oosting, R., Kirby, L., Santarelli, L., Beck, S., and Hen, R., $Serotonin_{1A}$ receptor acts during development to establish normal anxiety-like behaviour in the adult, *Nature,* 416, 396–400, 2002.
10. Bantick, R.A., Deakin, J.F., and Grasby, P.M., The 5-$HT_{1A}$ receptor in schizophrenia: a promising target for novel atypical neuroleptics, *J Psychopharmacol,* 15, 37–46, 2001.
11. Diaz-Mataix, L., Scorza, M.C., Bortolozzi, A., Toth, M., Celada, P., and Artigas, F., Involvement of 5-$HT_{1A}$ receptors in prefrontal cortex in the modulation of dopaminergic acitivity: role in atypical antipsychotic action, *J Neurosci,* 25, 10831–10843, 2005.
12. Celada, P., Puig, M.V., Amargós-Bosch, M., Adell, A., and Artigas, F., The therapeutic role of 5-$HT_{1A}$ and 5-HT2A receptors in depression, *J Psychiatr Neurosci,* 29, 252–265, 2004.
13. Santarelli, L., Saxe, M., Gross, C., Surget, A., Battaglia, F., Dulawa, S., Weisstaub, N., Lee, J., Duman, R., Aracio, O., Belzung, C., and Hen, R., Requirement of Hippocampal Neurogenesis for the Behavioral Effects of Antidepressants, *Science,* 301, 805–809, 2003.
14. Abdouh, M., Storring, J.M., Riad, M., Paquetter, Y., Albert, P.R., Drobetsky, E., and Kouassi, E., Transcriptional mechanisms for induction of 5-$HT_{1A}$ receptor mRNA and protein in activated B and T lymphocytes, *J Biol Chem,* 276, 4382–4388, 2001.
15. Abdouh, M., Albert, P.R., Drobetsky, E., Filep, J.G., and Kouassi, E., 5-$HT_{1A}$ mediated promotion of mitogen-activated T and B cell survival and proliferation is associated with increased translocation of NF-kB to the nucleus, *Brain Behav Immun,* 18, 24–34, 2004.

16. Husain, Z., Almeciga, I., Delgado, J.C., Clavijo, O.P., Castro, J.E., Belalcazar, V., Pinto, C., Zuniga, J., Romero, V., and Yunis, E.J., Increased FasL expression correlates with apoptotic changes in granulocytes cultured with oxidized clozapine, *Toxicol Appl Pharmacol,* 2006 (in press, corrected proof).
17. Schulte, P.F.J., Risk of clozapine-associated agranulocytosis and mandatory white blood cell monitoring, *Ann Pharmacother,* 40, 683–688, 2006.
18. Tiihonen, J., Fatal agranulocytosis 4 years after discontinuation of clozapine, *Am J Psychiatr,* 163, 161, 2006.
19. Santana, N,B.A., Serrats, J, Mengod, G., and Artigas, F., Expression of serotonin $_{1A}$ and serotonin 2A receptors in pyramidal and GABAergic neurons of the reat prefrontal cortex, *Cereb Cortex,* 14, 1100–1109, 2004.
20. Day, H.E., Greenwood, B.N., Hammack, S.E., Watkins, L.R., Fleshner, M., Maier, S.F., and Campeau, S., Differential expression of 5HT-1A, alpha 1b adrenergic, CRFR1, and CRF-R2 receptor mRNA in serotonergic, gamma-aminobutyric acidergic, and catecholaminergic cells of the rat dorsal raphe nucleus, *J Comp Neurol,* 474, 364–378, 2004.
21. Palchaudhuri, M. and Flugge, G., 5-HT$_{1A}$ receptor expression in pyramidal neurons of cortical and limbic brain regions, *Cell Tissue Res,* 321, 159–172, 2005.
22. Luna-Munguia, H., Manuel-Apolinar, L., Rocha, L., and Meneses, A., 5-HT$_{1A}$ receptor expression during memory formation, *Psychopharmacology (Berl),* 181, 309–318, 2005.
23. Aznavour, N., Rbah, L., Leger, L., Buda, C., Sastre, J.P., Imhof, A., Charnay, Y., and Zimmer, L., A comparison of *in vivo* and *in vitro* neuroimaging of 5-HT$_{1A}$ receptor binding sites in the cat brain, *J Chem Neuroanat,* 31, 226–232, 2006.
24. Geyer, S., Luppino, G., Ekamp, H., and Zilles, K., The macaque inferior parietal lobule: cytoarchitecture and distribution pattern of serotonin 5-HT$_{1A}$ binding sites, *Anat Embryol (Berl),* 210, 353–362, 2005.
25. Parsey, R.V., Arango, V., Olvet, D.M., Oquendo, M.A., Van Heertum, R.L., and Mann, J.J., Regional heterogeneity of 5-HT$_{1A}$ receptors in human cerebellum as assessed by positron emission tomography, *J Cereb Blood Flow Metab,* 25, 785–793, 2005.
26. Azmitia, E.C., Yu, I., Akbari, H.M., Kheck, N., Whitaker-Azmitia, P.M., and Marshak, D.R., Antipeptide antibodies against the 5-HT$_{1A}$ receptor, *J Chem Neuroanat,* 5, 289–298, 1992.
27. Santana, N., Bortolozzi, A., Serrats, J., Mengod, G., and Artigas, F., Expression of serotonin $_{1A}$ and serotonin 2A receptors in pyramidal and GABAergic neurons of the reat prefrontal cortex, *Cereb Cortex,* 14, 1100–1109, 2004.
28. Patel, T.D. and Zhou, F.C., Ontogeny of 5-HT$_{1A}$ receptor expression in the developing hippocampus, *Brain Res Dev Brain Res,* 157, 42–57, 2005.
29. Chen, Y. and Penington, N.J., Differential effects of protein kinase C activation on 5-HT$_{1A}$ receptor coupling to $Ca^{2+}$ and K+ currents in rat serotonergic neurones, *J Physiol,* 496, 129–137, 1996.
30. Jeong, H.-J., Han, S.-H., Min, B.-I., and Cho, Y.-W., 5-HT$_{1A}$ receptor-mediated activation of G-protein-gated inwardly rctifying $K^+$ current in rat periaqueductal gray neurons, *Neuropharmacology,* 41, 175–185, 2001.
31. Lei, Q., Talley, E.M., and Bayliss, D.A., Receptor-mediated inhibition of G protein-coupled inwardly rectifying potassium channels involves Gaq family subunits, phospholipase C, and a readily diffusible messenger, *J Biol Chem,* 276, 16720–16730, 2001.
32. Adayev, T., Ranasinghe, B., and Banerjee, P., Transmembrane signaling in the brain by serotonin, a key regulator of physiology and emotion, *Biosci Rep,* 25, 363–385, 2005.

33. Hensler, J.G., Differential regulation of 5-HT$_{1A}$ receptor-G protein interactions in brain following chronic antidepressant administration, *Neuropsychopharmacology,* 26, 565–573, 2002.
34. Riad, M., Zimmer, L., Rbah, L., Watkins, K.C., Hamon, M., and Descarries, L., Acute treatment with the antidepressant fluoxetine internalizes 5-HT$_{1A}$ autoreceptors and reduces the *in vivo* binding of the PET radioligand [18F]MPPF in the nucleus raphe dorsalis of rat, *J Neurosci,* 24, 5420–5426, 2004.
35. Blier, P. and Ward, N.M., Is there a role for 5-HT$_{1A}$ agonists in the treatment of depression?, *Biol Psychiatr,* 53, 193–203, 2003.
36. Albert, P.R., Lemonde, S., 5-HT$_{1A}$ receptors, gene repression, and depression: guilt by association, *Neuroscientist,* 10, 575–593, 2004.
37. Haddjeri, N., Ortemann, C., de Montigny, C., and Blier, P., Effect of sustained administration of the 5-HT receptor agonist 1A flesinoxan on rat 5-HT neurotransmission, *Eur Neuropsychopharmacol,* 9, 427–440, 1999.
38. Zimmer, L., Riad, M., Rbah, L., Belkacem-Kahlouli, A., Le Bars, D., Renaud, B., and Descarries, L., Toward brain imaging of serotonin 5-HT$_{1A}$ autoreceptor internalization, *Neuroimage,* 22, 1421–1426, 2004.
39. Adayev, T., El-Sherif, Y., Barua, M., and Banerjee, P., Agonist stimulation of the serotonin$_{1A}$ receptor causes supression of anoxia-induced apoptosis via mitogen-activated protein kinase in neuronal HN2-5 cells, *J Neurochem,* 72, 1489–1496, 1999.
40. Adayev, T., Ray, I., Sondhi, R., Sobocki, T., and Banerjee, P., The G protein-coupled 5-HT$_{1A}$ receptor causes suppression of caspase-3 through MAPK and protein kinase C, *Biochim Biophys Acta,* 1640, 85–96, 2003.
41. Mehta, M., Ahmed, Z., Cano-Sanchez, P., Fernando, S.S., Wieraszko, A., and Banerjee, P., The 5-HT1A receptor, brain development, and emotional disorders, *37th Annual Meeting of the American Society for Neurochemistry,* 96, 102, 2006.
42. Kushwaha, N. and Albert, N., Coupling of 5-HT$_{1A}$ autoreceptors to inhibition of mitogen-activated protein kinase activation via G beta gamma subunit signaling, *Eur J Neurosci,* 21, 721–732, 2005.
43. Banerjee, P., Berry-Kravis, E., Bonafede-Chhabra, D., and Dawson. G., Heterologous expression of the serotonin 5-HT$_{1A}$ receptor in neural and nonneural cell lines, *Biochem Biophys Res Commun,* 192, 104–110, 1993.
44. Burnet, P.W., Eastwood, S.L., and Harrison, P.J., [$^3$H]WAY-100635 for 5-HT$_{1A}$ receptor autoradiography in human brain: a comparison with [$^3$H]8-OH-DPAT and demonstration of increased binding in the frontal cortex in schizophrenia, *Neurochem Int,* 30, 565–574, 1997.
45. Lai, M.K., Tsang, S.W., Francis, P.T., Esiri, M.M., Keene, J., Hope, T., and Chen, C.P., Reduced serotonin 5-HT$_{1A}$ receptor binding in the temporal cortex correlates with aggressive behavior in Alzheimer disease, *Brain Res,* 974, 82–87, 2003.
46. Sheline, Y.I., Wang, P.W., Gado, M.H., Csernansky, J.G., and Vannier, M.W., Hippocampal atrophy in recurrent major depression, *Proc Natl Acad Sci USA,* 93, 3908–3913, 1996.
47. Pantel, J., Schroder, J., Essig, M., Schad, L.R., Popp, D., Dech, H., Knopp, M.V., Schad, L.R., Eysenbach, K., Backenstrass, M., and Friedlinger, M., Quantitative magnetic resonance imaging in geriartric depression and primary degenerative dementia, *J Affect Disord,* 42, 69–83, 1997.
48. Drevets, W.C., Price, J.L., Simpson, J.R. (Jr.), Todd, R.D., Reich, T., Vannier, M., and Raichle, M.E., Subgenual prefrontal abnormalities in mood disorders, *Nature,* 386, 824–827, 1997.

49. Ashtari, M., Greenwald, B.S., Kramer-Ginsberg, E., Hu, J., Wo, H., Patel, M., Aupperle, P., and Pollack, S., Hippocampal/amygdala volumes in geriatric depression, *Psychol Med,* 29, 629–638, 1999.
50. Weinstein, D., Magnuson, D., and Lee, J., Altered G-protein coupling of a frontal cortical low affinity [$^3$H]8-hydroxy-N.N-dipropyl-2-aminotetralin serotonergic binding site in Alzheimer's disease, *Behav Brain Res,* 73, 325–329, 1996.
51. Zhou, F.C., McKinzie, D.L., Patel, T.D., Lumeng, L., and Li, T.K., Additive reduction of alcohol drinking by 5-$HT_{1A}$ antagonist WAY 100635 and serotonin uptake blocker fluoxetine in alcohol-preferring P rats, *Alcohol Clin Exp Res,* 22, 266–269, 1998.
52. Hensler, J.G., Ladenheim, E.E., and Lyons, W.E., Ethanol consumption and serotonin-1A (5-$HT_{1A}$) receptor function in heterozygous BDNF (+/-) mice, *J Neurochem,* 85, 1139–1147, 2003.
53. Hofmann, C.E., Simms, W., Yu, W.K., and Weinberg, J., Prenatal ethanol exposure in rats alters serotonergic-mediated behavioral and physiological function, *Psychopharmacology,* 161, 379–386, 2002.
54. Schechter, L.E., Dawson, L.A., and Harder, J.A., The potential utility of 5-$HT_{1A}$ receptor antagonists in the treatment of cognitive dysfunction associated with Alzheimer's disease, *Curr Pharm Des,* 8, 139–145, 2002.
55. Popova, N.K. and Ivanova, E.A., 5-$HT_{1A}$ receptor antagonist p-MPPI attenuates acute ethanol effects in mice and rats, *Neurosci Lett, 322*, 1–4, 2002
56. Meltzer, H.Y., The role of serotonin in antipsychotic drug action, *Neuropsychopharmacology*, 21, 106S–115S, 1999.
57. Muller, C.P., De Souza, Silva M.A., DePalma, G., Tomaz, C., Carey, R.J., and Huston, J.P., The selective serotonin $_{1A}$-receptor antagonist WAY 100635 blocks behavioral stimulating effects of cocaine but not ventral striatal dopamine increase, *Behav Brain Res*, 134, 337–346, 2002.
58. Muller, C.P., Carey, R.J., De Souza, M.A., Jocham, G., and Huston, J.P., Cocaine increases serotonergic activity in the hippocampus and nucleus accumbens *in vivo*: 5$HT_{1A}$-receptor antagonism blocks behavioral but potentiates serotonergic activation, *Synapse,* 45, 67–77, 2002.
59. Ichikawa, J., Ishii, H., Bonaccorso, S., Fowler, W.L., O'Laughlin, A.A., and Meltzer, H.Y., 5-HT2A and D receptor blockade increases cortical DA release via 5-$HT_{1A}$ receptor activation: a possible mechanism of atypical antipsychotic-induced cortical dopamine release, *J Neurochem,* 76, 1521–1531, 2001.
60. Rollema, H., Lu, Y., Schmidt, A.W., and Zorn, S.H., Clozapine increase dopamine release in prefrontal cortex by 5-$HT_{1A}$ receptor activation, *Eur J Pharmacol,* 338, R3–R5, 1997.
61. Millan, M.J., Improving the treatment of schizophrenia: focus on serotonin $(5HT)_{1A}$ receptors, *J Pharmacol Exp Ther,* 295, 853–861, 2000.
62. Assie, M.B., Ravailhe, V., Faucillon, V., and Newman-Tancredi, A., Contrasting contribution of 5-hydroxytryptamine $_{1A}$ receptor activation to neurochemical profile of novel antipsychotics: frontocortical dopamine and hippocampal serotonin release in rat brain, *J Pharmacol Exp Ther,* 315, 265–272, 2005.
63. Ichikawa, J., Kuroki, T., Dai, J., and Meltzer, H.Y., Effect of antipsychotic drugs on extracellular serotonin levels in rat medial prefrontal cortex and nucleus accumbens, *Eur J Pharmacol, 351*, 163–171, 1998.
64. Baldomero, E.B., Ubago, J.G., Cercos, C.L., Ruiloba, J.V., Calvo, C.G., and Lopez, R.P., Venlafaxine extended release versus conventional antidepressants in the remission of depressive disorders after previous antidepressant failure: ARGOS study, *Depression Anxiety*, 22, 68–76, 2005.

65. Debonnel, G., Saint-Andre, E., Hebert, C., de Montigny, C., Lavoie, N., and Blier, P., Differential physiological effects of a low dose and high doses of venlafaxine in major depression, *Int J Neuropsychopharmacol*, 1–11, 2006 [Epub ahead of print].
66. Deecher, D.C., Beyer, C.E., Johnston, G., Bray, J., Shah, S., Abou-Gharbia, M., and Andree, T.H., Desvenlafaxine succinate: a new serotonin and norepinephrine reuptake inhibitor, *J Pharmacol Exp Ther,* 2006 [Epub ahead of print].
67. Dell'Osso, B., Nestadt, G., Allen, A., and Hollander, E., Serotonin-norepinephrine reuptake inhibitors in the treatment of obsessive-compulsive disorder: a critical review, *J Clin Psychiatr,* 67, 600–610, 2006.
68. Sir, A., D'Souza R.F., Uguz, S., George, T., Vahip, S., Hopwood, M., Martin, A.J., Lam, W., and Burt, T., Randomized trial of sertraline versus venlafaxine XR in major depression: efficacy and discontinuation symptoms, *J Clin Psychiatr,* 66, 1312–1320, 2005.
69. Thase, M.E., Treatment of anxiety disorders with venlafaxine XR, *Expert Rev Neurother,* 6, 269–282, 2006.
70. Iken, K., Chheng, S., Fargin, A., Goulet, A.-C., and Kouassi, E., Serotonin upregulates mitogen-stimulated B lymphocyte proliferation through 5-$HT_{1A}$ receptors, *Cell Immunol,* 163, 1–9, 1995.
71. Ferriere, F., Khan, N.A., Troutaud, D., and Deschaux, P., Serotonin modulation of lymphocyte proliferation via 5-$HT_{1A}$ receptors in rainbow trout (Oncorhynchus mykiss), *Dev Comp Immunol,* 20, 273–283, 1996.
72. Cloez-Tayarani, I., Kayyali, U.S., Fanburg, B.L., Cavaillon, J.M., 5-HT activates ERK MAP kinase in cultured-human peripheral blood mononuclear cells via 5-$HT_{1A}$ receptors, *Life Sci,* 76, 429–443, 2004.
73. Cattaneo, M.G., Fesce, R., and Vicentini, L.M., Mitogenic effect of serotonin in human small cell lung carcinoma cells via both 5-$HT_{1A}$ and 5-HT1D receptors, *Eur J Pharmacol,* 291, 209–211, 1995.
74. Dizeyi, N., Bjartell, A., Nilsson, E., Hansson, J., Gadaleanu, V., Cross, N., and Abrahamsson, P.-A., Expression of serotonin receptors and role of serotonin in human prostate cancer tissue and cell lines, *The Prostate,* 59, 328–336, 2004.
75. Rizk, A.N. and Hesketh, P.J., Antiemetics for cancer chemotherapy-induced nausea and vomiting. A review of agents in development, *Drugs R D,* 2, 229–235, 1999.
76. Siddiqui, E.J., Thompson, C.S., Mikhailidis, D.P., and Mumtaz, F.H., The role of serotonin in tumour growth (review), *Oncol Rep,* 14, 1593–1597, 2005.
77. Baik, S.-Y., Jung, K.H., Choi, M.-R., Yang, B.-H., Kim, S.-H., Lee, J.-S., Oh, D.-Y., Choi, I.-G., Chung, H., and Chai, Y.G., Fluoxetine-induced up-regulation of 14-33zeta and tryptophan hydroxylase levels in RBL-2H3 cells, *Neurosci Lett,* 374, 53–57, 2005.
78. Kim, S.W., Park, S.Y., and Hwang, O., Up-regulation of tryptophan hydroxylase expression and serotonin synthesis by sertraline, *Mol Pharmacol, 61*, 778–785, 2002.
79. Boularand, S., Darmon, M.C., Ravassard, P., and Mallet, J., Characterization of the human tryptophan hydroxylase gene promoter. Transcriptional regulation by cAMP requires a new motif distinct from the cAMP-responsive element, *J Biol Chem,* 270, 3757–3764, 1995.
80. Kim, D.-K., Tolliver, T.J., Huang, S.-J., Martin, B.J., Andrews, A.M., Wichems, C., Holmes, A., Lesch, K.-P., and Murphy, D.L., Altered serotonin synthesis, turnover and dynamic regulation in multiple brain regions of mice lacking the serotonin transporter, *Neuropharmacology*, 49, 798–810, 2005.
81. Park, D.H., Stone, D.M., Kim, K.S., and Joh, T.H., Characterization of recombinant mouse tryptophan hydroxylase expressed in *Escherichia coli*, *Mol Cell Neurosc,* 5, 87–93.

82. Barchas, J.D. and Altemus, M., Biochemical hypotheses of mood and anxiety disorders, in *Basic Neurochemistry: Molecular, Cellular, and Medical Aspects,* Siegel, G.J., Agranoff, B.W., Albers, R.W., Fisher, S.K., and Uhler, M.D., Eds., Lippincott-Raven, Philadelphia, PA, 1999, pp. 1077–1078.
83. Berton, O. and Nestler, E., New approaches to antidepressant drug discovery: beyond monoamines, *Nat Rev Neurosci,* 7, 137–151, 2006.
84. Kapur, S., Psychosis as a state of aberrant salience: a framework linking biology, phenomenology, and pharmacology in schizophrenia, *Am J Psychiatr,* 160, 13–23, 2003.
85. Winsauer, P.J., Rodriguez, F.H., Cha, A.E., and Moerschbaecher, J.M., Full and partial 5-$HT_{1A}$ receptor agonists disrupt learning and performance in rats, *J Pharmacol Exp Ther,* 288, 335–347, 1999.
86. Dao-Castellana, M.-H., Paillere-Martinot, M.-L., Hantraye, P., Attar-Levy, D., Remy, P., Crouzel, C., Artiges, E., Feline, A., Syrota, A., and Martinot, J.-L., Presynaptic dopaminergic function in the striatum of schizophrenic patients, *Schizophr Res,* 23, 167–174, 1997.
87. Reith, J., Benkelfat, C., Sherwin, A., Yasuhara, Y., Kuwabara, H., Andermann, F., Bachneff, S., Cumming, P., Diksic, M., Dyve, S.E., Etienne, P., Evans, A.C., Lal, S., Shevell, M., Savard, G., Wong, D.F., Chouinard, G., and Gjedde, A., Elevated dopa decarboxylase activity in living brain of patients with psychosis, *PNAS 91,* 11651–11654, 1994.
88. Lindstrom, L.H., Gefvert, O., Hagberg, G., Lundberg, T., Bergstrom, M., Hartvig, P., and Langstrom, B., Increased dopamine synthesis rate in medial prefrontal cortex and striatum in schizophrenia indicated by -([beta]-11C) DOPA and PET, *Biol Psychiatr,* 46, 681–688, 1999.
89. Laruelle, M., Abi-Dargham, A., van Dyck, C.H., Gil, R., D'Souza, C.D., Erdos, J., McCance, E., Rosenblatt, W., Fingado, C., Zoghbi, S.S., Baldwin, R.M., Seibyl, J.P., Krystal, J.H., Charney, D.S., and Innis, R.B., Single photon emission computerized tomography imaging of amphetamine-induced dopamine release in drug-free schizophrenic subjects, *PNAS*, 93, 9235–9240, 1996.
90. Abi-Dargham, A., Gil, R., Krystal, J., Baldwin, R.M., Seibyl, J.P., Bowers, M., van Dyck, C.H., Charney, D.S., Innis, R.B., and Laruelle, M., Increased striatal dopamine transmission in schizophrenia: confirmation in a second cohort, *Am J Psychiatr,* 155, 761–767, 1998.
91. Breier, A., Su, T.P., Saunders, R., Carson, R.E., Kolachana, B.S., de Bartolomeis, A., Weinberger, D.R., Weisenfeld, N., Malhotra, A.K., Eckelman, W.C., and Pickar, D., Schizophrenia is associated with elevated amphetamine-induced synaptic dopamine concentrations: evidence from a novel positron emission tomography method, *PNAS* 94, 2569–2574, 1997.
92. Gerber, D.J. and Tonegawa, S., Psychotomimetic effects of drugs —a common pathway to schizophrenia?, *N Engl J Med,* 350, 1047–1048, 2004.
93. Munafò, M.R., Thiselton, D.L., Clark, T.G., and Flint, J., Association of the NRG1 gene and schizophrenia: a meta-analysis, *Mol Psychiatr,* 11, 539–546, 2006.
94. Kristiansen, L., Beneyto, M., Haroutunian, V., and Meador-Woodruff, J.H., Changes in NMDA receptor subunits and interacting PSD proteins in dorsolateral prefrontal and anterior cingulate cortex indicate abnormal regional expression in schizophrenia, *Mol Psychiatr,* 11, 737–747, 2006.
95. Kristiansen, L., Beneyto, M., Haroutunian, V., and Meador-Woodruff, J.H., Altered NMDA receptor expression in schizophrenia, *Mol Psychiatr,* 11, 705, 2006.
96. Karin, M., The regulation of AP-1 by mitogen-activated protein kinases, *J Biol Chem,* 270, 16483–16486, 1995.

97. Shaulian, E., and Karin, M., AP-1 in cell proliferation and survival, *Oncogene,* 20, 2390–2400, 2001.
98. Bakiri, L., Lallemand, D., Bossy-Wetzel, E., and Yaniv, M., Cell cycle-dependent variations in c-Jun and JunB phosphorylation: a role in the control of cyclin D1 expression, *EMBO J,* 19, 2056–2068, 2000.
99. Stoppini, L., Buchs, P.-A., and Muller, D., A simple method for organotypic cultures of nervous tissue, *J Neurosci Methods,* 37, 173–182, 1991.
100. Xiang, Z., Hrabetova, S., Moskowitz, S.I., Casaccia-Bonnefil, P., Young, S.R., Nimmrich, V.C., Tiedge, H., Einheber, S., Karnup, S., Bianchi, R., and Bergold, P.J., Long-term maintenance of mature hippocampal slices *in vitro, J Neurosci Methods,* 98, 145–154, 2000.
101. Grimpe, B., Dong, S., Doller, C., Temple, K., Malouf, A.T., and Silver, J., The critical role of basement membrane-independent laminin 1 chain during axon regeneration in the CNS, *J Neurosci, 22,* 3144–3160, 2002.
102. Miner, L.H., Schroeter, S., Blakely, R.D., and Sesack, S.R., Ultrastructural localization of the serotonin transporter in superficial and deep layers of the rat prelimbic prefrontal cortex and its spatial relationship to dopamine terminals, *J Comp Neurol,* 427, 220–234, 2000.
103. Pickel, V.M. and Chan, J., Ultrastructural localization of the serotonin transporter in limbic and motor compartments of the nucleus accumbens, *J Neurosci,* 19, 7356–7366, 1999.

# 8 Do Limits of Neuronal Plasticity Represent an Opportunity for Mental Diseases, Such as Addiction to Food and Illegal Drugs? Use and Utilities of Serotonin Receptor Knock-Out Mice

*Valerie Compan*

## CONTENTS

Abstract ........ 158
Introduction ........ 158
Human 5-HT Receptor Gene Polymorphisms and Addictions to Food and Illegal Drugs ........ 160
    Polymorphism within Human 5-HT Receptor Genes and Eating Disorders ........ 161
    Polymorphism within Human 5-HT Receptor Genes and Addiction to Illegal Drugs ........ 161
Knock-Out of 5-HT Receptors Classically Required Environmental Challenges to Produce Maladaptive Behavior ........ 163
    5-HT Receptor Knock-Out Mice and Eating Disorders ........ 163
    5-HT Receptor Knock-Out Mice and Novelty-Induced Locomotion ........ 165
    5-HT Receptor Knock-Out Mice and Illegal Drugs such as Ecstasy and Cocaine ........ 167

5-HT Receptor Knock-Out Mice and Adaptive Changes in Central Serotoninergic Systems ........ 168
Conclusion ........ 168
Acknowledgment ........ 168
References ........ 170

## ABSTRACT

The properties of neuronal plasticity are not only studied after brain injury. Adaptive changes also take place in the neurons of genetically modified animals, which provide a rational for genetic studies in humans suffering from mental diseases. Following a brief summary describing the alterations in human serotonin (5-HT, 5-hydroxytryptamine) gene receptors found in one of the most complex of mental functions—motivation related to food or illegal drugs intake—we will describe that the only gene deficit, destroying 5-HT receptor, is not enough to produce apparent maladaptive behavior in rodents. However, when combined with a novel environmental challenge, modified genetic predisposition induces deviant behavior. Locomotion, feeding, and emotion-like state disorders (anxiety-like behavior, depression-like syndrome) are classically underlined by brain deficits and subsequently linked to adaptive neuronal changes. This review gathered evidence to postulate that following genetic alterations, the brain partly "burns" its adaptive resources over time, and lacks a sufficient flexibility to adapt during adulthood. The limits of adaptive changes in neurons in the face of an unexpected environmental context, may therefore represent an "open door" for mental-like disorders to emerge.

## INTRODUCTION

*Neuronal plasticity concept* (neuroplasticity) refers to neurobiological processes that include adaptive changes in the morphology and function of neuronal systems following environmental modifications and/or brain injury. The brain may then exert an adaptive control of behavior. Neuroplasticity persists during adulthood and confers the ability to perform adapted behavior in the face of environmental changes. In other words, over time during development (from conception to death), a "cerebral maturity" may occur, i.e., a progressive equilibrium in neuronal interactions may allow organisms to adapt behavior under environmental challenges, while taking into account internal factors. Surprisingly, behavioral states may appear adapted even when a neuronal imbalance between neuronal interactions exists (neuroplasticity). Subsequently, adaptive changes in neurons following neuronal lesion may represent a "neurobiological support" of adaptive behavioral strategies. In addition, adaptive changes are not equivalent to compensation. Neurons may effectively adapt but may not compensate all deficits. In other words, limits in neuronal adaptive changes exist.

The serotoninergic system is one of the most flexible neuronal systems,[1] and there is evidence for a contribution of this neuromodulator system to synaptic plasticity.[2] Again, in neuroscience history, major advances have taken advantage of studies on the functions of serotonin (5-hydroxytryptamine, 5-HT) in the brain and, in particular, of 5-HT receptors (5-HTR). This article synthesizes results from previous studies and suggests that the neuroplasticity confers to mouse models the ability to adapt in spite of an engineered genetic deficit. In other words, organisms maintained in their usual context (home cage) overcome a genetic deficit, as demonstrated for specific 5-HTR knock-out (KO) mouse.[3–12] In contrast, KO phenotypes are often, but not systematically, best revealed following an environmental or pharmacological challenge.

We have summarized results from previous studies focusing on motivational deficits related to feeding, novelty-induced locomotion, and addictive behavior to illustrate this point. Several studies have reported impairments in novelty-induced locomotion in 5-HTR KO mice: 5-$HT_{1A,}$[5] 5-$HT_{1B,}$[6] 5-$HT_{4,}$[10,11] and 5-$HT_{5A,}$[9] whereas these mutant mice did not exhibit any motor impairments in their home cage. These studies reinforce the view that (1) adaptive mechanisms are limited in these mice, and (2) serotoninergic systems modulate stress responsiveness in situations such as reactivity to novel environment. Likewise, locomotion induced by psychoactive stimulants (cocaine, 3,4-*N*-methylenedioxymethamphetamine or ecstasy) was altered in 5-$HT_{1B}$, 5-$HT_{2C,}$[7,8,13] and 5-$HT_4$R KO mice (Compan, unpublished). Along these lines, adult 5-$HT_{1B}$R and 5-$HT_4$R KO mice displayed maladaptive feeding behavior under unusual context,[14,10] where 5-$HT_{2C}$R KO mouse spontaneously displayed obesity.[15]

This review illustrates that deficits in adaptation to stressors (novel environment, psychostimulants, pharmacological substances) coexist with changes in the serotonergic system themselves in the absence of only one 5-HTR, as so far revealed in 5-$HT_{1A}$R and 5-$HT_4$R KO mice during adulthood but not in 5-$HT_{1B}$R-null adult mouse.[16–18]

Altered 5-HT neuromodulation may underlie a wide range of mental diseases, including anorexia, addiction to illegal drugs or food (bulimia), major depression (stress factor[19]), or anxiety.[20–24] Seven families of 5-HTRs with eighteen subtypes, without including mRNA editing and splice variant 5-HTR isoforms, are currently identified.[25,26] As will be summarized in the first section, a goal of many studies has been to find whether deficits in genes encoding 5-HTRs are or are not associated with mental disorders, predominantly with depression and schizophrenia, and less frequently with feeding disorders and addiction. However, as for most mental diseases, a genetic component combined with environmental determinism is likely. Such combinations of interactions between internal and external factors are, however, complex to study and are seldom investigated, though gene KO strategies may contribute to this knowledge. So far, ten 5-HTR KO mice lacking either 5-$HT_{1A}$R,[4,5] 5-$HT_{1B}$R,[27] 5-$HT_{2A}$,[12] 5-$HT_{2B}$,[28], 5-$HT_{2C}$,[15] 5-$HT_3$,[29] 5-$HT_4$,[10] 5-$HT_{5A}$,[9] 5-$HT_6$,[15] and 5-$HT_7$[30,31] receptors have been generated to study 5-HTR functions.

In this conceptual context, where environmental and biologic factors overlap, it is difficult to believe that only a therapeutic treatment would overcome mental diseases. Identifying effective therapeutic treatments from knowledge of specific 5-HTR functions and other related biological events (transduction systems, other neuronal and glial systems), may be promising but unpredictable. However, this article opens the possibility that a genetic deficit is not an absolute "un-overcome" because of neuronal adaptive changes. Extending the limits of adaptive changes in neurons may open the possibility to circumvent mental disease. It can be expected that the overall view of a "psynogenic puzzle"* will suggest possible combined therapies established on both pharmacological and psychological treatments. However, the potential for affecting biologic predispositions as they interact with environmental factors remains considerably unknown.

## HUMAN 5-HT RECEPTOR GENE POLYMORPHISMS AND ADDICTIONS TO FOOD AND ILLEGAL DRUGS

The neuronal processes involved in motivation, particularly for eating, have a high reward component. In humans, eating disorders are classified as mental disorders according to the Diagnostic and Statistical Manual of Mental Disorders (DSM IV-TR, APA, 2003). Bulimia is included in addictive behavior, and binge eating disorder is an excessive consumption of food, also observed in patients suffering from bulimia and sometimes from anorexia, and a few obese patients. The excess or absence of food intake is therefore not only a metabolic and/or endocrinological matter, but also, like in the consumption of drugs of abuse, a mental problem. One of the main principles of neural science is that disturbances of brain function underlie mental disorders. "Miscegenation" of neurobiology and psychology ("being in relation") should be further studied because stress can influence feeding disorders, which coexist with anxiety and depression.[32–34]

If the focus of the following association studies is almost systematically based on maladapted phenotypes of 5-HTR KO mouse, in turn, such analyses provide fundamental interest to both *in vitro* and *in vivo* pharmacological studies using animal models. Twelve different genes encode metabotropic 5-HTRs in humans (NCBI, Map Viewer at www.ncbi.nlm.nih.gov/mapview/) (e.g., gene structure in Reference 26) and, five genes encode the only 5-HTR ligand-gated ion channels, $5\text{-HT}_3$R (HTR3A and HTR3B[35]; $\text{HTR3C}_{1\text{-}4}$[36]; HTR3D, HTR3E[37]). Single nucleotide polymorphisms (SNP) may influence transduction systems (e.g., Reference 38); $B_{\text{max}}$ of $5\text{-HT}_{1A}$R,[39] $5\text{-HT}_{1B}$R,[40] and $5\text{-HT}_{2A}$R[41–43] and even affect 5-HT-stimulated intracellular calcium flux in cultured cells.[44,45] The substitution of cysteine for serine at codon 23 in the human HTR2C gene, though not systematically related to mental disorders (see below I.1 polymorphism within human 5-HTR genes and eating disorders and Table 2 in Reference 46 for review), may alter until the constitutive activity of the corresponding receptor.[47] Such an association may be of great importance as the constitutive activity of $5\text{-HT}_{2C}$R has recently been reported to influence the release of neuromediator (dopamine) within the striatum of freely moving rodents[48]; see also Reference 49). Each of these examples opens the obvious possibility that synonymous or nonsynonymous polymorphisms may exert an impact on specific neuronal response, including the efficacy of any therapeutic agents. Few association studies focused on the benefit of therapeutic drugs related to HTR5-HT gene polymorphisms, and did not establish a systematic positive relation. For instance, responses to clozapine, used to treat schizophrenia, failed to be linked to the T25N HTR2A SNP.[50]

* Psynogeny® groups neurobiology, psychology, and psychiatry (V. Compan, 2003).

## Polymorphism within Human 5-HT Receptor Genes and Eating Disorders

In humans suffering from eating disorders, the HTR2A gene, encoding 5-$HT_{2A}$R, has been the most frequently analyzed over the last decade. Both positive and absence of associations have been detected between HTR2A gene polymorphisms and eating disorders.[51–57] The 5-$HT_{2A}$R gene promoter polymorphism (1438A/G) is described to be associated with both anorexia and bulimia nervosa[51–53] or not.[54,55] Furthermore, no evidence of HTR2A gene polymorphism (1438A/G) has been related to binge eating disorders, whereas again such an association was revealed in patients suffering from anorexia or bulimia nervosa[56] or obesity.[58] In other words, positive or negative associations between HTR2A/2C genes and eating disorders appear to depend mainly on the studied polymorphism[46] and may likely depend on the complexity of psychiatric diagnosis (anorexia, bulimia, hyperphagia, hypophagia, binge eating disorders: e.g., DSM IV-TR, APA, 2003). Likewise, changes in HTR2C gene, encoding 5-HT2CR, are often but not systematically associated with obesity.[59–66] Indeed, the polymorphism of HTR2C gene at codon 23 (cys23ser) has not been associated with bulimia nervosa, binge eating disorders or obesity,[59,60] and its association with both weight loss and gain is unclear.[54,59,67] The association between the 759C/T polymorphism of HTR2C gene and weight gain (obesity, antipsychotic-induced body weight gain) appears more reproducible, as a positive link has been reported by at least five studies.[61–63,65,68] It has been further highlighted that defects in 5-$HT_{2C}$R pre-mRNA processing controlled by small nuclear RNA may contributes to the Prader-Willi syndrome, which includes hyperphagia and obesity.[66]

Less human studies have examined the association of polymorphism of some other human 5-HT receptor genes and feeding disorders. Hinney et al. (1999)[69] concluded that polymorphisms in the HTR1Dß (phe124cys: T371G) and HTR7 (pro279leu), encoding, respectively, 5-$HT_{1Dß}$R (rodent 5-$HT_{1B}$R analogue) and 5-$HT_7$R, are unlikely related in the regulation of body weight or anorexia nervosa. In contrast, the G861C polymorphism of the HTR1Dß gene was associated with minimum and maximum lifetime body mass indices in women suffering from bulimia[70] and could be related to their potential predisposition to obsessive–compulsive disorder.[71]

To our knowledge, no further studies have been conducted to relate any alteration within HTR1A, HTR3, HTR4, HTR5, HTR6, and HTR7 genes with eating disorders. Based on our previous studies, possible alterations within the HTR4 gene related to eating disorders is discussed in the 5-HTR KO and eating disorders section.

## Polymorphism within Human 5-HT Receptor Genes and Addiction to Illegal Drugs

Polymorphisms within human 5-HTR genes and alcoholism have been extensively studied focusing on HTR1Dß and HTR2A/2C gene polymorphisms. Only two studies have reported the absence of any influence of HTR1A gene polymorphism on alcoholism.[72,73]

Positive evidence exists to associate G861C polymorphism of the HTR1Dß gene with alcoholism, although the association resulted from either a high frequency in the HTR1Dß 861C allele[74,75] or the HTR1Dß 861G allele.[76] An association was also found between a lower frequency of the HTR1Dß 861C gene and alcoholism.[77] In divergence, no association with alcohol abuse has been related to the G861C or other polymorphism of the HTR1Dß gene.[78–82] In addition, the HTR1Dß A-161T polymorphism has been reported to be associated with alcohol dependence.[83] Such discrepancies have also been reported with the T/C 102 polymorphism within the HTR2A gene analyzed in patients suffering from alcoholism, with positive[84] and negative associations described.[85,86] The 1438 G/A HTR2A gene polymorphism has been reported, so far, to influence alcoholism development.[58,87,88] In sharp contrast, there exists a systematic absence of any impact of the cys23ser HTR2C polymorphism on alcohol dependence[85,89,90] associated or not with, other mental diseases such as attention-deficit hyperactivity disorder,[91] panic disorder, generalized anxiety disorder including narcolepsy[76] and, alcohol-withdrawal induced seizure.[92]

Again, other possible alterations of human 5-HTR genes have not yet been reported to constitute a risk factor for alcoholism. Only one study has described a possible protective influence of the allele 1180G within HTR1Dß gene from addiction to a drugs "cocktail" including heroin and cocaine,[93] whereas this is not systematically the case in subjects prominently addicted to cocaine and bearing the A1180G SNP.[78] Mutations within HTR2B gene further appear to influence abuse of illegal drugs,[94] but neither T102C nor G1438G HTR2A SNPs seem to exert an impact on heroin consumption.[95]

Such discrepancies between results from association studies are classically thought to come from differences in phenotypes of people suffering from eating disorders, and abuse of alcohol and illegal drugs,[82] as well as the number of subjects.[96] Mental disorder diagnosis is not limited to one aspect but is based on a large spectrum of maladaptive behavior; one given patient may suffer from both bulimia and anorexia, which both coexist with an elevated level of anxiety.[97] One must also consider the possibility that control subjects may later develop the disorders being studied, so they may not provide an appropriate control comparison. In addition, from results summarized by the above section, it is important to note that 1438 A/G HTR2A and G861C HTR1Dß, but not the cys23ser HTR2C, polymorphisms have been reported to be associated with both eating disorders and alcoholism. In light of the study of Drysdale et al.,[98] it appears of important relevance to further take into account, when possible, the impact of multiple single nucleotide polymorphisms (SNPs) within a haplotype (the set of SNPs within one chromosome) on one given phenotype.

Such a wide spectrum of human complexity may certainly explained why there is sometimes, in a first approximation, no reproducibility between phenotypes-related to HTR5 gene polymorphism and 5-HTR KO mice. However, even though we do not intend to anthropomorphize the mouse, increased body weight gain has been related to the 759C/T polymorphism of the HTR2C gene in humans[61–63,65,68] as well as, in 5-$HT_{2C}$R knockout mouse.[15]

# KNOCK-OUT OF 5-HT RECEPTORS CLASSICALLY REQUIRED ENVIRONMENTAL CHALLENGES TO PRODUCE MALADAPTIVE BEHAVIOR

## 5-HT Receptor Knock-Out Mice and Eating Disorders

In industrialized countries, anorexia nervosa is among the mental diseases having the highest mortality rates,[99] but no therapeutic strategy is effective to treat this disorder. There is as yet no animal model of this condition, either. If one defines anorexia as a self-imposed starvation despite an energy demand, a similar behavior can be produced in animals and humans following increase in 5-HT neuromodulation. Among drugs increasing the synaptic levels of 5-HT, fenfluramine inhibits the consumption of food in human and rodent.[100–102] Likewise, drugs of abuse such as amphetamine classically interrupt the physiological drive to eat. In particular, the psychogenic compound Ecstasy, 3,4-N-methylenedioxymethamphetamine (MDMA), diminishes both the consumption of food in humans[103] and rats[104] and reduces starvation-induced eating in mice.[105] Recently, 5-$HT_{1B}$R agonists and fenfluramine have been found to reduce food intake by reciprocally regulating melacortin neurons at the level of the arcuate nucleus of hypothalamus.[106] Pharmacological studies have shown that 5-$HT_{2C}$R is also importantly involved in 5-HT-mediated hypophagia.[107–109] The contribution of 5-$HT_{2C}$R in both fenfluramine and MDMA-induced anorexia-like behavior has been demonstrated.[105,110,111] 5-$HT_{2C}$R KO mice display a lower sensitivity to fenfluramine,[110] as described in 5-$HT_{1B}$R KO mice (Lucas et al., 1998[14]). An attenuated function of 5-$HT_{2C}$R may account for the decreased sensitivity of 5-$HT_{1B}$R KO mice to fenfluramine.[112] The common downstream target of both 5-$HT_{1B}$R and 5-$HT_{2C}$R activation may be the melacortin 4 receptor activation in the hypothalamus.[106,113] Appetite suppressants remain effective in the absence of 5-$HT_{1B}$R. Indeed, 5-$HT_{1B}$R KO mice display a normal response to MDMA-induced anorexia-like behavior.[105] Our yet unpublished findings clearly indicate that an elevated function of 5-$HT_4$R may account for the insensitivity of 5-$HT_{1B}$R KO mice to the anorectic effect of MDMA. Indeed, the intraperitoneal injection of RS39604, a 5-$HT_4$R antagonist, has clearly suppressed MDMA-induced anorexia in starved 5-$HT_{1B}$R KO compared to wild-type mice (Compan, unpublished). Along these lines, MDMA-treated 5-$HT_4$R KO mice consume a higher amount of food than wild-type mice (Compan, unpublished).

The obesity associated with overeating in 5-$HT_{2C}$R KO mice further supports that 5-$HT_{2C}$R influences feeding behavior.[15] Both 5-$HT_{1A}$R and 5-$HT_{1B}$R KO are also heavier when born from homozygote lines,[6,105,114–116] while the loss of 5-$HT_{2A}$R, 5-$HT_{3A}$R, 5-$HT_4$R, as well as 5-$HT_{5A}$R function did not induce any change in body weight under basal conditions.[9,10,12,117]

In particular, overweight in the absence of 5-$HT_{1A}$R has been reported in male,[116] but not in female mutant mice.[114] To our knowledge, no information on the amount of food consumed over time during development has been revealed in the absence of mHTR1A gene under basal condition. In contrast to all other 5-HTR agonists tested so far, the pharmacological stimulation of 5-HT$_{1A}$R augments the intake of food.[118,119] 5-HT$_{1A}$R agonist-induced hyperphagia is classically associated with their negative influence on serotonergic neurons activity[17,120–126] (cf. section titled "5-HT Receptor Knock-Out Mice and Adaptive Changes in Central Serotoninergic Systems"). In other words, the stimulation of 5-HT$_{1A}$R induced increases in food intake probably because the activation of 5-HT$_{1A}$R inhibits the electrophysiological activity of serotonergic neurons. On the contrary, 5-HT$_{1A}$R KO mice display increased spontaneous firing of DRN 5-HT neurons[126,127] (cf. section titled "5-HT Receptor Knock-Out Mice and Adaptive Changes in Central Serotoninergic Systems"). It may then be expected that 5-HT$_{1A}$R KO mice would eat a lesser amount of food than their wild-type congeners and would be underweight. Subsequently, the possible adaptive changes that certainly overcome such hypophagia and low body weight can be suspected to highly counterbalance the loss of 5-HT$_{1A}$R function.

Two earlier studies have reported that male 5-HT$_{1B}$R KO mice are overweight and consume a higher amount of food than wild-type mice.[6,115,116] In addition, female 5-HT$_{1B}$R KO mice are also overweight.[116] However, results became the subject of an interesting debate because animals were the offspring of homozygote breeders, suggesting an incidence of parental care.[115] Homozygote offspring from heterozygote breeding pairs were then used[105] in order to avoid indirect effects of parental care behavioral responses, as discussed by Bouwknecht et al.[115] No differences in either baseline food intake or body weight were detected between wild-type and 5-HT$_{1B}$R KO mice, even following a 24 h period of food deprivation,[105] as previously reported.[14,128,129] Whether animals of identical genotype were housed together or not may also likely influence any state (I. Seif, personal observation), including feeding behavior in mutant mice that may display elevated aggressiveness (5-HT$_{1B}$R KO[27]), hyperanxiety-like state (5-HT$_{1A}$R KO[4,5]), hyposensitivity to stress (5-HT$_4$R KO[10]). Less investigated is the phenomenon of stress-induced hypophagia. Hypothalamic-pituitary-adrenal axis hormones have been, however, suggested to be involved in this stress response, at least partially, via the release of 5-HT in the medial prefrontal cortex, nucleus accumbens, amygdala and dorsal hippocampus.[130–132]

One study proposed that 5-HT$_{2A/2C}$Rs could be implicated in this cascade of events, using a pharmacological approach.[133] 5-HT$_{2C}$R KO mice display opposite feeding responses to mild stress, depending on age, whereas only old mutant mice displayed hypersensitivity to repeated stress, which is associated with increased levels in stress hormones.[134]

To date, 5-HT$_4$Rs KO mouse is the only known animal model displaying less sensitivity to stress-induced anorexia.[10,11] Such a maladaptive feeding response to restraint stress in 5-HT$_4$R KO mouse is likely due to deficiencies in the activity of serotonergic systems (see Reference 18; cf. section titled "5-HT Receptor Knock-Out Mice and Adaptive Changes in Central Serotoninergic Systems") rather than a direct consequence of the hyperactivity of hypothalamo-pituitary axis.[10]

Only a few sparse results related to food intake have been reported so far in other 5-HTR KO mice under stressful context such as the open-field. In the novelty suppressed feeding paradigm, starved mice face a dual conflict between the physiological drive to explore a novel environment, eat, and find safety (see Reference 12 and Reference 135). The latency to eat in this novel space is interpreted as an index of anxiety-like state, when possible bias in locomotor and exploratory activity. The latency to eat is enhanced in starved 5-$HT_{1A}$R KO,[135] and reduced in 5-$HT_{2A}$R KO mice,[12] which were related to their respective hyper- and hypo-anxiety-like state. Along these lines, feeding disorders in 5-$HT_4$R KO mice might be related to a higher level of anxiety-like behavior following stress.[10]

Altogether (Table 8.1), these results suggest the presence of at least two modes of action of 5-HT to regulate feeding behavior. Under baseline conditions, the body weight appears to be mainly regulated via the 5-$HT_{2C}$R whereas after an unusual and strong stressful event, 5-$HT_4$R may become involved. How other 5-HTRs contribute to the ability of stress to decrease food intake remains unexplored, although the loss of 5-$HT_{1A}$R function, as well as the loss of 5-$HT_4$R function, appears to be associated with overeating when mice were frightened in the open field. It is the inverse in the absence 5-$HT_{2A}$R. All in all, these examples, with the exception of 5-$HT_{2C}$R KO mice, indicate that feeding disorders associated with a genetic deficit are best revealed in a stressful context or following a pharmacological challenge.

As stressful context appears as a determinant factor in human suffering from anorexia that coexists with anxiety and depression,[32–34] polymorphisms within HTR4 gene may contribute to eating disorders. We have recently raised the possibility that 5-$HT_4$R may represent a new therapeutic target of depression.[18,136,137] 5-$HT_4$R-gene polymorphism has been associated with the bipolar syndrome (manic depression) in humans,[138] and increased 5-$HT_4$R density has recently been described in patients suffering from depression leading to suicide.[139] Along these lines, using a range of molecular, biochemical, and physiological *in vivo* studies, we have gathered evidence indicating that the intra-accumbal activation of the 5-$HT_4$Rs/cAMP/PKA pathway reduced eating and elevated the mRNA expression of an important satiety factor: cocaine- and amphetamine-regulated transcript (CART) into the nucleus accumbens (Compan, unpublished). Neuronal disturbances in the nucleus accumbens are associated with anxiety,[140] abusive consumption of illegal drugs,[141] and ecstasy-induced anorexia (Compan, unpublished results). Interestingly, one of the HTR4 introns encodes the adrenaline ß2 receptor, a polymorphism of which is related to obesity. Both 5-$HT_4$R and ß2R lead to their heterodimerization.[142] CART gene polymorphisms have also been identified in obese patients (for review see Reference 143). As multiple SNPs may account for any given mental disease,[98] it is interesting to note that both HTR4 and CART genes are also located on the same human chromosome; htR4 gene: 5q31-33,[144] CART gene: 5q13-14.[144–146]

## 5-HT Receptor Knock-Out Mice and Novelty-Induced Locomotion

Central serotoninergic systems are widely suspected to modulate stress responsiveness in situations such as reactivity to a novel environment.[147] Several studies have further revealed a relationship between specific 5-HTR and novelty-induced locomotion using KO mice: 5-$HT_{1A}$,[5] 5-$HT_{1B}$,[6] 5-$HT_{2C}$,[8] 5-$HT_4$,[10] and 5-$HT_{5A}$.[9] In the novel open-field test, rodents face a conflict between the physiological drive to explore a novel environment and safety. In addition, under specific environmental conditions, wild-type mice explored the center part less than the periphery of the open field.

## TABLE 8.1
## Serotonin (5-hydroxytryptamine, 5-HT) Receptor Knock-Out Mice Display Maladapted Behaviors under an Environmental or Pharmacological Challenge

| KNOCK-OUT | HOME CAGE | | | STRESSORS | | | | | | | |
|---|---|---|---|---|---|---|---|---|---|---|---|
| | | | | OPEN-FIELD | | | OPEN-FIELD FOLLOWING ILLEGAL DRUGS | | | FOOD INTAKE UNDER CHALLENGES | ANXIETY-LIKE STATE |
| | Activity | Body weight | Food intake | Total path | Center path | Center time | MDMA | Cocaine | Other | | "fear-state" |
| 5-HT$_{1A}$R | Unchanged[19,20] | Increased[10,12] | ? | Decreased[19,29] | Decreased[19,20] | Decreased[19,20] | ? | ? | ? | Latency to eat in the open-field: increased[14] | Enhanced[14,18,19] |
| 5-HT$_{1B}$R | Unchanged[28] | Increased/ no change [3,4,8,11,12,17] | Unchanged[8,17] | Increased No change[4,24,28] | Increased[28] | Increased[28] | Reduced[23] | Enhanced[21] | ? | Fenfluramine-induced anorexia: suppressed[17]<br>MDMA-induced anorexia: maintained[8] | Decreased[4,28] |
| 5-HT$_{2A}$R | Unchanged[27] | Unchanged[27] | ? | No change[27] | Increased[27] | ? | ? | ? | ? | Latency to eat in the open-field: decreased[27] | Decreased[27] |
| 5-HT$_{2C}$R | Increased[25] | Increased[18] | Increased[18] | Increased[25] | ? | ? | ? | Enhanced[22] | ? | ± Hyper-responsive after repeated stress (age)[5]<br>Fenfluramine-induced anorexia: reduced[26]<br>MDMA-induced anorexia: ?-decreased post-RS102221[8] | ? |
| 5-HT$_3$R | ? | Unchanged[16] | Unchanged[16] | No change[1,15,16] | ? | ? | ? | ? | No change post-ethanol[15] | ? | Decreased[16] |
| 5-HT$_4$R | Unchanged[6] | Unchanged[6] | Unchanged[6] | Decreased[6] | No change[6] | Decreased[6] | Reduced[7] | Enhanced[7] | ? | Hyposensitive to stress-induced anorexia[6]<br>MDMA-induced anorexia: reduced[7]<br>Cocaine-induced anorexia: reduced[7] | Enhanced post-stress[6] |
| 5-HT$_{5A}$R | Unchanged[13] | Unchanged[13] | ? | Increased[13] | No change[13] | ? | ? | ? | Decreased post-LSD[13] | ? | Unchanged[13] |
| 5-HT$_6$R | Unchanged[2] | ? | ? | No change[2] | No change[2] | No change[2] | ? | ? | Enhanced post-ethanol[2] | ? | Unchanged[2] |
| 5-HT$_7$R | ? | ? | ? | ? | ? | ? | ? | ? | ? | ? | ? |

1. Bhatnagar et al., 2004; 2. Bonasera et al., 2006; 3. Bouwknecht et al., 1999; 4. Brunner et al., 1999; 5. Chou-green et al., 2003; 6. Compan et al., 2004; 7. Compan, unpublished; 8. Conductier et al., 2005; 9. Conductier et al., 2006; 10. Dirks et al., 2001; 11. Dulawa et al., 1997; 12. Dulawa et al., 2000; 13. Grailhe et al., 1999; 14. Gross et al., 2002; 15. Hodge et al., 2004; 16. Kelley et al., 2003; 17. Lucas et al., 1998; 18. Nonogaki et al., 2003; 19. Parks et al., 1998; 20. Ramboz et al., 1998; 21. Rocha et al., 1998; 22. Rocha et al., 2002; 23. Scearce-Levie et al., 1999; 24. Saudou et al., 1994; 25. Tecott et al., 1995; 26. Vickers et al., 1999; 27. Weisstaub et al., 2006; 28. Zhuang et al., 1999

5-$HT_{1A}$R and 5-$HT_4$R KO mice are less reactive to novel environments, whereas their motor activity was not altered in their home cages.[4,5,10] These results suggest that the absence of 5-$HT_{1A}$R or 5-$HT_4$R results in an attenuation of the motor responses induced by novel environments. In contrast, it is the inverse for 5-$HT_{1B}$, 5-$HT_{2C}$ or 5-$HT_{5A}$R KO mice, which display an elevated horizontal activity in the open field (5-$HT_{1B}$[6,148]; not systematically see Reference 27; 5-$HT_{2C}$R[27]; 5-$HT_{5A}$R[9]). In contrast, loss of 5-$HT_{2A}$R,[12] 5-$HT_{3/3A}$R[117,149,150] or, 5-$HT_6$R[151] function has no effect on novelty-induced locomotion (5-$HT_{2A}$R[12]; 5-$HT_{3/3A}$R[117,149,150]; 5-$HT_6$R[151]). As well as for 5-$HT_{1A}$R or 5-$HT_4$R KO, all other generated 5-HTR KO mice did not exhibit any change in home cage activity (5-$HT_{1B}$R[6,148]; 5-$HT_{2A}$R[12]; 5-$HT_4$R[10]; 5-$HT_{5A}$R[9]; 5-$HT_6$R[151]), with the exception of 5-$HT_{2C}$R KO that display hyperactivity.[152] To our knowledge, no studies have reported whether the loss of 5-$HT_3$R or 5-$HT_7$R function may alter home cage activity or not.

These data are summarized in Table 8.1 and suggest that a permanent absence of a gene encoding one of the 5-HTR provokes either a hypo- (5-$HT_{1A}$R and 5-$HT_4$R), hyper- (5-$HT_{1B}$R, 5-$HT_{2C}$R, 5-$HT_{5A}$R) or control-reactivity (5-$HT_{2A}$R, 5-$HT_3$R, 5-$HT_6$R) to novelty. As novelty-seeking behavior is often related to abuse of illegal drugs,[153,154] the goal of several studies was to test the effects of illegal drugs in the absence of gene encoding 5-HTR. These results are discussed in the following section.

## 5-HT Receptor Knock-Out Mice and Illegal Drugs such as Ecstasy and Cocaine

Stress may enhance the effects of illegal drugs such as cocaine, which are associated with changes on dopaminergic systems.[155] It has been further demonstrated that cocaine effects involve 5-HTRs.[24] It is unlikely that one given physiological function does not depend on multiple neuronal systems, even if multiple studies have focused on isolated mechanisms. Subsequently, numerous studies have tested whether 5-HTRs may also contribute to both neuronal and behavioral effects of illegal drugs such as cocaine and MDMA (Ecstasy).

MDMA is a "substrate-type 5-HT releaser"[156] and is an amphetamine-like stimulant. MDMA (10 mg/kg) increases the levels of 5-HT but can also increase dopamine (DA) levels, especially at higher doses.[157] Although using different mechanisms of action, cocaine is also well known to target monoamine (DA, 5-HT, NA) transporters.[158]

MDMA is a psychoactive substance, first described as an appetite-suppressant in humans, inducing side effects and even death. As summarized in Table 8.1, we have recently shown that MDMA-induced anorexia involves both 5-$HT_{2C}$R[105] and 5-$HT_4$R (Compan, unpublished). In contrast, the feeding responses to MDMA treatment were maintained in 5-$HT_{1B}$R KO mice or in animals treated with the 5-$HT_{1B/1D}$R antagonist GR127935.[105] A specific dose of a 5-$HT_4$R antagonist (RS39604) suppressed the hypophagic effect of MDMA in 5-$HT_{1B}$R KO mice, suggesting that 5-$HT_4$R functions adapt over time during development when mice are 5-$HT_{1B}$R deprived (Compan, unpublished).

MDMA also elicits a dramatic increase in locomotion,[159–161] which in mice involves 5-$HT_{1B}$R, 5-$HT_{2C}$R, and 5-$HT_4$R[13,105] (Compan, unpublished). Conversely, the loss of 5-$HT_{1B}$R, 5-$HT_{2C}$,[7,8] as well as 5-$HT_4$R (Compan, unpublished), function has clearly enhanced cocaine-induced locomotion in the novel open field test.

## 5-HT RECEPTOR KNOCK-OUT MICE AND ADAPTIVE CHANGES IN CENTRAL SEROTONINERGIC SYSTEMS

Over the last two decades, studies have demonstrated that 5-$HT_{1A}$ autoreceptors exert a negative feedback influence on 5-HT neuron activity and 5-HT levels in the dorsal (DRN) and medial (MRN) raphe nuclei.[17,120–126] A similar conclusion has been reached following studies with 5-$HT_{1A}$R KO mice, which display increased spontaneous firing of DRN 5-HT neurons (Figure 8.1).[126,127]

Reduced 5-$HT_{1A}$R density and sensitivity may also coexist with decreased DRN 5-HT neuron activity. This is observed in monoamine oxidase-A (catabolism enzyme of 5-HT) null mice and interpreted as an overcompensation for the excessive level of 5-HT in these mice.[16] 5-HT neuron firing and the level of 5-HT in DRN are also negatively regulated by 5-$HT_{1B}$ autoreceptors, depending on the species examined.[16,162] In 5-$HT_{1B}$R null adult mice, subtle adaptive changes are suspected to overcome the loss-of-function mutation in 5-$HT_{1B}$R, because 5-HT neuron activity and 5-HT levels are unchanged in the raphe nuclei (Figure 8.1).[16,17] Indeed, 5-HT neuron firing and the level of 5-HT in DRN are negatively regulated by 5-$HT_{1B}$ autoreceptors, depending on the species examined.[16,162,163] Although decreased 5-HT neurotransmission may exist in depressed humans,[21] there is as yet no genetically modified animal for any G protein-coupled receptor that displays decreased 5-HT neuron activity and 5-HT content in the raphe nuclei. We have recently provided such an animal model, lacking 5-$HT_4$R (Figure 8.1).[18]

## CONCLUSION

It is surprising that such dramatic changes in the activity of the serotonergic system as detected in the absence of mHtr1A and mHtr4 genes did not produce more apparent behavioral deficiencies under basal conditions. The neuroplasticity of the brain over the course of development is fascinating and suggests that such adaptive neuronal mechanisms may be able to circumvent genetic deficits.

In contrast, stress-induced adaptive behavior is impaired in genetically modified animals lacking 5-HTR (see Table 8.1), suggesting some limitations on the extent of plasticity in the brain. To an extent, the brain may adapt to underline adapted behavior but appears to "burn" its adaptive resources over time under stress. To date, genetic deficits related to 5-HTR mostly produced less sensitivity to the anorectic effects of stressors, with different changes in reactivity to novelty or illegal drugs and fear states. These responses provide a focus for our efforts to further examine neural mechanisms whereby serotonin systems may take part to a mixed pattern that underline mental disease.

## ACKNOWLEDGMENT

I greatly appreciate the assistance of Kerri Holick in editing this text.

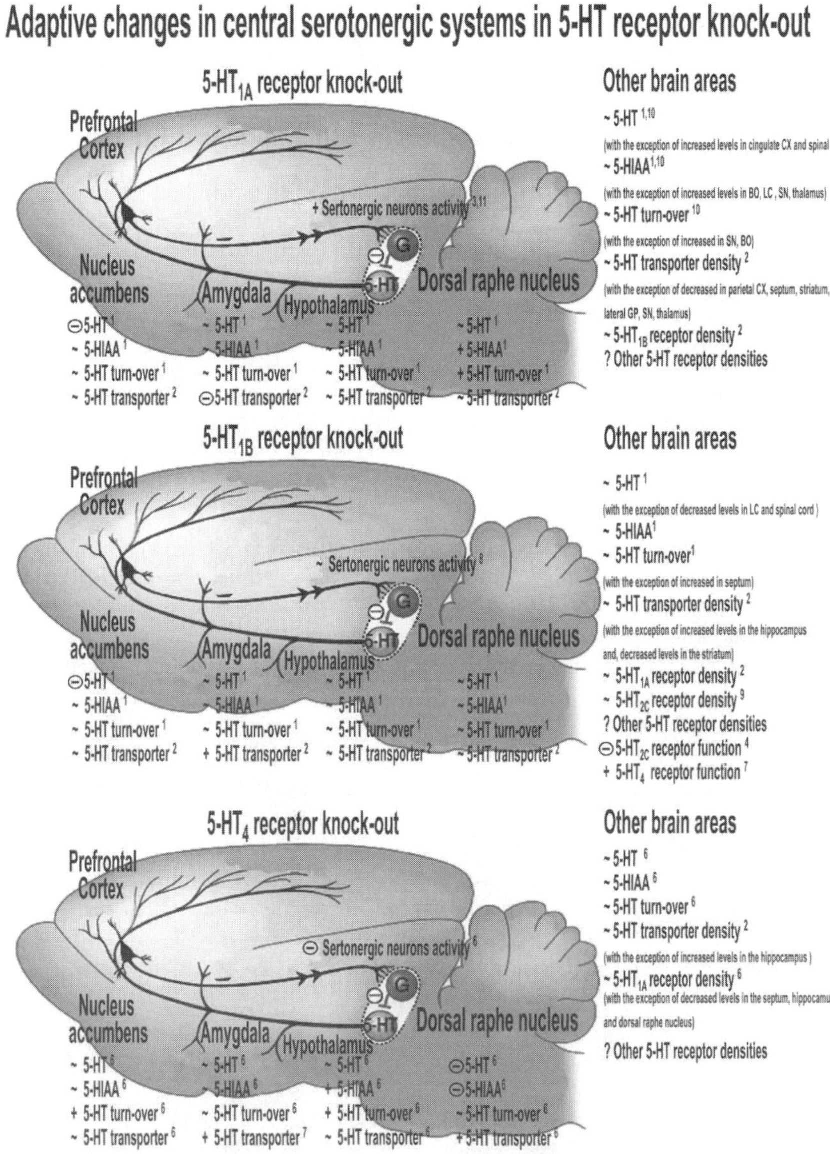


**FIGURE 8.1 (Color figure follows p. 110.)** Adaptive changes in serotonergic neurons in 5-HT receptor knock-out mice. BO: olfactory bulb, CX: cortex, GP: globus pallidus, LC: locus coeruleus, SN: substantia nigra. (From 1. Ase, A.R. et al., *J Neurochem* 75(6), 2415–26, 2000; 2. Ase, A. R. et al., *J Neurochem*, 78(3), 619–630; 3. Bortolozzi, A. et al., *J Neurochem,* 88(6), 1373–1379, 2004; 4. Clifton, P.G. et al., *Eur J Neurosci,* 17(1), 185–190, 2003; 5. Compan, unpublished; 6. Conductier, G., et al., *Eur J Neurosci,* 24(4), 1053–1062, 2006; 8. Evrard, A. et al., *Eur J Neurosci,* 11(11), 3823–3831, 1999; 9. Lee, M. D. et al., *Psychopharmacology* 176(1), 39–49,2004; 10. Ramboz, S. et al., *Proc Natl Acad Sci USA,* 95(24), 14476–14481, 1998; Richer, M. et al., *Eur J Pharmacol,* 435(2–3), 195–203, 2002. With permission.)

## REFERENCES

1. Azmitia, E.C. and Whitaker-Azmitia, P.M., Awakening the sleeping giant: anatomy and plasticity of the brain serotonergic system, *J Clin Psychiatr,* 52(Suppl.), 4–16, 1991.
2. Byrne, J.H. and Kandel, E.R., Presynaptic facilitation revisited: state and time dependence, *J Neurosci,* 16(2), 425–435, 1996.
3. Heisler, L.K., Chu, H.M., Brennan, T.J., Danao, J.A., Bajwa, P., Parsons, L.H., and Tecott, L.H., Elevated anxiety and antidepressant-like responses in serotonin 5-HT1A receptor mutant mice, *Proc Natl Acad Sci USA,* 95(25), 15049–15054, 1998.
4. Parks, C.L., Robinson, P.S., Sibille, E., Shenk, T., and Toth, M., Increased anxiety of mice lacking the serotonin1A receptor, *Proc Natl Acad Sci USA,* 95(18), 10734–10739, 1998.
5. Ramboz, S., Oosting, R., Amara, D.A., Kung, H.F., Blier, P., Mendelsohn, M., Mann, J.J., Brunner, D., and Hen, R., Serotonin receptor 1A knockout: an animal model of anxiety-related disorder, *Proc Natl Acad Sci USA,* 95(24), 14476–14481, 1998.
6. Brunner, D., Buhot, M.C., Hen, R., and Hofer, M., Anxiety, motor activation, and maternal-infant interactions in 5HT1B knockout mice, *Behav Neurosci,* 113(3), 587–601, 1999.
7. Rocha, B.A., Scearce-Levie, K., Lucas, J.J., Hiroi, N., Castanon, N., Crabbe, J.C., Nestler, E.J., and Hen, R., Increased vulnerability to cocaine in mice lacking the serotonin-1B receptor, *Nature,* 393(6681), 175–178, 1998.
8. Rocha, B.A., Goulding, E.H., O'Dell, L.E., Mead, A.N., Coufal, N.G., Parsons, L.H., and Tecott, L.H., Enhanced locomotor, reinforcing, and neurochemical effects of cocaine in serotonin 5-hydroxytryptamine 2C receptor mutant mice, *J Neurosci,* 22(22), 10039–10045, 2002.
9. Grailhe, R., Waeber, C., Dulawa, S.C., Hornung, J.P., Zhuang, X., Brunner, D., Geyer, M.A., and Hen, R., Increased exploratory activity and altered response to LSD in mice lacking the 5-HT(5A) receptor, *Neuron,* 22(3), 581–591, 1999.
10. Compan, V., Zhou, M., Grailhe, R., Gazzara, R.A., Martin, R., Gingrich, J., Dumuis, A., Brunner, D., Bockaert, J., and Hen, R., Attenuated response to stress and novelty and hypersensitivity to seizures in 5-HT4 receptor knock-out mice, *J Neurosci,* 24(2), 412–419, 2004.
11. Compan, V., Charnay, Y., Dusticier, N., Daszuta, A., Hen, R., and Bockaert, J., Feeding disorders in 5-HT4 receptor knockout mice, *J Soc Biol,* 198(1), 37–49, 2004.
12. Weisstaub, N.V., Zhou, M., Lira, A., Lambe, E., Gonzalez-Maeso, J., Hornung, J.P., Sibille, E., Underwood, M., Itohara, S., Dauer, W.T., Ansorge, M.S., Morelli, E., Mann, J.J., Toth, M., Aghajanian, G., Sealfon, S.C., Hen, R., and Gingrich, J.A., Cortical 5-HT2A receptor signaling modulates anxiety-like behaviors in mice, *Science,* 313(5786), 536–540, 2006.
13. Scearce-Levie, K., Viswanathan, S.S., and Hen, R., Locomotor response to MDMA is attenuated in knockout mice lacking the 5-HT1B receptor, *Psychopharmacology (Berl),* 141(2), 154–161, 1999.
14. Lucas, J.J., Yamamoto, A., Scearce-Levie, K., Saudou, F., and Hen, R., Absence of fenfluramine-induced anorexia and reduced c-Fos induction in the hypothalamus and central amygdaloid complex of serotonin 1B receptor knock-out mice, *J Neurosci,* 18(14), 5537–5544, 1998.
15. Tecott, L.H., Sun, L.M., Akana, S.F., Strack, A.M., Lowenstein, D.H., Dallman, M.F., and Julius, D., Eating disorder and epilepsy in mice lacking 5-HT2c serotonin receptors, *Nature,* 374(6522), 542–546, 1995.

16. Evrard, A., Laporte, A.M., Chastanet, M., Hen, R., Hamon, M., and Adrien, J., 5-HT1A and 5-HT1B receptors control the firing of serotoninergic neurons in the dorsal raphe nucleus of the mouse: studies in 5-HT1B knock-out mice, *Eur J Neurosci,* 11(11), 3823–3831, 1999.
17. Ase, A.R., Reader, T.A., Hen, R., Riad, M., and Descarries, L., Altered serotonin and dopamine metabolism in the CNS of serotonin 5-HT(1A) or 5-HT(1B) receptor knockout mice, *J Neurochem,* 75(6), 2415–2426, 2000.
18. Conductier, G., Dusticier, N., Lucas, G., Cote, F., Debonnel, G., Daszuta, A., Dumuis, A., Nieoullon, A., Hen, R., Bockaert, J., and Compan, V., Adaptive changes in serotonin neurons of the raphe nuclei in 5-HT(4) receptor knock-out mouse, *Eur J Neurosci,* 24(4), 1053–1062, 2006.
19. Lopez, J.F., Akil, H., and Watson, S.J., Neural circuits mediating stress, *Biol Psychiatr,* 46(11), 1461–1471, 1999.
20. Chaouloff, F., Serotonin, stress and corticoids, *J Psychopharmacol,* 14(2), 139–151, 2000.
21. Nemeroff, C.B. and Owens, M.J., Treatment of mood disorders, *Nat Neurosci,* 5(Suppl.), 1068–1070, 2002.
22. Vaswani, M. and Kalra, H., Selective serotonin re-uptake inhibitors in anorexia nervosa, *Expert Opin Invest Drugs,* 13(4), 349–357, 2004.
23. Leonardo, E.D. and Hen, R., Genetics of affective and anxiety disorders, *Annu Rev Psychol,* 57, 117–137, 2006.
24. Muller, C.P. and Huston, J.P., Determining the region-specific contributions of 5-HT receptors to the psychostimulant effects of cocaine, *Trends Pharmacol Sci,* 27(2), 105–112, 2006.
25. Hoyer, D., Hannon, J.P., and Martin, G.R., Molecular, pharmacological and functional diversity of 5-HT receptors, *Pharmacol Biochem Behav,* 71(4), 533–554, 2002.
26. Bockaert, J., Claeysen, S., Becamel, C., Dumuis, A., and Marin, P., Neuronal 5-HT metabotropic receptors: fine-tuning of their structure, signaling, and roles in synaptic modulation, *Cell Tissue Res,* 326, 553–572, 2006.
27. Saudou, F., Amara, D.A., Dierich, A., LeMeur, M., Ramboz, S., Segu, L., Buhot, M.C., and Hen, R., Enhanced Aggressive Behavior in Mice Lacking 5-HT1B Receptor, *Science,* 265(5180), 1875–1878, 1994.
28. Nebigil, C.G., Choi, D.S., Dierich, A., Hickel, P., Le Meur, M., Messaddeq, N., Launay, J.M., and Maroteaux, L., Serotonin 2B receptor is required for heart development, *Proc Natl Acad Sci USA,* 97(17), 9508–9513, 2000.
29. Zeitz, K.P., Guy, N., Malmberg, A.B., Dirajlal, S., Martin, W.J., Sun, L., Bonhaus, D.W., Stucky, C.L., Julius, D., and Basbaum, A.I., The 5-HT3 subtype of serotonin receptor contributes to nociceptive processing via a novel subset of myelinated and unmyelinated nociceptors, *J Neurosci,* 22(3), 1010–1019, 2002.
30. Guscott, M.R., Egan, E., Cook, G.P., Stanton, J.A., Beer, M.S., Rosahl, T.W., Hartmann, S., Kulagowski, J., McAllister, G., Fone, K.C., and Hutson, P.H., The hypothermic effect of 5-CT in mice is mediated through the 5-HT7 receptor, *Neuropharmacology,* 44(8), 1031–1037, 2003.
31. Hedlund, P.B., Danielson, P.E., Thomas, E.A., Slanina, K., Carson, M.J., and Sutcliffe, J.G., No hypothermic response to serotonin in 5-HT7 receptor knockout mice, *Proc Natl Acad Sci USA,* 100(3), 1375–1380, 2003.
32. Donohoe, T.P., Stress-induced anorexia: implications for anorexia nervosa, *Life Sci,* 34(3), 203–218, 1984.
33. Casper, R.C., Depression and eating disorders, *Depression Anxiety,* 8(Suppl. 1), 96–104, 1998.

34. Godart, N.T., Flament, M.F., Lecrubier, Y., and Jeammet, P., Anxiety disorders in anorexia nervosa and bulimia nervosa: co-morbidity and chronology of appearance, *Eur Psychiatr,* 15(1), 38–45, 2000.
35. Miyake, A., Mochizuki, S., Takemoto, Y., and Akuzawa, S., Molecular cloning of human 5-hydroxytryptamine3 receptor: heterogeneity in distribution and function among species, *Mol Pharmacol,* 48(3), 407–416, 1995.
36. Karnovsky, A.M., Gotow, L.F., McKinley, D.D., Piechan, J.L., Ruble, C.L., Mills, C.J., Schellin, K.A., Slightom, J.L., Fitzgerald, L.R., Benjamin, C.W., and Roberds, S.L., A cluster of novel serotonin receptor 3-like genes on human chromosome 3, *Gene,* 319, 137–148, 2003.
37. Niesler, B., Frank, B., Kapeller, J., and Rappold, G.A., Cloning, physical mapping and expression analysis of the human 5-HT3 serotonin receptor-like genes HTR3C, HTR3D and HTR3E, *Gene,* 310, 101–111, 2003.
38. Serretti, A., Benedetti, F., Zanardi, R., and Smeraldi, E., The influence of serotonin transporter promoter polymorphism (SERTPR) and other polymorphisms of the serotonin pathway on the efficacy of antidepressant treatments, *Prog Neuropsychopharmacol Biol Psychiatr,* 29(6), 1074–1084, 2005.
39. Del Tredici, A.L., Schiffer, H.H., Burstein, E.S., Lameh, J., Mohell, N., Hacksell, U., Brann, M.R., and Weiner, D.M., Pharmacology of polymorphic variants of the human 5-HT1A receptor, *Biochem Pharmacol,* 67(3), 479–490, 2004.
40. Huang, Y.Y., Grailhe, R., Arango, V., Hen, R., and Mann, J.J., Relationship of psychopathology to the human serotonin1B genotype and receptor binding kinetics in postmortem brain tissue, *Neuropsychopharmacology,* 21(2), 238–246, 1999.
41. Hazelwood, L.A. and Sanders-Bush, E., His452Tyr polymorphism in the human 5-HT2A receptor destabilizes the signaling conformation, *Mol Pharmacol,* 66(5), 1293–1300, 2004.
42. Khait, V.D., Huang, Y.Y., Zalsman, G., Oquendo, M.A., Brent, D.A., Harkavy-Friedman, J.M., and Mann, J.J., Association of serotonin 5-HT2A receptor binding and the T102C polymorphism in depressed and healthy Caucasian subjects, *Neuropsychopharmacology,* 30(1), 166–172, 2005.
43. Davies, M.A., Setola, V., Strachan, R.T., Sheffler, D.J., Salay, E., Hufeisen, S.J., and Roth, B.L., Pharmacologic analysis of non-synonymous coding h5-HT2A SNPs reveals alterations in atypical antipsychotic and agonist efficacies, *Pharmacogenomics J,* 6(1), 42–51, 2006.
44. Ozaki, N., Manji, H., Lubierman, V., Lu, S.J., Lappalainen, J., Rosenthal, N.E., and Goldman, D., A naturally occurring amino acid substitution of the human serotonin 5-HT2A receptor influences amplitude and timing of intracellular calcium mobilization, *J Neurochem,* 68(5), 2186–2193, 1997.
45. Reist, C., Mazzanti, C., Vu, R., Fujimoto, K., and Goldman, D., Inter-relationships of intermediate phenotypes for serotonin function, impulsivity, and a 5-HT2A candidate allele: His452Tyr, *Mol Psychiatr,* 9(9), 871–878, 2004.
46. Sanders-Bush, E., Fentress, H., and Hazelwood, L., Serotonin 5-ht2 receptors: molecular and genomic diversity, *Mol Interv,* 3(6), 319–330, 2003.
47. Okada, M., Northup, J.K., Ozaki, N., Russell, J.T., Linnoila, M., and Goldman, D., Modification of human 5-HT(2C) receptor function by Cys23Ser, an abundant, naturally occurring amino-acid substitution, *Mol Psychiatr,* 9(1), 55–64, 2004.
48. De Deurwaerdere, P., Navailles, S., Berg, K.A., Clarke, W.P., and Spampinato, U., Constitutive activity of the serotonin2C receptor inhibits *in vivo* dopamine release in the rat striatum and nucleus accumbens, *J Neurosci,* 24(13), 3235–3241, 2004.

49. Navailles, S., De Deurwaerdere, P., and Spampinato, U., Clozapine and haloperidol differentially alter the constitutive activity of central serotonin2C receptors *in vivo*, *Biol Psychiatr,* 59(6), 568–575, 2006.
50. Nothen, M.M., Rietschel, M., Erdmann, J., Oberlander, H., Moller, H.J., Nober, D., and Propping, P., Genetic variation of the 5-HT2A receptor and response to clozapine, *Lancet,* 346(8979), 908–909, 1995.
51. Nacmias, B., Ricca, V., Tedde, A., Mezzani, B., Rotella, C.M., and Sorbi, S., 5-HT2A receptor gene polymorphisms in anorexia nervosa and bulimia nervosa, *Neurosci Lett,* 277(2), 134–136, 1999.
52. Ziegler, A., Hebebrand, J., Gorg, T., Rosenkranz, K., Fichter, M., Herpertz-Dahlmann, B., Remschmidt, H., and Hinney, A., Further lack of association between the 5-HT2A gene promoter polymorphism and susceptibility to eating disorders and a meta-analysis pertaining to anorexia nervosa, *Mol Psychiatr,* 4(5), 410–412, 1999.
53. Nishiguchi, N., Matsushita, S., Suzuki, K., Murayama, M., Shirakawa, O., and Higuchi, S., Association between 5HT2A receptor gene promoter region polymorphism and eating disorders in Japanese patients, *Biol Psychiatr,* 50(2), 123–128, 2001.
54. Karwautz, A., Rabe-Hesketh, S., Hu, X., Zhao, J., Sham, P., Collier, D.A., and Treasure, J.L., Individual-specific risk factors for anorexia nervosa: a pilot study using a discordant sister-pair design, *Psychol Med,* 31(2), 317–329, 2001.
55. Gorwood, P., Ades, J., Bellodi, L., Cellini, E., Collier, D.A., Di Bella, D., Di Bernardo, M., Estivill, X., Fernandez-Aranda, F., Gratacos, M., Hebebrand, J., Hinney, A., Hu, X., Karwautz, A., Kipman, A., Mouren-Simeoni, M.C., Nacmias, B., Ribases, M., Remschmidt, H., Ricca, V., Rotella, C.M., Sorbi, S., and Treasure, J., The 5-HT(2A) -1438G/A polymorphism in anorexia nervosa: a combined analysis of 316 trios from six European centres, *Mol Psychiatr,* 7(1), 90–94, 2002.
56. Ricca, V., Nacmias, B., Cellini, E., Di Bernardo, M., Rotella, C.M., and Sorbi, S., 5-HT2A receptor gene polymorphism and eating disorders, *Neurosci Lett,* 323(2), 105–108, 2002.
57. Bruce, K.R., Steiger, H., Joober, R., Ng Ying Kin, N.M., Israel, M., and Young, S.N., Association of the promoter polymorphism -1438G/A of the 5-HT2A receptor gene with behavioral impulsiveness and serotonin function in women with bulimia nervosa, *Am J Med Genet B Neuropsychiatr Genet,* 137(1), 40–44, 2005.
58. Aubert, R., Betoulle, D., Herbeth, B., Siest, G., and Fumeron, F., 5-HT2A receptor gene polymorphism is associated with food and alcohol intake in obese people, *Int J Obes Relat Metab Disord,* 24(7), 920–924, 2000.
59. Lentes, K.U., Hinney, A., Ziegler, A., Rosenkranz, K., Wurmser, H., Barth, N., Jacob, K., Coners, H., Mayer, H., Grzeschik, K.H., Schafer, H., Remschmidt, H., Pirke, K.M., and Hebebrand, J., Evaluation of a Cys23Ser mutation within the human 5-HT2C receptor gene: no evidence for an association of the mutant allele with obesity or underweight in children, adolescents and young adults, *Life Sci,* 61(1), PL9–PL16, 1997.
60. Burnet, P.W., Smith, K.A., Cowen, P.J., Fairburn, C.G., and Harrison, P.J., Allelic variation of the 5-HT2C receptor (HTR2C) in bulimia nervosa and binge eating disorder, *Psychiatr Genet,* 9(2), 101–104, 1999.
61. Yuan, X., Yamada, K., Ishiyama-Shigemoto, S., Koyama, W., and Nonaka, K., Identification of polymorphic loci in the promoter region of the serotonin 5-HT2C receptor gene and their association with obesity and type II diabetes, *Diabetologia,* 43(3), 373–376, 2000.
62. Reynolds, G.P., Zhang, Z.J., and Zhang, X.B., Association of antipsychotic drug-induced weight gain with a 5-HT2C receptor gene polymorphism, *Lancet,* 359(9323), 2086–2087, 2002.

63. Reynolds, G.P., Zhang, Z., and Zhang, X., Polymorphism of the promoter region of the serotonin 5-HT(2C) receptor gene and clozapine-induced weight gain, *Am J Psychiatr,* 160(4), 677–679, 2003.
64. McCarthy, S., Mottagui-Tabar, S., Mizuno, Y., Sennblad, B., Hoffstedt, J., Arner, P., Wahlestedt, C., and Andersson, B., Complex HTR2C linkage disequilibrium and promoter associations with body mass index and serum leptin, *Hum Genet,* 117(6), 545–557, 2005.
65. Pooley, E.C., Fairburn, C.G., Cooper, Z., Sodhi, M.S., Cowen, P.J., and Harrison, P.J., A 5-HT2C receptor promoter polymorphism (HTR2C — 759C/T) is associated with obesity in women, and with resistance to weight loss in heterozygotes, *Am J Med Genet B Neuropsychiatr Genet,* 126(1), 124–127, 2004.
66. Kishore, S. and Stamm, S., The snoRNA HBII-52 regulates alternative Splicing of the serotonin receptor 2C, *Science,* 311(5758), 230–232, 2006.
67. Westberg, L., Bah, J., Rastam, M., Gillberg, C., Wentz, E., Melke, J., Hellstrand, M., and Eriksson, E., Association between a polymorphism of the 5-HT2C receptor and weight loss in teenage girls, *Neuropsychopharmacology,* 26(6), 789–793, 2002.
68. Miller del, D., Ellingrod, V.L., Holman, T.L., Buckley, P.F., and Arndt, S., Clozapine-induced weight gain associated with the 5HT2C receptor -759C/T polymorphism, *Am J Med Genet B Neuropsychiatr Genet,* 133(1), 97–100, 2005.
69. Hinney, A., Herrmann, H., Lohr, T., Rosenkranz, K., Ziegler, A., Lehmkuhl, G., Poustka, F., Schmidt, M.H., Mayer, H., Siegfried, W., Remschmidt, H., and Hebebrand, J., No evidence for an involvement of alleles of polymorphisms in the serotonin1Dbeta and 7 receptor genes in obesity, underweight or anorexia nervosa, *Int J Obes Relat Metab Disord,* 23(7), 760–763, 1999.
70. Levitan, R.D., Kaplan, A.S., Masellis, M., Basile, V.S., Walker, M.L., Lipson, N., Siegel, G.I., Woodside, D.B., Macciardi, F.M., Kennedy, S.H., and Kennedy, J.L., Polymorphism of the serotonin 5-HT1B receptor gene (HTR1B) associated with minimum lifetime body mass index in women with bulimia nervosa, *Biol Psychiatr,* 50(8), 640–643, 2001.
71. Levitan, R.D., Kaplan, A.S., Masellis, M., Basile, V.S., Richter, M.A., and Kennedy, J.L., The serotonin-1Dbeta receptor gene and severity of obsessive-compulsive disorder in women with bulimia nervosa, *Eur Neuropsychopharmacol,* 16(1), 1–6, 2006.
72. Hammoumi, S., Payen, A., Favre, J.D., Balmes, J.L., Benard, J.Y., Husson, M., Ferrand, J.P., Martin, J.P., and Daoust, M., Does the short variant of the serotonin transporter linked polymorphic region constitute a marker of alcohol dependence?, *Alcohol,* 17(2), 107–112, 1999.
73. Himei, A., Kono, Y., Yoneda, H., Sakai, T., Koh, J., Sakai, J., Inada, Y., and Imamichi, H., An association study between alcoholism and the serotonergic receptor genes, *Alcohol Clin Exp Res,* 24(3), 341–342, 2000.
74. Lappalainen, J., Long, J.C., Eggert, M., Ozaki, N., Robin, R.W., Brown, G.L., Naukkarinen, H., Virkkunen, M., Linnoila, M., and Goldman, D., Linkage of antisocial alcoholism to the serotonin 5-HT1B receptor gene in 2 populations, *Arch Gen Psychiatr,* 55(11), 989–994, 1998.
75. Hasegawa, Y., Higuchi, S., Matsushita, S., and Miyaoka, H., Association of a polymorphism of the serotonin 1B receptor gene and alcohol dependence with inactive aldehyde dehydrogenase-2, *J Neural Transm,* 109(4), 513–521, 2002.
76. Fehr, C., Szegedi, A., Anghelescu, I., Klawe, C., Hiemke, C., and Dahmen, N., Sex differences in allelic frequencies of the 5-HT2C Cys23Ser polymorphism in psychiatric patients and healthy volunteers: findings from an association study, *Psychiatr Genet,* 10(2), 59–65, 2000.

77. Soyka, M., Preuss, U.W., Koller, G., Zill, P., and Bondy, B., Association of 5-HT1B receptor gene and antisocial behavior in alcoholism, *J Neural Transm,* 111(1), 101–109, 2004.
78. Cigler, T., LaForge, K.S., McHugh, P.F., Kapadia, S.U., Leal, S.M., and Kreek, M.J., Novel and previously reported single-nucleotide polymorphisms in the human 5-HT(1B) receptor gene: no association with cocaine or alcohol abuse or dependence, *Am J Med Genet,* 105(6), 489–497, 2001.
79. Gorwood, P., Aissi, F., Batel, P., Ades, J., Cohen-Salmon, C., Hamon, M., Boni, C., and Lanfumey, L., Reappraisal of the serotonin 5-HT(1B) receptor gene in alcoholism: of mice and men, *Brain Res Bull,* 57(1), 103–107, 2002.
80. Kranzler, H.R., Hernandez-Avila, C.A., and Gelernter, J., Polymorphism of the 5-HT1B receptor gene (HTR1B): strong within-locus linkage disequilibrium without association to antisocial substance dependence, *Neuropsychopharmacology,* 26(1), 115–122, 2002.
81. Huang, Y.Y., Oquendo, M.A., Friedman, J.M., Greenhill, L.L., Brodsky, B., Malone, K.M., Khait, V., and Mann, J.J., Substance abuse disorder and major depression are associated with the human 5-HT1B receptor gene (HTR1B) G861C polymorphism, *Neuropsychopharmacology,* 28(1), 163–169, 2003.
82. Sinha, R., Cloninger, C.R., and Parsian, A., Linkage disequilibrium and haplotype analysis between serotonin receptor 1B gene variations and subtypes of alcoholism, *Am J Med Genet B Neuropsychiatr Genet,* 121(1), 83–88, 2003.
83. Sun, H.F., Chang, Y.T., Fann, C.S., Chang, C.J., Chen, Y.H., Hsu, Y.P., Yu, W.Y., and Cheng, A.T., Association study of novel human serotonin 5-HT(1B) polymorphisms with alcohol dependence in Taiwanese Han, *Biol Psychiatr,* 51(11), 896–901, 2002.
84. Hwu, H.G. and Chen, C.H., Association of 5HT2A receptor gene polymorphism and alcohol abuse with behavior problems, *Am J Med Genet,* 96(6), 797–800, 2000.
85. Schuckit, M.A., Mazzanti, C., Smith, T.L., Ahmed, U., Radel, M., Iwata, N., and Goldman, D., Selective genotyping for the role of 5-HT2A, 5-HT2C, and GABA alpha 6 receptors and the serotonin transporter in the level of response to alcohol: a pilot study, *Biol Psychiatr,* 45(5), 647–651, 1999.
86. Fehr, C., Schleicher, A., Szegedi, A., Anghelescu, I., Klawe, C., Hiemke, C., and Dahmen, N., Serotonergic polymorphisms in patients suffering from alcoholism, anxiety disorders and narcolepsy, *Prog Neuropsychopharmacol Biol Psychiatr,* 25(5), 965–982, 2001.
87. Nakamura, T., Matsushita, S., Nishiguchi, N., Kimura, M., Yoshino, A., and Higuchi, S., Association of a polymorphism of the 5HT2A receptor gene promoter region with alcohol dependence, *Mol Psychiatr,* 4(1), 85–88, 1999.
88. Galeeva, A.R., Iur'ev, E.B., and Khusnutdinova, E.K., Polymorphism of the serotonin 2A receptor gene in populations from the Volga-Ural region, *Genetika,* 37(11), 1581–1584, 2001.
89. Lappalainen, J., Long, J.C., Virkkunen, M., Ozaki, N., Goldman, D., and Linnoila, M., HTR2C Cys23Ser polymorphism in relation to CSF monoamine metabolite concentrations and DSM-III-R psychiatric diagnoses, *Biol Psychiatr,* 46(6), 821–826, 1999.
90. Wodarz, N., Lange, K., Laufkotter, R., and Johann, M., ADHD and alcohol dependence: a common genetic predisposition?, *Psychiatr Prax,* 31(Suppl. 1), S111–S113, 2004.
91. Johann, M., Bobbe, G., Putzhammer, A., and Wodarz, N., Comorbidity of alcohol dependence with attention-deficit hyperactivity disorder: differences in phenotype with increased severity of the substance disorder, but not in genotype (serotonin transporter and 5-hydroxytryptamine-2c receptor), *Alcohol Clin Exp Res,* 27(10), 1527–1534, 2003.

92. Samochowiec, J., Smolka, M., Winterer, G., Rommelspacher, H., Schmidt, L.G., and Sander, T., Association analysis between a Cys23Ser substitution polymorphism of the human 5-HT2c receptor gene and neuronal hyperexcitability, *Am J Med Genet,* 88(2), 126–130, 1999.
93. Proudnikov, D., LaForge, K.S., Hofflich, H., Levenstien, M., Gordon, D., Barral, S., Ott, J., and Kreek, M.J., Association analysis of polymorphisms in serotonin 1B receptor (HTR1B) gene with heroin addiction: a comparison of molecular and statistically estimated haplotypes, *Pharmacogenet Genomics,* 16(1), 25–36, 2006.
94. Lin, Z., Walther, D., Yu, X.Y., Drgon, T., and Uhl, G.R., The human serotonin receptor 2B: coding region polymorphisms and association with vulnerability to illegal drug abuse, *Pharmacogenetics,* 14(12), 805–811, 2004.
95. Li, T., Liu, X., Zhao, J., Hu, X., Ball, D.M., Loh el, W., Sham, P.C., and Collier, D.A., Allelic association analysis of the dopamine D2, D3, 5-HT2A, and GABA(A)gamma2 receptors and serotonin transporter genes with heroin abuse in Chinese subjects, *Am J Med Genet,* 114(3), 329–335, 2002.
96. Uhl, G.R., Liu, Q.R., Walther, D., Hess, J., and Naiman, D., Polysubstance abuse-vulnerability genes: genome scans for association, using 1,004 subjects and 1,494 single-nucleotide polymorphisms, *Am J Hum Genet,* 69(6), 1290–1300, 2001.
97. Jones-Chesters, M.H., Monsell, S., and Cooper, P.J., The disorder-salient stroop effect as a measure of psychopathology in eating disorders, *Int J Eat Disord,* 24(1), 65–82, 1998.
98. Drysdale, C.M., McGraw, D.W., Stack, C.B., Stephens, J.C., Judson, R.S., Nandabalan, K., Arnold, K., Ruano, G., and Liggett, S.B., Complex promoter and coding region beta 2-adrenergic receptor haplotypes alter receptor expression and predict *in vivo* responsiveness, *Proc Natl Acad Sci USA,* 97(19), 10483–10488, 2000.
99. Sullivan, P.F., Mortality in anorexia nervosa, *Am J Psychiatr,* 152(7), 1073–1074, 1995.
100. Meguid, M.M., Fetissov, S.O., Varma, M., Sato, T., Zhang, L., Laviano, A., and Rossi-Fanelli, F., Hypothalamic dopamine and serotonin in the regulation of food intake, *Nutrition,* 16(10), 843–857, 2000.
101. Naqvi, T.Z. and Gross, S.B., Anorexigen-induced cardiac valvulopathy and female gender, *Curr Womens Health Rep,* 3(2), 116–125, 2003.
102. O'Connor, K.A. and Roth, B.L., Finding new tricks for old drugs: an efficient route for public-sector drug discovery, *Nat Rev Drug Discov,* 4(12), 1005–1014, 2005.
103. Rochester, J.A. and Kirchner, J.T., Ecstasy (3,4-methylenedioxymethamphetamine): history, neurochemistry, and toxicology, *J Am Board Fam Pract,* 12(2), 137–142, 1999.
104. Frith, C.H., Chang, L.W., Lattin, D.L., Walls, R.C., Hamm, J., and Doblin, R., Toxicity of methylenedioxymethamphetamine (MDMA) in the dog and the rat, *Fundam Appl Toxicol,* 9(1), 110–119, 1987.
105. Conductier, G., Crosson, C., Hen, R., Bockaert, J., and Compan, V., 3,4-N-methlenedioxymethamphetamine-induced hypophagia is maintained in 5-HT1B receptor knockout mice, but suppressed by the 5-HT2C receptor antagonist RS102221, *Neuropsychopharmacology,* 30(6), 1056–1063, 2005.
106. Heisler, L.K., Jobst, E.E., Sutton, G.M., Zhou, L., Borok, E., Thornton-Jones, Z., Liu, H.Y., Zigman, J.M., Balthasar, N., Kishi, T., Lee, C.E., Aschkenasi, C.J., Zhang, C.Y., Yu, J., Boss, O., Mountjoy, K.G., Clifton, P.G., Lowell, B.B., Friedman, J.M., Horvath, T., Butler, A.A., Elmquist, J.K., and Cowley, M.A., Serotonin reciprocally regulates melanocortin neurons to modulate food intake, *Neuron,* 51(2), 239–249, 2006.
107. Kennett, G.A. and Curzon, G., Evidence that hypophagia induced by mCPP and TFMPP requires 5-HT1C and 5-HT1B receptors; hypophagia induced by RU 24969 only requires 5-HT1B receptors, *Psychopharmacology (Berl),* 96(1), 93–100, 1988.

108. Klodzinska, A. and Chojnacka-Wojcik, E., Anorexia induced by M-trifluorometh-ylphenylpiperazine (TFMPP) in rats, *Pol J Pharmacol Pharm,* 42(1), 13–17, 1990.
109. Yamada, J., Sugimoto, Y., Yoshikawa, T., Noma, T., and Horisaka, K., The effects of peripheral serotonin2 receptor agonist on food intake of rats, *Adv Exp Med Biol,* 398, 555–557, 1996.
110. Vickers, S.P., Clifton, P.G., Dourish, C.T., and Tecott, L.H., Reduced satiating effect of d-fenfluramine in serotonin 5-HT(2C) receptor mutant mice, *Psychopharmacology (Berl),* 143(3), 309–314, 1999.
111. Vickers, S.P., Dourish, C.T., and Kennett, G.A., Evidence that hypophagia induced by d-fenfluramine and d-norfenfluramine in the rat is mediated by 5-HT2C receptors, *Neuropharmacology,* 41(2), 200–209, 2001.
112. Clifton, P.G., Lee, M.D., Somerville, E.M., Kennett, G.A., and Dourish, C.T., 5-HT1B receptor knockout mice show a compensatory reduction in 5-HT2C receptor function, *Eur J Neurosci,* 17(1), 185–190, 2003.
113. Heisler, L.K., Cowley, M.A., Tecott, L.H., Fan, W., Low, M.J., Smart, J.L., Rubinstein, M., Tatro, J.B., Marcus, J.N., Holstege, H., Lee, C.E., Cone, R.D., and Elmquist, J.K., Activation of central melanocortin pathways by fenfluramine, *Science,* 297(5581), 609–611, 2002.
114. Dulawa, S.C., Gross, C., Stark, K.L., Hen, R., and Geyer, M.A., Knockout mice reveal opposite roles for serotonin 1A and 1B receptors in prepulse inhibition, *Neuropsychopharmacology,* 22(6), 650–659, 2000.
115. Bouwknecht, J.A., van der Gugten, J., Hijzen, T.H., Maes, R.A., Hen, R., and Olivier, B., Male and female 5-HT(1B) receptor knockout mice have higher body weights than wildtypes, *Physiol Behav,* 74(4–5), 507–516, 2001.
116. Dirks, A., Pattij, T., Bouwknecht, J.A., Westphal, T.T., Hijzen, T.H., Groenink, L., van der Gugten, J., Oosting, R.S., Hen, R., Geyer, M.A., and Olivier, B., 5-HT1B receptor knockout, but not 5-HT1A receptor knockout mice, show reduced startle reactivity and footshock-induced sensitization, as measured with the acoustic startle response, *Behav Brain Res,* 118(2), 169–178, 2001.
117. Kelley, S.P., Bratt, A.M., and Hodge, C.W., Targeted gene deletion of the 5-HT3A receptor subunit produces an anxiolytic phenotype in mice, *Eur J Pharmacol,* 461(1), 19–25, 2003.
118. Aulakh, C.S., Hill, J.L., and Murphy, D.L., A comparison of feeding and locomotion responses to serotonin agonists in three rat strains, *Pharmacol Biochem Behav,* 31(3), 567–571, 1988.
119. Bendotti, C. and Samanin, R., The role of putative 5-HT1A and 5-HT1B receptors in the control of feeding in rats, *Life Sci,* 41(5), 635–642, 1987.
120. Sprouse, J.S. and Aghajanian, G.K., Electrophysiological responses of serotoninergic dorsal raphe neurons to 5-HT1A and 5-HT1B agonists, *Synapse,* 1(1), 3–9, 1987.
121. Haj-Dahmane, S., Hamon, M., and Lanfumey, L., K+ channel and 5-hydroxytryptamine1A autoreceptor interactions in the rat dorsal raphe nucleus: an *in vitro* electrophysiological study, *Neuroscience,* 41(2–3), 495–505, 1991.
122. Le Poul, E., Lima, L., Laporte, A.M., Even, C., Doucet, E., Fattaccini, C.M., Laaris, N., Hamon, M., and Lanfumey, L., Central serotonin receptors and chronic treatment with selective serotonin reuptake inhibitors in the rat: comparative effects of fluoxetine and paroxetine, *Encephale,* 21(2), 123–132, 1995.
123. Casanovas, J.M., Lesourd, M., and Artigas, F., The effect of the selective 5-HT1A agonists alnespirone (S-20499) and 8-OH-DPAT on extracellular 5-hydroxytryptamine in different regions of rat brain, *Br J Pharmacol,* 122(4), 733–741, 1997.

124. Casanovas, J.M., Berton, O., Celada, P., and Artigas, F., *In vivo* actions of the selective 5-HT1A receptor agonist BAY x 3702 on serotonergic cell firing and release, *Naunyn Schmiedebergs Arch Pharmacol,* 362(3), 248–254, 2000.
125. Celada, P., Puig, M.V., Casanovas, J.M., Guillazo, G., and Artigas, F., Control of dorsal raphe serotonergic neurons by the medial prefrontal cortex: involvement of serotonin-1A, GABA(A), and glutamate receptors, *J Neurosci,* 21(24), 9917–9929, 2001.
126. Bortolozzi, A., Amargos-Bosch, M., Toth, M., Artigas, F., and Adell, A., *In vivo* efflux of serotonin in the dorsal raphe nucleus of 5-HT1A receptor knockout mice, *J Neurochem,* 88(6), 1373–1379, 2004.
127. Richer, M., Hen, R., and Blier, P., Modification of serotonin neuron properties in mice lacking 5-HT1A receptors, *Eur J Pharmacol,* 435(2–3), 195–203, 2002.
128. Dulawa, S.C., Hen, R., Scearce-Levie, K., and Geyer, M.A., Serotonin1B receptor modulation of startle reactivity, habituation, and prepulse inhibition in wild-type and serotonin1B knockout mice, *Psychopharmacology (Berl),* 132(2), 125–134, 1997.
129. Boutrel, B., Franc, B., Hen, R., Hamon, M., and Adrien, J., Key role of 5-HT1B receptors in the regulation of paradoxical sleep as evidenced in 5-HT1B knock-out mice, *J Neurosci,* 19(8), 3204–3212, 1999.
130. Inoue, T., Tsuchiya, K., and Koyama, T., Regional changes in dopamine and serotonin activation with various intensity of physical and psychological stress in the rat brain, *Pharmacol Biochem Behav,* 49(4), 911–920, 1994.
131. Ge, J., Barnes, N.M., Costall, B., and Naylor, R.J., Effect of aversive stimulation on 5-hydroxytryptamine and dopamine metabolism in the rat brain, *Pharmacol Biochem Behav,* 58(3), 775–783, 1997.
132. Konstandi, M., Johnson, E., Lang, M.A., Malamas, M., and Marselos, M., Noradrenaline, dopamine, serotonin: different effects of psychological stress on brain biogenic amines in mice and rats, *Pharmacol Res,* 41(3), 341–346, 2000.
133. Grignaschi, G., Mantelli, B., and Samanin, R., The hypophagic effect of restraint stress in rats can be mediated by 5-HT2 receptors in the paraventricular nucleus of the hypothalamus, *Neurosci Lett,* 152(1–2), 103–106, 1993.
134. Chou-Green, J.M., Holscher, T.D., Dallman, M.F., and Akana, S.F., Repeated stress in young and old 5-HT(2C) receptor knockout mice, *Physiol Behav,* 79(2), 217–226, 2003.
135. Gross, C., Zhuang, X., Stark, K., Ramboz, S., Oosting, R., Kirby, L., Santarelli, L., Beck, S., and Hen, R., Serotonin1A receptor acts during development to establish normal anxiety-like behaviour in the adult, *Nature,* 416(6879), 396–400, 2002.
136. Lucas, G., Compan, V., Charnay, Y., Neve, R.L., Nestler, E.J., Bockaert, J., Barrot, M., and Debonnel, G., Frontocortical 5-HT4 receptors exert positive feedback on serotonergic activity: viral transfections, subacute and chronic treatments with 5-HT4 agonists, *Biol Psychiatr,* 57(8), 918–925, 2005.
137. Holick, K.A., Dulawa, S., Compan, V., Hen, R., The contribution of 5-HT1A and 5-HT4 receptors to the effects of fluoxetine in the mouse, Society for Neuroscience Online Abstract viewer/Itinerary Planner (Program N°567.14.), Washington, D.C., 2005.
138. Ohtsuki, T., Ishiguro, H., Detera-Wadleigh, S.D., Toyota, T., Shimizu, H., Yamada, K., Yoshitsugu, K., Hattori, E., Yoshikawa, T., and Arinami, T., Association between serotonin 4 receptor gene polymorphisms and bipolar disorder in Japanese case-control samples and the NIMH genetics initiative bipolar pedigrees, *Mol Psychiatr,* 7(9), 954–961, 2002.
139. Rosel, P., Arranz, B., Urretavizcaya, M., Oros, M., San, L., and Navarro, M.A., Altered 5-HT2A and 5-HT4 postsynaptic receptors and their intracellular signalling systems IP3 and cAMP in brains from depressed violent suicide victims, *Neuropsychobiology,* 49(4), 189–195, 2004.

140. Barrot, M., Wallace, D.L., Bolanos, C.A., Graham, D.L., Perrotti, L.I., Neve, R.L., Chambliss, H., Yin, J.C., and Nestler, E.J., Regulation of anxiety and initiation of sexual behavior by CREB in the nucleus accumbens, *Proc Natl Acad Sci USA,* 102(23), 8357–8362, 2005.
141. Carlezon, W.A., Jr., Duman, R.S., and Nestler, E.J., The many faces of CREB, *Trends Neurosci,* 28(8), 436–445, 2005.
142. Berthouze, M., Ayoub, M., Russo, O., Rivail, L., Sicsic, S., Fischmeister, R., Berque-Bestel, I., Jockers, R., and Lezoualc'h, F., Constitutive dimerization of human serotonin 5-HT4 receptors in living cells, *FEBS Lett,* 579(14), 2973–2980, 2005.
143. Hunter, R.G., Philpot, K., Vicentic, A., Dominguez, G., Hubert, G.W., and Kuhar, M.J., CART in feeding and obesity, *Trends Endocrinol Metab,* 15(9), 454–459, 2004.
144. Cichon, S., Kesper, K., Propping, P., and Nothen, M.M., Assignment of the human serotonin 4 receptor gene (HTR4) to the long arm of chromosome 5 (5q31–q33), *Mol Membr Biol,* 15(2), 75–78, 1998.
145. Echwald, S.M., Sorensen, T.I., Andersen, T., Hansen, C., Tommerup, N., and Pedersen, O., Sequence variants in the human cocaine and amphetamine-regulated transcript (CART) gene in subjects with early onset obesity, *Obes Res,* 7(6), 532–536, 1999.
146. Douglass, J. and Daoud, S., Characterization of the human cDNA and genomic DNA encoding CART: a cocaine- and amphetamine-regulated transcript, *Gene,* 169(2), 241–245, 1996.
147. Lowry, C.A. and Moore, F.L., Regulation of behavioral responses by corticotropin-releasing factor, *Gen Comp Endocrinol,* 146(1), 19–27, 2006.
148. Zhuang, X., Gross, C., Santarelli, L., Compan, V., Trillat, A.C., and Hen, R., Altered emotional states in knockout mice lacking 5-HT1A or 5-HT1B receptors, *Neuropsychopharmacology,* 21(2 Suppl.), 52S–60S, 1999.
149. Hodge, C.W., Kelley, S.P., Bratt, A.M., Iller, K., Schroeder, J.P., and Besheer, J., 5-HT(3A) receptor subunit is required for 5-HT3 antagonist-induced reductions in alcohol drinking, *Neuropsychopharmacology,* 29(10), 1807–1813, 2004.
150. Bhatnagar, S., Nowak, N., Babich, L., and Bok, L., Deletion of the 5-HT3 receptor differentially affects behavior of males and females in the Porsolt forced swim and defensive withdrawal tests, *Behav Brain Res,* 153(2), 527–535, 2004.
151. Bonasera, S.J., Chu, H.M., Brennan, T.J., and Tecott, L.H., A null mutation of the serotonin 6 receptor alters acute responses to ethanol, *Neuropsychopharmacology,* 31(8), 1801–1813, 2006.
152. Nonogaki, K., Abdallah, L., Goulding, E.H., Bonasera, S.J., and Tecott, L.H., Hyperactivity and reduced energy cost of physical activity in serotonin 5-HT(2C) receptor mutant mice, *Diabetes,* 52(2), 315–320, 2003.
153. Bardo, M.T., Donohew, R.L., and Harrington, N.G., Psychobiology of novelty seeking and drug seeking behavior, *Behav Brain Res,* 77(1–2), 23–43, 1996.
154. Bowirrat, A. and Oscar-Berman, M., Relationship between dopaminergic neurotransmission, alcoholism, and reward deficiency syndrome, *Am J Med Genet B Neuropsychiatr Genet,* 132(1), 29–37, 2005.
155. Piazza, P.V. and Le Moal, M., The role of stress in drug self-administration, *Trends Pharmacol Sci,* 19(2), 67–74, 1998.
156. Rothman, R.B. and Baumann, M.H., Therapeutic and adverse actions of serotonin transporter substrates, *Pharmacol Ther,* 95(1), 73–88, 2002.
157. Colado, M.I., O'Shea, E., and Green, A.R., Acute and long-term effects of MDMA on cerebral dopamine biochemistry and function, *Psychopharmacology (Berl),* 173(3–4), 249–263, 2004.

158. Ritz, M.C., Cone, E.J., and Kuhar, M.J., Cocaine inhibition of ligand binding at dopamine, norepinephrine and serotonin transporters: a structure-activity study, *Life Sci,* 46(9), 635–645, 1990.
159. Geyer, M.A., Callaway, C.W., Behavioral pharmacology of ring-substituted amphetamine analogs, in *Amphetamine and Its Analogs*, Cho, A.R. and Segal, O.S., Eds., Academic Press, New York, 1994, pp. 177–201.
160. McNamara, M.G., Kelly, J.P., and Leonard, B.E., Some behavioural and neurochemical aspects of subacute (+/-)3,4-methylenedioxymethamphetamine administration in rats, *Pharmacol Biochem Behav,* 52(3), 479–484, 1995.
161. Ball, K.T., Budreau, D., and Rebec, G.V., Context-dependent behavioural and neuronal sensitization in striatum to MDMA (ecstasy) administration in rats, *Eur J Neurosci,* 24(1), 217–228, 2006.
162. Adell, A., Celada, P., and Artigas, F., The role of 5-HT1B receptors in the regulation of serotonin cell firing and release in the rat brain, *J Neurochem,* 79(1), 172–182, 2001.
163. Hamon, M., The main features of central 5-HT1A receptors. Serotoninergic Neurons and 5-HT Receptors in the CNS, in *Handbook of Experimental Pharmacology*, Baumgarten, H.G. and Göthert, M., Eds., Vol. 129, 1997, pp. 239–268.

# 9 Use of Mice with Targeted Genetic Inactivation in the Serotonergic System for the Study of Anxiety

*Miklos Toth*

## CONTENTS

Role of the 5-HT System in Modulating Anxiety States ..... 181
Mice with Targeted Genetic Inactivation in the Serotonergic System for the Study of Anxiety ..... 182
Behavioral Tests for Anxiety ..... 183
Mice with Genetic Inactivation of 5-HT Receptors and the 5-HT Transporter ..... 183
$5\text{-HT}_{1A}$ Receptor Knockout Mice ..... 184
$5\text{-HT}_{1B}$ Receptor Knockout Mice ..... 186
$5\text{-HT}_{2}$ Receptor Knockout Mice ..... 186
$5\text{-HT}_{3}$ Receptor Knockout Mice ..... 187
$5\text{-HT}_{4-7}$ Receptor Knockout Mice ..... 187
5-HT Transporter Knockout Mice ..... 188
Summary ..... 189
References ..... 189

## ROLE OF THE 5-HT SYSTEM IN MODULATING ANXIETY STATES

Anxiety is a normal reaction to threatening situations, and it represents a physiological protective function. Anxiety is often manifested as avoidance of threatening situations and is also characterized by overt sympathetic reactions. Pathological anxiety is a level of anxiety that is disproportionate to the threat and can be manifested even in the absence of threat. Individuals seem to have a consistent level of anxiety over their lifetime suggesting the importance of genetic factors. DSM-IV (American Psychiatric Association, 1994) and the International Classification of

Diseases [ICD-10] (World Health Organization, 1992), two categorical systems, set the boundary at which a particular level of anxiety becomes an anxiety disorder. These boundaries are often based on the number and the duration of symptoms. DSM-IV provides diagnostic criteria for anxiety disorders, including panic disorder (PD), specific and social phobias, obsessive compulsive disorder (OCD), posttraumatic stress disorder (PTSD), and generalized anxiety disorder (GAD) (American Psychiatric Association, 1994; Noyes, 2004).

The 5-HT system has been implicated in the modulation of anxiety levels; some components of the 5-HT system promote anxiety whereas others reduce its symptoms. However, the 5-HT is only one of the many systems that have been implicated in anxiety disorders and GABA, norepinephrine, dopamine, and neuropeptides including corticotropin-releasing hormone, cholecystokinin, and neuropeptide Y have been shown to modulate anxiety.

The level of 5-HT is regulated by both the 5-HT transporter and the 5-$HT_{1A}$ autoreceptor (in the serotonergic raphe nuclei) (Blier et al., 1998; Inoue et al., 2004). Inhibiting the 5-HT transporter by selective serotonin reuptake inhibitors (SSRIs) is an effective treatment for certain anxiety disorders and depression (Blier et al., 1998; Inoue et al., 2004), whereas partial 5-$HT_{1A}$ receptor agonists such as buspirone have an anxiolytic effect (Goldberg et al., 1983). The 5-HT-mediated anxiolytic actions of these drugs do not in any way support the role of the 5-HT system in the pathogenesis of anxiety disorders. More compelling are the findings of recent genetic studies demonstrating a relatively small but significant increase in neuroticism in individuals who carry the *s/s* (short promoter repeat) alleles of the 5-HT transporter as compared to individuals with *s/l* (long) or *l/l* alleles (Lesch et al., 1996). Further studies and a recent meta-analysis of these studies found a moderate but significant association between the *s* allele and measures of anxiety (Sen et al., 2004; Munafo et al., 2004). A more robust effect of the *s* allele was seen in neuroimaging studies when amygdala activation was measured to fearful and angry human facial expressions (Hariri et al., 2002). These data are somewhat surprising because the *s* allele is associated with decreased transporter activity (and presumably more synaptic 5-HT), whereas pharmacological inhibition of the 5-HT transporter by SSRIs results in an anxiolytic effect. It has been hypothesized that a constitutive increase in 5-HT levels, specifically during development, leads to the increase in anxiety.

The involvement of the 5-HT system in the pathogenesis of anxiety disorders is also supported by the association between reduced 5-$HT_{1A}$ receptor levels in the anterior cingulate, posterior cingulate and raphe in PD and PTSD (Lesch et al., 1992; Lopez et al., 1998; Mann, 1999; Lemonde et al., 2003; Neumeister et al., 2004).

## MICE WITH TARGETED GENETIC INACTIVATION IN THE SEROTONERGIC SYSTEM FOR THE STUDY OF ANXIETY

Although the association studies mentioned above have the potential of discovering disease-predisposing genes, the small contribution of individual genes and the variability of their effects make the discovery of these genes rare and difficult. In contrast, knockout technology allows the creation of genetic abnormalities in a large number of animals with the same genetic background (where variability in gene–gene interaction is not a factor). These animals can be tested for behavioral alterations under identical environmental conditions that is also important in finding small effects. A further advantage is that mouse knockout models are amenable for mechanistic studies to reveal the molecular and cellular pathways underlying a specific behavior.

## BEHAVIORAL TESTS FOR ANXIETY

Anxiety cannot be easily reproduced in animals, but it is possible to create conditions resembling anxiety disorders, especially with genetically modified mice. Fearful situations in animals elicit avoidance and the unwillingness to be exposed to novel situations are also common symptoms of anxiety disorders including GAD, PD, phobias, and PTSD. Therefore, fear reaction can conceptually be interpreted as anxiety-like behavior (Rodgers, 1997; Crawley, 1999; Weiss et al., 2000). As in humans, fear/anxiety in mice can be viewed as a behavior serving a protective function and an increase in the normal level of fear reaction as a result of a genetic manipulation may be interpreted as conditions similar to anxiety disorders.

Fear and anxiety-like behavior can be unconditioned or conditioned. Unconditioned tests often measure the natural conflict experienced by animals as they avoid or explore a novel, potentially dangerous environment for food, water, or social reward. Measurements of avoidant behaviors, such as decreased activity in a particular region of the testing apparatus, compared to overall activity, provide a quantifiable measure to assess the level of anxiety. Novel stressful situations include high platforms (elevated plus maze) (Lister, 1987) and a brightly lit area (open field, light–dark box) (Crawley, 1981; Treit and Fundytus, 1988). Animal models of conditioned fear on the other hand measure anxiety in environments associated with prior or present stress (Davis, 1990; Geller et al., 1962; Vogel et al., 1971). Similar to avoidance of unconditioned situations, avoidance of conditioned situations is also a characteristic symptom of PD, PTSD, and in some degree of all anxiety disorders. As unconditioned and conditioned anxiety tests are straightforward and relatively simple to conduct and interpret, they are frequently used in behavioral studies. More complex methods are also available, including social interaction tests that measure a different aspect of avoidant behavior that is highly relevant to some anxiety disorders (for example social phobias) (File, 1985). Additional information on anxiety-related tests can be found in several excellent reviews (Griebel, 1995; Rodgers, 1997; Rodgers et al., 1997; Weiss et al., 2000; Uys et al., 2003).

## MICE WITH GENETIC INACTIVATION OF 5-HT RECEPTORS AND THE 5-HT TRANSPORTER

5-HT receptors mediate the pre- and postsynaptic actions of 5-HT and are classified into seven groups (5-$HT_{1-7}$), comprising a total of at least 14 structurally and pharmacologically distinct mammalian receptor subtypes (Hoyer et al., 1994). At the molecular level, most 5-HT receptors are seven transmembrane-spanning, G-protein-coupled metabotropic receptors (Barnes and Sharp, 1999). The 5-HT$_3$ receptor is the only member of the 5-HT receptor family that is a ligand-gated ion channel (Derkach et al., 1989).

The 5-HT transporter is central in the regulation of 5-HT activity in the brain. Following its release, 5-HT is taken up by presynaptically located 5-HT transporters for recycling or metabolic degradation. Under normal conditions, 5-HT transporter activity is the principal mechanism for clearing 5-HT from the synapse and extracellular space that consequently determines the duration and intensity of the 5-HT signal.

## 5-HT$_{1A}$ Receptor Knockout Mice

5-HT$_{1A}$ receptor knockout mice were generated on Swiss-Webster (SW), C57Bl6, and 129sv genetic backgrounds (Heisler et al., 1998; Parks et al., 1998; Ramboz et al., 1998). Independently of the genetic background, mutant mice exhibited anxiety-like behaviors (in open field, elevated plus, and zero maze) and reduced immobility in the forced swim test (Parks et al., 1998) or tail suspension test (Heisler et al., 1998; Ramboz et al., 1998). Later studies with 5-HT$_{1A}$ receptor knockout mice on the 129sv background also showed anxiety-like behavior in the novelty-induced suppression of feeding test and in fear conditioning (Gross et al., 2000; Klemenhagen et al., 2006). It was also shown that 5-HT$_{1A}$ receptor knockout mice have reduced locomotor activity, another sign of increased anxiety-like behavior (Gross et al., 2002). Increased autonomic arousal, a typical symptom of anxiety disorders, was also observed in these mice. For example, heart rate and body temperature was increased following footshock and saline injection in 5-HT$_{1A}$ receptor knockout mice (Gross et al., 2000; Pattij et al., 2002). These data demonstrate that 5-HT$_{1A}$ receptor knockout mice have abnormalities in three important measures of anxiety: increased avoidance, decreased locomotor activity, and increased autonomic arousal. Consequently, these studies firmly established the 5-HT$_{1A}$ receptor as a modulator of anxiety, at least in mice.

Altered sleep pattern is also a characteristic abnormality in anxiodepressive diseases, and 5-HT$_{1A}$ receptor knockout mice have been shown to have altered sleep–wake regulation. Specifically, rapid eye movement (REM) sleep was enhanced in 5-HT$_{1A}$ receptor knockout mice during both the light and the dark phases, whereas slow wave sleep was unaltered (Boutrel et al., 2002).

In addition to anxiety-related phenotypes, data indicate a hippocampal-dependent learning and memory deficit in 5-HT$_{1A}$ receptor knockout mice. Behavior of mutant mice on the SW background is impaired in the hidden platform (spatial) version of the Morris water maze and the delayed version of the Y maze (Sarnyai et al., 2000). In contrast, nonhippocampal memory tasks such as the visible platform (nonspatial) version of the Morris water maze, the immediate version of the Y maze, and the spontaneous-alternation test of working memory were normal in 5-HT$_{1A}$ receptor knockout mice. 5-HT$_{1A}$ receptor knockout mice on the 129sv genetic background showed a similar phenotype but the abnormality was apparent in older (22-month-old) but not in young (3-month-old) mice (Wolff et al., 2004). As the mice in the Sarnyai et al. study (Sarnyai et al., 2000) were 4–6 months old, it may take a certain degree of aging when the learning and memory defect becomes apparent in 5-$HT_{1A}$ receptor knockout mice. Alternatively, there are baseline differences among mouse strains in Morris water maze performance (Clapcote and Roder, 2004) that could influence the onset of the learning and memory deficit in 5-$HT_{1A}$ receptor knockout mice.

It is important to note that acute pharmacological inhibition of the 5-$HT_{1A}$ receptor does not result in an anxiety-like phenotype (Millan, 2003). Consistent with this, knockout mice with an adult-specific receptor loss showed no anxiety-like phenotype. Only when receptor expression was inactivated during early postnatal life mice developed anxiety that persisted through life (Gross et al., 2002). The development-specific effects of receptor inactivation, however, does not apply to the 5-$HT_{1A}$ receptor mediated regulation of REM sleep because pharmacological blockade of the receptor by the selective antagonist WAY 100635 promoted REM sleep similar to the effect of the genetic inactivation (Boutrel et al., 2002). This indicates a direct role of the 5-$HT_{1A}$ receptor in the regulation of REM sleep.

The brain regions associated with the development of the anxiety-like phenotype in 5-$HT_{1A}$ receptor knockout mice are important to determine if we want to understand the neuronal circuitries underlying this phenotype. Although hippocampal (in particular dorsal hippocampal) dysfunction has traditionally been associated with spatial cognitive tasks, data also indicate a ventral hippocampal involvement in anxiety-like behavior (Moser and Moser, 1998; Kjelstrup et al., 2002). Consistent with the behavioral abnormalities that can be linked to the hippocampus, 5-$HT_{1A}$ receptor knockout mice show electrophysiological alterations in the hippocampus. Paired-pulse facilitation and paired pulse inhibition were impaired in the dentate gyrus and CA1 region of the hippocampus of SW 5-$HT_{1A}$ receptor knockout mice, respectively (Sarnyai et al., 2000; Sibille et al., 2000). Another study demonstrated an increase in theta oscillations in 5-$HT_{1A}$ receptor knockout mice on the 129sv genetic background during the exploration of the open arm of the elevated plus maze (Gordon et al., 2005). Theta oscillations are 4–12 Hz waves present in local field potentials recorded throughout the hippocampus during navigation (Buzsaki, 2002). Also, 5-$HT_{1A}$ receptor knockout mice displayed higher limbic excitability manifested as lower seizure threshold and higher lethality in response to kainic acid administration (Sibille et al., 2000). Another brain region that may be involved in the anxiety-like phenotype of 5-$HT_{1A}$ receptor knockout mice is the prefrontal cortex (PFC). PFC malfunction was detected in anxiety disorders (Malizia et al., 1998; Bremner et al., 2000; Bystritsky et al., 2001; Dilger et al., 2003) and increased neuronal activity was associated with anxiety in animals (Shah and Treit, 2003; Singewald et al., 2003; Shah and Treit, 2004). Reductions in $GABA_A$ receptor expression were reported in 5-$HT_{1A}$ receptor knockout mice on the SW background (Bailey and Toth, 2004) while changes in glutamate and GABA uptake were found in knockout mice on the C57Bl6 background (Bruening et al., 2006). This demonstrates that the inactivation of the 5-$HT_{1A}$ receptor elicits different and genetic-background-dependent perturbations in the prefrontal cortex GABA/glutamate system. These perturbations can result in a change in the balance between excitation and inhibition on both genetic backgrounds and may contribute to the anxiety phenotype of 5-$HT_{1A}$ receptor knockout mice.

The electrophysiological and biochemical studies summarized above are consistent with a role for 5-$HT_{1A}$ receptors at the postsynaptic sites in anxiety-related behaviors. 5-$HT_{1A}$ receptors are expressed both presynaptically on 5-HT neurons in the raphe nuclei as somatodendritic autoreceptors and postsynaptically in 5-HT target areas (such as hippocampus and cortex). Although it seems plausible that lack of the presynaptic receptors results in an increase in 5-HT leading to the anxiety phenotype, several lines of evidence suggest that it is rather the absence of postsynaptic 5-$HT_{1A}$ receptors that can be linked to the anxiety phenotype. First, basal 5-HT levels are not altered, as measured by *in vivo* microdialysis, in 5-$HT_{1A}$ receptor knockout mice (He et al., 2001; Bortolozzi et al., 2004). Second, expression of 5-$HT_{1A}$ receptors in forebrain regions (at the postsynaptic sites) rescued the anxiety phenotype of 5-$HT_{1A}$ receptor knockout mice (Gross et al., 2002).

In summary, experiments with 5-$HT_{1A}$ receptor knockout mice clearly demonstrate that the expression of the receptor is essential for the establishment and/or maintenance of a normal level of anxiety. Data indicate the involvement of a 5-$HT_{1A}$ receptor-dependent early postnatal developmental process in the anxiety-like phenotype of 5-$HT_{1A}$ receptor knockout mice, but the actual neurobiological process is still unknown.

## 5-$HT_{1B}$ Receptor Knockout Mice

In contrast to the genetic inactivation of the 5-$HT_{1A}$ receptor, knockout of the 5-$HT_{1B}$ receptor resulted in a reduced anxiety level in an open field test and in the novelty-induced suppression of feeding test (Zhuang et al., 1999). However, the light–dark box and elevated plus maze tests showed no significant change in anxiety-like behavior in the 5-$HT_{1B}$ receptor knockout mice (Malleret et al., 1999; Phillips et al., 1999). Moreover, the reduced anxiety phenotype of 5-$HT_{1B}$ receptor knockout mice could not be reproduced in various laboratories, even using the same supply of mice (Phillips et al., 1999) indicating that the phenotype is not robust and/or there are confounding, such as environmental, factors that may strongly influence the phenotype. The 5-$HT_{1B}$ receptor may also be involved in regulating sleep because its inactivation increased REM sleep (Boutrel et al., 1999). The rather limited anxiety-related phenotype of 5-$HT_{1B}$ receptor knockout mice is surprising because the receptor is widely expressed in both presynaptic and postsynaptic locations (Gothert, 1990; Boschert et al., 1994). It has been proposed that the lack of a robust anxiety-related phenotype in 5-$HT_{1B}$ receptor knockout mice is the result of adaptive changes in the coupling of 5-$HT_{1A}$ and 5-$HT_{2C}$ receptors (Knobelman et al., 2001; Ase et al., 2002; Clifton et al., 2003).

## 5-$HT_2$ Receptor Knockout Mice

Similarly to 5-$HT_{1A}$ receptor specific drugs, compounds that target the 5-$HT_{2A}$ receptors are widely used in the treatment of depression and schizophrenia. Also, 5-$HT_{2A}$ receptor antagonists have an anxiolytic profile in a number of behavioral paradigms (Stutzmann et al., 1991; Motta et al., 1992; Costall and Naylor, 1995). Consistent with the pharmacological data, global disruption of 5-$HT_{2A}$ receptors in mice resulted in a reduced level of anxiety in various conflict-based tests, including the elevated plus maze, open field and the light–dark shuttle test (Weisstaub et al., 2006). These data indicate that it is the direct effect of the receptor loss rather than the absence of the receptor during developmental that leads to the knockout phenotype. The genetic inactivation of the receptor did not alter fear-conditioning and depression-related (forced swim and tail suspension) behaviors. This finding is surprising because antisense-mediated downregulation of the 5-$HT_{2A}$ receptor decreased immobility in the forced swim test (Sibille et al., 1997). Similarly, pharmacological blockade of the receptor decreases immobility (Patel et al., 2004), and treatment with antidepressants downregulates the 5-$HT_{2A}$ receptor and results in less immobility. The different behavioral outcomes of the genetic inactivation and the antisense knockdown/pharmacological blockade of the receptor suggest that developmental changes in knockout mice blunt the effect of the receptor deficit on depression-related behavior.

Selective cortical re-expression of the 5-$HT_{2A}$ receptor rescued the reduced anxiety-like behavior of 5-$HT_{2A}$ receptor knockout mice, indicating a role for cortical 5-$HT_{2A}$ receptors in the modulation of conflict based anxiety-related behavior (Weisstaub et al., 2006).

Genetic inactivation of another member of the 5-$HT_2$ receptor family, the 5-$HT_{2B}$ receptor, results in embryonic and neonatal death caused by heart defects (Nebigil et al., 2000), and therefore a possible role for the receptor in anxiety is impossible to study by conventional knockout. Genetic inactivation of still another member of the family, the 5-$HT_{2C}$ receptor, indicated the involvement of the receptor in food intake (Chou-Green et al., 2003b, a; Tecott and Abdallah, 2003). It was proposed that these mice may model weight gain seen in depression and compulsive disorders.

## 5-$HT_3$ Receptor Knockout Mice

The only ionotropic 5-HT receptor, the 5-$HT_3$ receptor, has also been linked to anxiety-like behavior by using targeted mutagenesis. 5-$HT_3$ receptor deficient mice exhibit reduced anxiety-like behavior in the elevated plus maze and light dark shuttle box (Kelley et al., 2003). As pharmacological blockade of the receptor results in an anxiolytic profile in behavioral tests (Costall and Naylor, 1992), the receptor seems to directly regulate anxiety levels in mice. The reduced anxiety-like phenotype of 5-$HT_3$ receptor knockout mice is similar to that of the 5-$HT_{2A}$ receptor knockout mice and opposite to that of the 5-$HT_{1A}$ receptor-deficient mice, indicating that some 5-HT receptors increase while others reduce anxiety levels.

## 5-$HT_{4–7}$ Receptor Knockout Mice

Genetic inactivation of the 5-$HT_4$ receptor resulted in reduced locomotor activity in novel environment but increased feeding following stress (Compan et al., 2004). It was concluded that 5-$HT_4$ receptors modulate feeding and activity in stressful situations and are perhaps involved in eating disorders.

Of the two subtypes of 5-$HT_5$ receptor, 5-$HT_{5A}$ and 5-$HT_{5B}$, only the former is present in human CNS (Grailhe et al., 1999; Grailhe et al., 2001). 5-$HT_{5A}$ knockout mice exhibit increased exploratory activity when exposed to novel environment but show no change in anxiety-related behaviors in several behavioral paradigms (Grailhe et al., 1999; Grailhe et al., 2001).

Deletion of the 5-HT$_6$ receptor resulted in no abnormal behavioral phenotype tested in numerous behavioral assays including models of anxiety (Bonasera et al., 2006). The only demonstrated abnormality of 5-HT$_6$ receptor knockout mice is a reduced sensitivity to ethanol (Bonasera et al., 2006).

Pharmacological studies showed no clear evidence for the involvement of the 5-HT$_7$ receptor in anxiety-related behavior but suggested a role for the receptor in sleep. Consistent with these pharmacological data, 5-HT$_7$ receptor knockout mice spent less time in, and had less frequent episodes of, REM sleep, behaviors opposite to that observed in depression (Hedlund et al., 2005). 5-HT$_7$ receptor knockout mice have less immobility in the forced swim and tail suspension tests that also indicate that the receptor modulates behavior related to depressive states. This behavior could also be reproduced by the selective pharmacological blockade of the receptor. This indicates that both the regulation of sleep pattern and depression-related behaviors may be directly linked to the activation of the 5-HT$_7$ receptor.

## 5-HT Transporter Knockout Mice

Although individuals with the *s/s* 5–HT transporter alleles seems to have an increased susceptibility to neuroticism and anxiety (see Introduction), initial studies with 5-HT transporter knockout mice indicated no anxiety-like phenotype (Bengel et al., 1998). A later study showed anxiety, at least in females, in 5-HT transporter knockout mice (Murphy et al., 2001). A more recent analysis of these mice demonstrated anxiety in some (latency to feed in novel environment) but not other behavioral paradigms (the open field and elevated plus maze tests) (Lira et al., 2003). Another study however reported increased anxiety-like behavior of the 5-HT transporter knockout mice in the elevated plus maze and light–dark transition test (Holmes et al., 2003). Although the difference in the genetic background may have accounted for this discrepancy, these studies nevertheless indicate that the 5-HT transporter null phenotype is relatively weak. This is not entirely surprising considering the small contribution of the *s* allele in neuroticism and anxiety in humans (Lesch et al., 1996).

Mice null for the 5-HT transporter have secondary adaptive alterations in the 5-HT system. Specifically, the density and expression, but not G-protein coupling, of 5-HT$_{1A}$ receptor were reduced in 5-HT transporter knockout mice (Li et al., 2000). This finding is consistent with the reduced 5-HT$_{1A}$ receptor binding in individuals with the s allele of the 5-HT transporter (David et al., 2005). The reduced 5-HT$_{1A}$ receptor level in 5-HT transporter knockout mice may actually contribute to the anxiety-like phenotype as inactivation of this receptor results in a robust, anxiety-like phenotype (see above). Indeed, it has been shown that expression of 5-HT$_{1A}$ receptors by adenovirus-mediated gene transfer in the medial hypothalamus rescues the increased stress response (measured as exaggerated adrenocorticotropin responses) and reduced locomotor activity of 5-HT transporter knockout mice (Li et al., 2004). It is not known if limbic (for example, amygdalar and hippocampal) expression of the receptor would rescue the anxiety-like phenotype of 5-HT transporter knockout mice.

The anxiety-like phenotype of 5-HT transporter knockout mice was more robust when BDNF was also genetically inactivated (Ren-Patterson et al., 2005). In addition to a higher level of anxiety-like behavior, double knockout mice, as compared to 5-HT transporter knockout, BDNF heterozygote knockout and wild-type mice, displayed significantly reduced levels of 5-HT and 5-hydroxyindole acetic acid in the hippocampus and hypothalamus, and greater increases in plasma ACTH after a stressful stimulus. These data are consistent with the notion that anxiety disorders are polygenic characterized by individually small contributions from several genes.

As pharmacological blockade of the 5-HT transporter has an anxiolytic profile, the increased anxiety-like behavior of 5-HT transporter knockout mice was presumed to be related to the absence of 5-HT transporter during development. This notion received support recently because the pharmacological blockade of 5-HT transporter during early postnatal life also elicited anxiety-like behavior in adults (Ansorge et al., 2004). This scenario is similar to that observed with the 5-$HT_{1A}$ receptor because receptor loss resulted in an anxiety-like phenotype only if absent during early postnatal life.

## SUMMARY

The first knockout of a 5-HT receptor was published more than 10 years ago and since then most of the 5-HT receptors have been genetically inactivated in mice. Reviewing the anxiety-related behavior of individual receptor knockout lines, it is apparent that genetic inactivation of some receptors, including the 5-$HT_{2A}$ and 5-$HT_3$ receptors, reproduced the phenotype elicited by the acute pharmacological blockade of the receptor. This indicates that signaling through some 5-HT receptors is directly involved in the regulation of anxiety states. In other cases, however, the genetic inactivation and pharmacological blockade of the receptor resulted in different and sometimes opposing behaviors. For example, the genetic inactivation of the 5-HT transporter results in increased anxiety whereas chronic pharmacological blockade of the 5-HT transporter by SSRIs leads to reduced anxiety. Mice null for the 5-$HT_{1A}$ receptor show increased anxiety, while selective pharmacological inhibitors of the receptor do not have a measurable effect on anxiety in adult mice. These are interesting findings because the genetic inactivation of the receptor could be similar to inherited conditions characterized by a loss of function mutation of the receptor/transporter. Indeed, 5-HT transporter knockout in mice may reproduce the *s* polymorphism in humans. Also, certain individuals have a reduced expression of the 5-$HT_{1A}$ receptor that, if present from early life, is similar to the null mutation of the receptor in mice. Therefore, the 5-$HT_{1A}$ receptor and 5-HT transporter knockout mouse strains may represent disease models that can be used to understand how a receptor or transporter deficiency leads to an anxiety-like phenotype.

## REFERENCES

American-Psychiatric-Association, *Diagnostic and Statistical Manual of Mental Disorders*, 4th ed., American Psychiatric Association, Washington, D.C., 1994.

Ansorge, M.S., Zhou, M., Lira, A., Hen, R., and Gingrich, J.A., Early-Life Blockade of the 5-HT Transporter Alters Emotional Behavior in Adult Mice, *Science*, 306, 879, 2004.

Ase, A.R., Senecal, J., Reader, T.A., Hen, R., and Descarries, L., Decreased G-protein coupling of serotonin 5-HT(1A) receptors in the brain of 5-HT(1B) knockout mouse, *Neuropharmacology*, 42, 941, 2002.

Bailey, S.J. and Toth, M., Variability in the benzodiazepine response of serotonin 5-$HT_{1A}$ receptor null mice displaying anxiety-like phenotype: evidence for genetic modifiers in the 5-HT-mediated regulation of GABA(A) receptors, *J Neurosci*, 24, 6343, 2004.

Barnes, N.M. and Sharp, T., A review of central 5-HT receptors and their function, *Neuropharmacology*, 38, 1083, 1999.

Bengel, D., Murphy, D.L., Andrews, A.M., Wichems, C.H., Feltner, D., Heils, A., Mossner, R., Westphal, H., and Lesch, K.P., Altered brain serotonin homeostasis and locomotor insensitivity to 3, 4-methylenedioxymethamphetamine ("Ecstasy") in serotonin transporter-deficient mice, *Mol Pharmacol*, 53, 649, 1998.

Blier, P., Pineyro, G., el Mansari, M., Bergeron, R., and de Montigny, C., Role of somatodendritic 5-HT autoreceptors in modulating 5-HT neurotransmission, *Ann NY Acad Sci.*, 86, 204, 1998.

Bonasera, S.J., Chu, H.M., Brennan, T.J., and Tecott, L.H., A null mutation of the serotonin 6 receptor alters acute responses to ethanol, *Neuropsychopharmacology*, 31, 1801, 2006.

Bortolozzi, A., Amargos-Bosch, M., Toth, M., Artigas, F., and Adell, A., *In vivo* efflux of serotonin in the dorsal raphe nucleus of 5-$HT_{1A}$ receptor knockout mice, *J Neurochem*, 88, 1373, 2004.

Boschert, U., Amara, D.A., Segu, L., and Hen, R., The mouse 5-hydroxytryptamine1B receptor is localized predominantly on axon terminals, *Neuroscience*, 58, 167, 1994.

Boutrel, B., Franc, B., Hen, R., Hamon, M., and Adrien, J., Key role of 5-$HT_{1B}$ receptors in the regulation of paradoxical sleep as evidenced in 5-$HT_{1B}$ knock-out mice, *J Neurosci*, 19, 3204, 1999.

Boutrel, B., Monaca, C., Hen, R., Hamon, M., and Adrien, J., Involvement of 5-$HT_{1A}$ receptors in homeostatic and stress-induced adaptive regulations of paradoxical sleep: studies in 5-$HT_{1A}$ knock-out mice, *J Neurosci*, 22, 4686, 2002.

Bremner, J.D., Innis, R.B., White, T., Fujita, M., Silbersweig, D., Goddard, A.W., Staib, L., Stern, E., Cappiello, A., Woods, S., Baldwin, R., and Charney, D.S., SPECT [I-123]iomazenil measurement of the benzodiazepine receptor in panic disorder, *Biol Psychiatr*, 47, 96, 2000.

Bruening, S., Oh, E., Hetzenauer, A., Escobar-Alvarez, S., Westphalen, R. I., Hemmings, H. C. Jr., Singewald, N., Shippenberg, T., and Toth, M., The anxiety-like phenotype of 5-HT receptor null mice is associated with genetic background-specific pertubations in the prefrontal cortex GABA-glutamate system. *J Neurochem* 99(3), 892, 2006.

Buzsaki, G., Theta oscillations in the hippocampus, *Neuron*, 33, 325–340, 2002.

Bystritsky, A., Pontillo, D., Powers, M., Sabb, F.W., Craske, M.G., and Bookheimer, S.Y., Functional MRI changes during panic anticipation and imagery exposure, *Neuroreport*, 12, 3953, 2001.

Chou-Green, J.M., Holscher, T.D., Dallman, M.F., and Akana, S.F., Compulsive behavior in the 5-$HT_{2C}$ receptor knockout mouse, *Physiol Behav*, 78, 641, 2003a.

Chou-Green, J.M., Holscher, T.D., Dallman, M.F., and Akana, S.F., Repeated stress in young and old 5-HT(2C) receptor knockout mice, *Physiol Behav*, 79, 217, 2003b.

Clapcote, S.J. and Roder, J.C., Survey of embryonic stem cell line source strains in the water maze reveals superior reversal learning of 129S6/SvEvTac mice, *Behav Brain Res*, 152, 35, 2004.

Clifton, P.G., Lee, M.D., Somerville, E.M., Kennett, G.A., and Dourish, C.T., 5-$HT_{1B}$ receptor knockout mice show a compensatory reduction in 5-HT2C receptor function, *Eur J Neurosci*, 17, 185, 2003.

Compan, V., Zhou, M., Grailhe, R.Gazzara, R.A., Martin, R., Gingrich, J., Dumuis, A., Brunner, D., Bockaert, J., and Hen, R., Attenuated response to stress and novelty and hypersensitivity to seizures in 5-HT4 receptor knock-out mice, *J Neurosci*, 24, 412, 2004.

Costall, B. and Naylor, R.J., Anxiolytic potential of 5-HT3 receptor antagonists, *Pharmacol Toxicol*, 70, 157, 1992.

Costall, B. and Naylor, R.J., Behavioural interactions between 5-hydroxytryptophan, neuroleptic agents and 5-HT receptor antagonists in modifying rodent responding to aversive situations, *Br J Pharmacol*, 116, 2989, 1995.

Crawley, J.N., Neuropharmacologic specificity of a simple animal model for the behavioral actions of benzodiazepines, *Pharmacol Biochem Behav*, 15, 695, 1981.

Crawley, J.N., Behavioral phenotyping of transgenic and knockout mice: experimental design and evaluation of general health, sensory functions, motor abilities, and specific behavioral tests, *Brain Res*, 835, 18, 1999.

David, S.P., Murthy, N.V., Rabiner, E.A., Munafo, M.R., Johnstone, E.C., Jacob, R.Walton, R.T., and Grasby, P.M., A functional genetic variation of the serotonin (5-HT) transporter affects 5-$HT_{1A}$ receptor binding in humans, *J Neurosci*, 25, 2586, 2005.

Davis, M., Animal models of anxiety based on classical conditioning: the conditioned emotional response (CER) and the fear-potentiated startle effect, *Pharmacol Ther*, 47, 147, 1990.

Derkach, V., Surprenant, A., and North, R.A., 5-HT3 receptors are membrane ion channels, *Nature*, 339, 706, 1989.

Dilger, S., Straube, T., Mentzel, H.J., Fitzek, C., Reichenbach, J.R., Hecht, H., Krieschel, S., Gutberlet, I., and Miltner, W.H., Brain activation to phobia-related pictures in spider phobic humans: an event-related functional magnetic resonance imaging study, *Neurosci Lett*, 348, 29, 2003.

File, S.E., Animal models for predicting clinical efficacy of anxiolytic drugs: social behaviour, *Neuropsychobiology*, 13, 55, 1985.

Geller, I., Kulak, J.T., Jr., and Seifter, J., The effects of chlordiazepoxide and chlorpromazine on a punishment discrimination, *Psychopharmacologia*, 3, 374, 1962.

Goldberg, M.E., Salama, A.I., Patel, J.B., and Malick, J.B., Novel non-benzodiazepine anxiolytics, *Neuropharmacology*, 22, 1499, 1983.

Gordon, J.A., Lacefield, C.O., Kentros, C.G., and Hen, R., State-dependent alterations in hippocampal oscillations in serotonin 1A receptor-deficient mice, *J Neurosci*, 25, 6509, 2005.

Gothert, M., Presynaptic serotonin receptors in the central nervous system, *Ann NY Acad Sci*, 604, 102, 1990.

Grailhe, R., Grabtree, G.W., and Hen, R., Human 5-HT(5) receptors: the 5-HT(5A) receptor is functional but the 5-HT(5B) receptor was lost during mammalian evolution, *Eur J Pharmacol*, 418, 157, 2001.

Grailhe, R., Waeber, C., Dulawa, S.C., Hornung, J.P., Zhuang, X., Brunner, D., Geyer, M.A., and Hen, R., Increased exploratory activity and altered response to LSD in mice lacking the 5-HT(5A) receptor, *Neuron*, 22, 581, 1999.

Griebel, G., 5-Hydroxytryptamine-interacting drugs in animal models of anxiety disorders: more than 30 years of research, *Pharmacol Ther*, 65, 319, 1995.

Gross, C., Santarelli, L., Brunner, D., Zhuang, X., and Hen, R., Altered fear circuits in 5-HT(1A) receptor KO mice, *Biol Psychiatr*, 48, 1157, 2000.

Gross, C., Zhuang, X., Stark, K., Ramboz, S., Oosting, R., Kirby, L., Santarelli, L., Beck, S., and Hen, R., Serotonin1A receptor acts during development to establish normal anxiety-like behaviour in the adult, *Nature*, 416, 396, 2002.

Hariri, A.R., Mattay, V.S., Tessitore, A., Kolachana, B., Fera, F., Goldman, M.F., Egan, M.F., and Weinberger, D.R., Serotonin Transporter Genetic Variation and the Response of the Human Amygdala, *Science*, 297, 400, 2002.

He, M., Sibille, E., Benjamin, D., Toth, M., and Shippenberg, T., Differential effects of 5-$HT_{1A}$ receptor deletion upon basal and fluoxetine-evoked 5-HT concentrations as revealed by *in vivo* microdialysis, *Brain Res*, 902, 11, 2001.

Hedlund, P.B., Huitron-Resendiz, S.Henriksen, S.J., and Sutcliffe, J.G., 5-HT7 receptor inhibition and inactivation induce antidepressantlike behavior and sleep pattern, *Biol Psychiatr*, 58, 831, 2005.

Heisler, L.K., Chu, H.M., Brennan, T.J., Danao, J.A., Bajwa, P., Parsons, L.H., and Tecott, L.H., Elevated anxiety and antidepressant-like responses in serotonin 5-$HT_{1A}$ receptor mutant mice, *Proc Natl Acad Sci USA*, 95, 15049, 1998.

Holmes, A., Lit, Q., Murphy, D.L., Gold, E., and Crawley, J.N., Abnormal anxiety-related behavior in serotonin transporter null mutant mice: the influence of genetic background, *Genes Brain Behav*, 2, 365, 2003.

Hoyer, D., Clarke, D.E., Fozard, J.R., Hartig, P.R., Martin, G.R., Mylecharane, E.J., Saxena, P.R., and Humphrey, P.P., International union of pharmacology classification of receptors for 5-hydroxytryptamine (Serotonin), *Pharmacol Rev*, 46, 157, 1994.

Inoue, T., Izumi, T., Li, X.B., Huang, J.Z., Kitaichi, Y., Nakagawa, S., and Koyama, T., Biological basis of anxiety disorders and serotonergic anxiolytics, *Nihon Shinkei Seishin Yakurigaku Zasshi*, 24, 125, 2004.

Kelley, S.P., Bratt, A.M., and Hodge, C.W., Targeted gene deletion of the 5-HT3A receptor subunit produces an anxiolytic phenotype in mice, *Eur J Pharmacol*, 461, 19, 2003.

Kjelstrup, K.G., Tuvnes, F.A., Steffenach, H.A., Murison, R., Moser, E.I., and Moser, M.B., Reduced fear expression after lesions of the ventral hippocampus, *Proc Natl Acad Sci USA*, 99, 10825, 2002.

Klemenhagen, K.C., Gordon, J.A., David, D.J., Hen, R., and Gross, C.T., Increased fear response to contextual cues in mice lacking the 5-$HT_{1A}$ receptor, *Neuropsychopharmacology*, 31, 101, 2006.

Knobelman, D.A., Hen, R., Blendy, J.A., and Lucki, I., Regional patterns of compensation following genetic deletion of either 5-hydroxytryptamine(1A) or 5-hydroxytryptamine(1B) receptor in the mouse, *J Pharmacol Exp Ther*, 298, 1092, 2001.

Lemonde, S., Turecki, G., Bakish, D., Du, L., Hrdina, P.D., Bown, C.D., Sequeira, A., Kushwaha, N., Morris, S.J., Basak, A., Ou, X.M., and Albert, P.R.. Impaired repression at a 5-hydroxytryptamine 1A receptor gene polymorphism associated with major depression and suicide, *J Neurosci*, 23, 8788, 2003.

Lesch, K.P., Aulakh, C.S., Wolozin, B.L., and Murphy, D.L., Serotonin (5-HT) receptor, 5-HT transporter and G protein-effector expression: implications for depression, *Pharmacol Toxicol*, 71(Suppl. 1), 49, 1992.

Lesch, K.P., Bengel, D., Heils, A., Sabol, S.Z., Greenberg, B.D., Petri, S., Benjamin, J., Muller, C.R., Hamer, D.H., and Murphy, D.L., Association of Anxiety-Related Traits with a Polymorphism in the Serotonin Transporter Gene Regulatory Region, *Science*, 274, 1527, 1996.

Li, Q., Holmes, A., Ma, L.Van de Kar, L.D., Garcia, F., and Murphy, D.L., Medial hypothalamic 5-hydroxytryptamine (5-HT)1A receptors regulate neuroendocrine responses to stress and exploratory locomotor activity: application of recombinant adenovirus containing 5-$HT_{1A}$ sequences, *J Neurosci*, 24, 10868, 2004.

Li, Q., Wichems, C., Heils, A., Lesch, K.P., and Murphy, D.L., Reduction in the density and expression, but not G-protein coupling, of serotonin receptors (5-$HT_{1A}$) in 5-HT transporter knock-out mice: gender and brain region differences, *J Neurosci*, 20, 7888, 2000.

Lira, A., Zhou, M., Castanon, N., Ansorge, M.S., Gordon, J.A., Francis, J.H., Bradley-Moore, M., Lira, J., Underwood, M.D., Arango, V., Kung, H.F., Hofer, M.A., Hen, R., and Gingrich, J.A., Altered depression-related behaviors and functional changes in the dorsal raphe nucleus of serotonin transporter-deficient mice, *Biol Psychiatr*, 54, 960, 2003.

Lister, R.G., The use of a plus-maze to measure anxiety in the mouse, *Psychopharmacology (Berl)*, 92, 180, 1987.

Lopez, J.F., Chalmers, D.T., Little, K.Y., and Watson, S.J., A.E. Bennett Research Award. Regulation of serotonin1A, glucocorticoid, and mineralocorticoid receptor in rat and human hippocampus: implications for the neurobiology of depression, *Biol Psychiatr*, 43, 547, 1998.

Malizia, A.L., Cunningham, V.J., Bell, C.J., Liddle, P.F., Jones, T., and Nutt, D.J., Decreased brain GABA(A)-benzodiazepine receptor binding in panic disorder: preliminary results from a quantitative PET study, *Arch Gen Psychiatr*, 55, 715, 1998.

Malleret, G., Hen, R., Guillou, J.L., Segu, L., and Buhot, M.C., 5-$HT_{1B}$ receptor knock-out mice exhibit increased exploratory activity and enhanced spatial memory performance in the Morris water maze, *J Neurosci*, 19, 6157, 1999.

Mann, J.J., Role of the serotonergic system in the pathogenesis of major depression and suicidal behavior, *Neuropsychopharmacology*, 21, 99S, 1999.

Millan, M.J., The neurobiology and control of anxious states, *Prog Neurobiol*, 70, 83, 2003.

Moser, M.B. and Moser, E.I., Functional differentiation in the hippocampus, *Hippocampus*, 8, 608, 1998.

Motta, V., Maisonnette, S., Morato, S., Castrechini, P., and Brandao, M.L., Effects of blockade of 5-HT2 receptors and activation of 5-$HT_{1A}$ receptors on the exploratory activity of rats in the elevated plus-maze, *Psychopharmacology (Berl)*, 107, 135, 1992.

Munafo, M.R., Clark, T., and Flint, J., Does measurement instrument moderate the association between the serotonin transporter gene and anxiety-related personality traits? A meta-analysis, *Mol Psychiatr*, 10, 415, 2004.

Murphy, D.L., Li, Q., Engel, S., Wichems, C., Andrews, A., Lesch, K.P., and Uhl, G., Genetic perspectives on the serotonin transporter, *Brain Res Bull*, 56, 487, 2001.

Nebigil, C.G., Choi, D.S., Dierich, A., Hickel, P., Le Meur, M., Messaddeq, N., Launay, J.M., and Maroteaux, L., Serotonin 2B receptor is required for heart development, *Proc Natl Acad Sci USA*, 97, 9508, 2000.

Neumeister, A., Bain, E., Nugent, A.C., Carson, R.E., Bonne, O., Luckenbaugh, D.A., Eckelman, W., Herscovitch, P., Charney, D.S., and Drevets, W.C., Reduced serotonin type 1A receptor binding in panic disorder, *J Neurosci*, 24, 589, 2004.

Noyes, R. and Hoehn-Saric, R., *The Anxiety Disorders*, Cambridge University Press, Cambridge, 2004.

Parks, C.L., Robinson, P.S., Sibille, E., Shenk, T., and Toth, M., Increased anxiety of mice lacking the serotonin1A receptor, *Proc Natl Acad Sci USA*, 95, 10734, 1998.

Patel, J.G., Bartoszyk, G.D., Edwards, E., and Ashby, C.R., Jr., The highly selective 5-hydroxytryptamine (5-HT)2A receptor antagonist, EMD 281014, significantly increases swimming and decreases immobility in male congenital learned helpless rats in the forced swim test, *Synapse*, 52, 73, 2004.

Pattij, T., Groenink, L., Hijzen, T.H., Oosting, R.S., Maes, R.A., van der Gugten, J., and Olivier, B., Autonomic changes associated with enhanced anxiety in 5-HT(1A) receptor knockout mice, *Neuropsychopharmacology*, 27, 380, 2002.

Phillips, T.J., Hen, R., and Crabbe, J.C., Complications associated with genetic background effects in research using knockout mice, *Psychopharmacology (Berl)*, 147, 5, 1999.

Ramboz, S., Oosting, R., Amara, D.A., Kung, H.F., Blier, P., Mendelsohn, M., Mann, J.J., Brunner, D., and Hen, R., Serotonin receptor 1A knockout: an animal model of anxiety-related disorder, *Proc Natl Acad Sci USA*, 95, 14476, 1998.

Ren-Patterson, R.F., Cochran, L.W., Holmes, A., Sherrill, S., Huang, S.J., Tolliver, T., Lesch, K.P., Lu, B., and Murphy, D.L., Loss of brain-derived neurotrophic factor gene allele exacerbates brain monoamine deficiencies and increases stress abnormalities of serotonin transporter knockout mice, *J Neurosci Res*, 79, 756, 2005.

Rodgers, R.J., Animal models of "anxiety": where next?, *Behav Pharmacol*, 8, 477, 1997.

Rodgers, R.J., Cao, B.J., Dalvi, A., and Holmes, A., Animal models of anxiety: an ethological perspective, *Braz J Med Biol Res*, 30, 289, 1997.

Sarnyai, Z., Sibille, E.L., Pavlides, C., Fenster, R.J., McEwen, B.S., and Toth, M., Impaired hippocampal-dependent learning and functional abnormalities in the hippocampus in mice lacking serotonin(1A) receptors, *Proc Natl Acad Sci USA*, 97, 14731, 2000.

Sen, S., Burmeister, M., and Ghosh, D., Meta-analysis of the association between a serotonin transporter promoter polymorphism (5-HTTLPR) and anxiety-related personality traits, *Am J Med Genet B Neuropsychiatr Genet*, 127, 85, 2004.

Shah, A.A. and Treit, D., Excitotoxic lesions of the medial prefrontal cortex attenuate fear responses in the elevated-plus maze, social interaction and shock probe burying tests, *Brain Res*, 969, 183, 2003.

Shah, A.A. and Treit, D., Infusions of midazolam into the medial prefrontal cortex produce anxiolytic effects in the elevated plus-maze and shock-probe burying tests, *Brain Res*, 996, 31, 2004.

Sibille, E., Pavlides, C., Benke, D., and Toth, M., Genetic inactivation of the Serotonin(1A) receptor in mice results in downregulation of major GABA(A) receptor alpha subunits, reduction of GABA(A) receptor binding, and benzodiazepine-resistant anxiety, *J Neurosci*, 20, 2758, 2000.

Sibille, E., Sarnyai, Z., Benjamin, D., Gal, J., Baker, H., and Toth, M., Antisense inhibition of 5-hydroxytryptamine2a receptor induces an antidepressant-like effect in mice, *Mol Pharmacol*, 52, 1056, 1997.

Singewald, N., Salchner, P., and Sharp, T., Induction of c-Fos expression in specific areas of the fear circuitry in rat forebrain by anxiogenic drugs, *Biol Psychiatr*, 53, 275, 2003.

Stutzmann, J.M., Eon, B., Darche, F., Lucas, M., Rataud, J., Piot, O., Blanchard, J.C., and Laduron, P.M., Are 5-HT2 antagonists endowed with anxiolytic properties in rodents?, *Neurosci Lett*, 128, 4, 1991.

Tecott, L.H. and Abdallah, L., Mouse genetic approaches to feeding regulation: serotonin 5-HT2C receptor mutant mice, *CNS Spectr*, 8, 584, 2003.

Treit, D. and Fundytus, M., Thigmotaxis as a test for anxiolytic activity in rats, *Pharmacol Biochem Behav*, 31, 959, 1988.

Uys, J.D., Stein, D.J., Daniels, W.M., and Harvey, B.H., Animal models of anxiety disorders, *Curr Psychiatr Rep*, 5, 274, 2003.

Vogel, J.R., Beer, B., and Clody, D.E., A simple and reliable conflict procedure for testing anti-anxiety agents, *Psychopharmacologia*, 21, 1, 1971.

Weiss, S.M., Lightowler, S., Stanhope, K.J., Kennett, G.A., and Dourish, C.T., Measurement of anxiety in transgenic mice, *Rev Neurosci*, 11, 59, 2000.

Weisstaub, N.V., Zhou, M., Lira, A., Lambe, E., Gonzalez-Maeso, J., Hornung, J.P., Sibille, E., Underwood, M., Itohara, S., Dauer, W.T., Ansorge, M.S., Morelli, E., Mann, J.J., Toth, M., Aghajanian, G., Sealfon, S.C., Hen, R., and Gingrich, J.A., Cortical 5-$HT_{2A}$ Receptor Signaling Modulates Anxiety-Like Behaviors in Mice, *Science*, 313, 536, 2006.

Wolff, M., Costet, P., Gross, C., Hen, R., Segu, L., and Buhot, M.C., Age-dependent effects of serotonin-1A receptor gene deletion in spatial learning abilities in mice, *Brain Res Mol Brain Res*, 130, 39, 2004.

World Health Organization, *The ICD-10 Classification of Mental and Behavioral Disorders — Clinical Description and Diagnostic Guidelines,* Vol. 14, Geneva, 1992, pp. 281–299.

Zhuang, X., Gross, C., Santarelli, L., Compan, V., Trillat, A.C., and Hen, R., Altered emotional states in knockout mice lacking 5-$HT_{1A}$ or 5-$HT_{1B}$ receptors, *Neuropsychopharmacology,* 21, 52S, 1999.

# Index


2-arachidonylglycerol (2-AG) signaling, with 5-HT$_{2A}$ receptor stimulation, 111
5-HT$_{4-7}$ receptor knockout mice, 187–188
5-HT$_1$ receptor, calmodulin and, 70
5-HT receptor knockout mice, 157, 183–184
  5-HT$_{4-7}$ receptors, 187–188
  5-HT$_2$ receptor, 186–187
  5-HT$_3$ receptor, 187
  5-HT$_{1A}$ receptor, 184–186
  5-HT$_{1B}$ receptor, 186
  adaptive changes in central serotonergic systems in, 168–169
  anxiety studies in, 181
  eating disorders in, 163–165
  engineered genetic deficits in, 159
  and illegal drug use, 167–169
  and maladaptive behavior with environmental challenges, 163–168
  and novelty-induced locomotion, 165, 167
  reduced reactivity to novel environments, 167
5-HT$_2$ receptor knockout mice, 186–187
5-HT$_3$ receptor knockout mice, 187
5-HT receptors, 62, 133
  and calmodulin, 67–73
  calmodulin binding sites on miscellaneous, 72–73
  eating disorders and polymorphisms, 161
  interactions with calmodulin, 61
  polymorphisms and addictions to food and illegal drugs, 160–162
  receptor desensitization by calmodulin, 66
  role of CaM in signaling by, 65–66
  seven families of, 62
5-HT$_2$ receptors, calmodulin and, 70
5-HT reuptake transporter (5-HTT), 83
5-HT transporter knockout mice, 183–184, 188–189
5-HT$_{1A}$ agonists, mechanisms of action, 141–142
5-HT$_{2A}$ protein expression, methods of ascertaining, 107
5-HT$_{1A}$ receptor. *See also* Serotonin receptors; Serotonin$_{1A}$ receptor
  aberrant expression and diseases, 137–138
  abundance in brain, 83
  antidepressant effect of agonists, 136
  cell proliferation effects in cancer, 140
  cellular distribution, 135
  coexpression with HvCNG channels, 26
  desensitization and acceleration of antidepressant action, 83, 136
  and early mouse brain development, 143–144
  example studies, 142–145
  experimental procedures, 146–148
  future studies in signaling mechanisms, 140–142
  helical wheel projections of putative CaM-binding domains, 69
  hyperpolarizing effect of, 136
  interactions with calmodulin, 67–70
  involvement in clozapine-evoked neuronal activity, 144–145
  mitogenic effects in cancer, 140
  ontogeny, 135–136
  overweight in absence of, 163–164
  presynaptic and postsynaptic signaling effects, 136–137
  role in cancer, 140
  role in immune system, 140
  as signaling hub in emotional balance, 133–135
  therapeutic agents regulating signaling of, 138–139
  transcription regulation of, 82
5-HT$_{2A}$ receptor
  activation studies, 109–111
  agonist-induced internalization of, 117
  amino acid composition in human, rat, and mouse, 107
  calmodulin binding to, 66
  caveolin-1 binding to, 64
  cell-type specific activation of, 113
  desensitization and resensitization, 106, 111
  functional selectivity, 111–115, 119
  helical wheel projections of putative CaM-binding domains, 71
  implication of brain localization in diseases, 108–109
  initial identification by hybridization, 106
  internalization and recycling of, 112, 116, 117
  intracellular localization, 115–116
  intracellular signaling cascades activated by, 109
  intracellular trafficking, 116–120
  IP$_3$ activation, 111
  ligand-dependent mechanisms and pathways, 105–106
  localization in brain, 106–109
  long-term regulation of, 122–124

oligomerization probes, 114–115
paradoxical regulation of, 123
phosphorylation and desensitization by CaM, 72
proteins associating with, 121
role in mental disorders, 106
subcellular localization, 116
5-$HT_{7A}$ receptor
IQ motif in human, 73
potential CaM-binding domains, 73
5-$HT_{1A}$ receptor knockout mice, 184–186
5-$HT_{2A}$ receptor oligomerization probes, 114–115
5-$HT_{1A}$ receptor signaling
future studies in mechanisms of, 140–142
mechanisms of action, 141–142
5-$HT_{2A}$ receptor subtypes, cloning and identification of, 106
5-$HT_{1B}$ receptor knockout mice, 186
5-$HT_{1D}$ receptor, potential CaM-binding domains, 70
5-$HT_{1E}$ receptor, potential CaM-binding domains, 70
5-$HT_{1F}$ receptor, potential CaM-binding domains, 70
5-$HT_4$R, elevated function in knockout mice, 163
8-OH-DPAT, inhibition of *Plasmodium falciparum* by, 44
5-RACE method, 83
for identifying transcriptional regulators, 85–86
5 RT-PCR method, for identifying 5-$HT_{1A}$ receptor transcriptional initiation region, 84

## A

Activation studies, for 5-$HT_{2A}$ receptor, 109–111
Adaptive changes
in 5-HT receptor knockout mice, 168, 169
impaired stress response in 5-HT receptor knockout mice, 168
in serotonergic neurons in 5-HT receptor knockout mice, 169
in spite of engineered genetic defects, 159
Adenylyl cyclases (AC)
cAMP level dependence on, 20
dependence of signal transduction on mobile fractions of proteins in reticulocyte plasma membranes, 49
overexpression and enhancement of beta-adrenergic receptor mediation, 49
Agranulocytosis, with clozapine use, 135
Alcoholism
altered signaling of 5-$HT_{1A}$ receptor in, 138
HTR1A gene polymorphisms and, 161, 162
role of 5-$HT_{1A}$ receptor in therapeutic agents for, 138
Alzheimer's disease
5-$HT_{2A}$ receptor in, 109
role of 5-$HT_{1A}$ receptor agents in, 139
Amino acid composition
5-$HT_{2A}$ receptor, 107
differences in 5-$HT_{2A}$ receptor between humans and rats, 120
Amphetamine-induced serotonin release, 16
Angiotensin II $AT_{1A}$ receptor, calmodulin interactions with, 63, 66
Anorexia nervosa, 163
5-HT receptor knockout mice insensitivity to stress-induced, 164
association of 5-$HT_{2A}$ gene promoter polymorphism with, 161
decreased 5-$HT_{2A}$ receptor binding in, 109
Antibody identification, in CHIP assays, 95
Antimalarial drug research, and serotonin$_{1A}$ receptor agonists, 44
Antipsychotic drugs, first- and second-generation, 142
Antisense probes, for identifying transcription factor activity, 97–98
Anxiety disorders
in 5-HT receptor knockout mice, 158, 164–165, 166
5-$HT_{1A}$ gene and, 82
5-$HT_{2A}$ receptor in, 108
association with HTR2C polymorphisms, 162
behavioral tests for, 183
with disruption of 5-$HT_{1A}$ gene in mice, 138
role of 5-HT system in modulating, 181–182
serotonin$_{1A}$ receptor studies and, 43–44
studies in 5-HT receptor knockout mice, 181
Arachidonic acid (AA) signaling, with 5-$HT_{2A}$ receptor stimulation, 111
Area bleaching, *vs.* point bleaching, 51
Arrestin, role in 5-$HT_{2A}$ receptor internalization, 117, 118
Attention deficit hyperactivity disorder (ADHD)
5-$HT_{2A}$ receptor in, 108–109
and HTR2C polymorphisms, 162
Atypical antipsychotics, 142
and 5-$HT_{1A}$ receptor signaling, 139

## B

Background fluorescence, in membrane dynamics studies, 52
BD Biosciences Marathon-Ready cDNA library, 85
Beam size, 3

Beam waist, 3
Behavior modification
and transcriptional modulation, 82
by transcriptional regulators, 82–83
Beta$_2$-adrenergic receptor, interactions with GPCR RIPs, 64
Binding reaction, in EMSA studies, 91–92
Binge eating disorder, 160
Bioinformatic analysis, of promoters, 87
Bioluminescence resonance energy transfer (BRET), 68, 71, 114
Bipolar disorders, 5-HT$_{2A}$ receptor in, 109
BLAST search, 97
Bleach duration, 51, 53
BLOCK-iT Fluorescent Oligo, 97
Body weight gain
in 5-HT receptor knockout mice, 165
and HTR2C gene polymorphism, 162
Brain
implication of 5-HT$_{2A}$ receptor localization in diseases, 108–109
localization of 5-HT$_{2A}$ receptor in, 106
Brain-derived neurotrophin (BDNF) promoter, 82
Brain development
and 5-HT$_{1A}$ receptor, 143–144
role of 5-HT$_{1A}$ receptor in, 134, 142
Bulimia, 160
polymorphisms of HTR2C gene associated with, 161

## C

Calcineurin, inhibition of 5-HT$_{2C}$ receptor desensitization by, 72
Calcium imaging techniques, cAMP measurement using, 21
Calcium-regulated transcription factor (CaRF), 82
Calmodulin (CaM), 64–65
as 5-HT receptor interacting and regulatory protein, 61
and 5-HT$_2$ receptors, 70–72
and 5-HT$_{1A}$ receptor, 67–70
antagonism with phosphorylation, 68–69
blunting of receptor phosphorylation by, 67
closed and open conformations, 64
G protein coupling blunting by, 67
interaction with G-protein coupled receptor-interacting proteins, 66–67
and miscellaneous 5-HT$_1$ receptors, 70
as modifier of GPCRs, 61
potential binding sites on 5-HT$_4$, 5-HT$_5$, 5-HT$_6$, 5-HT$_7$ receptors, 72–73
receptor desensitization by, 66
role in 5-HT receptor signaling, 65–66
cAMP levels
and activation of PKA and CNG, 27
and activation time constant, 26
Boltzmann fit of steady-state HvCNG activation curve, 23
calculating for each time constant, 23
cell perfusion, 24
and channel activation time constant, 23
electrophysiological approach to monitoring, 21–27
excitation pathway and emission light for measuring, 32
FRET-based sensor for monitoring, 27–36
Hill equation for dose-response curve, 23
measurement after agonist-induced receptor activation, 26
measurement in living cells, 20
measuring decreases in, 25
monitoring receptor-mediated changes in, 19–20
spatial and temporal resolution problems in measurement, 20
and time course of cell perfusion, 24
Cancer, role of 5-HT$_{1A}$ receptor in, 140
Catecholamine detection, with serotonin autofluorescence, 16–17
Caveolin-I, interactions with GPCR RIPs, 62
cDNA libraries, 85
Cell damage, minimizing with three-photon microscopy, 7
Cell line transfection, for promoter characterization, 88–89
Cerebellum, 5-HT$_{2A}$ receptor activity in developing, 108
Chromatin immunoprecipitation (CHIP) assay, identifying DNA binding proteins via, 94–95
Clozapine, 142, 143
and change in excitability of PFC neurons, 144–145
involvement of 5-HT$_{1A}$ receptor in neuronal activity of, 144–145
CNG-channel, 22
Cocaine abuse
and 5-HT receptor knock-out mice, 167–169
in 5-HT receptor knock-out mice, 167
altered 5-HT$_{1A}$ receptor expression in, 138
and HTR1DB gene polymorphism, 162
role of 5-HT$_{1A}$ receptor agonists and antagonists in, 139
Coimmunoprecipitation, 68, 114
Collimation, 8
Colocalization, in live neurons, 14
Confocal microscopy, in 5-HT$_{1A}$ receptor experimental procedures, 148

Consensus element, identification for transcriptional regulators, 87
CREB transcription factor, 82
Cyclic AMP (cAMP), as key second messenger for serotonin, 20
Cyclic nucleotide binding domain (CNBD), 22
Cyclic nucleotide gated channels (CNG), 22, 27
Cyclin D1, 5-$HT_{1A}$ receptor-mediated stimulation of, 144
Cytoskeleton
  as barrier to free protein diffusion, 54
  binding to alpha-$_2$ adrenergic receptors, 63
  role in 5-$HT_{2A}$ receptor trafficking, 118

## D

Dansyl chloride spectroscopy, 68, 71
DEAE-dextran, DNA delivery via, 89
Depolarization, and BDNF promoter sites, 82
Depression
  5-$HT_{1A}$ gene and, 82
  5-$HT_{1A}$ receptor agonists in treatment of, 139
  5-$HT_{2A}$ receptor in, 108
  reduced hippocampal 5-$HT_{1A}$ receptor volumes in, 138
  serotonin$_{1A}$ receptor studies and, 43–44
  SNRIs in treatment of, 139
Detergent insolubility, of serotonin$_{1A}$ receptor, 44, 47–48
Detergent resistant membranes (DRMs), detection of proteins in, 45
Developmental disorders, serotonin$_{1A}$ receptor uses for, 43
Diagnostic and Statistical Manual of Mental Disorders (DSM IV-TR), 160
Discontinuous single-electrode voltage-clamp amplifier SEC-051, 21
Distance calibration, in multiphoton imaging, 12
DNA binding proteins, 92
  CHIP assay method for identifying, 94–96
  cloning, 93
  yeast one-hybrid analysis of, 93–94
DNA delivery methods, 89
DNA elements
  EMSA gel retardation assay method for identifying, 90–92
  identifying for transcriptional regulators, 90
  supershift EMSA method for identifying, 92
DNA methylation, 82
Dopamine $D_2$ receptor
  and 5-$HT_{1A}$ receptor signaling, 139
  CaM binding to, 66
Dopamine imaging, in live neurons, 17
Dopaminergic neurons, in striatum and ventral tegmental area (VTA), 144
Drug development, and membrane organization studies, 43
Drug discovery, transcriptional approach to, 98
Dyskinesia, as side effect of first-generation antipsychotics, 142

## E

Eating disorders, 160
  in 5-HT receptor knock-out mice, 163–165, 164
  and human 5-HT polymorphisms, 161
EBP50, interactions with GPCR RIPs, 63
EcoPro, 91
Ecstasy, 163
  as appetite suppressant in humans, 167
  increase in locomotion with, 167
  use in 5-HT receptor knock-out mice, 167–169
Effectors, 134
Electrophoresis mobility shift assay (EMSA), 82
  binding reaction, 91–92
  identifying DNA elements by, 90
  probe design, 90–91
  recombinant protein sources, 91
Electrophysiological approach, 20
  examples, 25
  experimental system, setup, and data analysis, 21–25
  to monitoring receptor-mediated cAMP levels, 21
Electrostatic interactions, and cell membrane pore opening and closing, 70
EMCCD chips, 31, 33
Emission light, in electrophysical method of cAMP level measurement, 32
Emotional status
  5-$HT_{1A}$ receptor and, 133–135
  and serotonin system, 82
Enhanced yellow fluorescent protein (EYFP)
  detergent insolubility, 48
  serotonin$_{1A}$ receptor fusing with, 45
Enhancer regions, 87
Epac-construct, 29
  conformational change in, 28
Epi collection, setup for multiphoton microscopy, 9–10
Epifluorescent microscope, 31
Excitation, testing for order of, 13

Excitation efficiency, dependence on pulse width for two- and three-photon excitation, 4
Excitation intensity, near focus spot, 5
Excitation light, in electrophysical method of cAMP level measurement, 32
Excitation shielding, for non-epi collection, 10
Excitation wavelength
  in multiphoton microscopy (MPM), 3
  selection for multiphoton microscopy, 5–7
  setup for multiphoton microscopy, 8–9

## F

Femtoseconds (fs), 2
Fixation protocols, for CHIP assays, 95
Fluid mosaic model, for cell membranes, 42
Fluorescence detection, setup for multiphoton microscopy, 9
Fluorescence microscopy, membrane organization and dynamics of serotonin$_{1A}$ receptor monitored with, 41
Fluorescence recovery after photobleaching (FRAP), 43
  estimating diffusion properties in cells with, 50
  of serotonin$_{1A}$-EYFP receptors in CHP cells, 54
Fluorescence recovery plot, 54
Fluorescence resonance energy transfer (FRET), 114
Fluorescent proteins, monitoring receptor-mediated changes in cAMP levels with, 19–20
Fluorescent receptor tagging, 119
Fluoxetine, 141
Focusing requirements, for multiphoton microscopy, 4–5
Food addictions, and human 5-HT receptor gene polymorphisms, 160–162
Förster, Theodor, 28
Förster resonance energy transfer (FRET), 20, 27. *See also* FRET-based cAMP sensors
  acceptor photobleaching method of measurement, 30
  algorithm for calculating, 30
  methods for analyzing, 29
Free diffusion, plasma membrane cytoskeleton as barrier to, 54
FRET-based cAMP sensors, 27–28
  analysis results, 34
  camera setup, 33
  examples, 33–35
  experimental system using Epac1 sensor, 31–33
  FRET principles and FRET-based analysis, 28–31
  half-images from CPF and YFP channels, 35
  light sources, 31
  optics setup, 31–32
  outlook, 35–36
FRET efficiency, measuring, 35–36
Functional selectivity
  in 5-HT$_{2A}$ receptor, 111–115
  in 5-HT$_{2A}$ receptor activation and trafficking, 118
  in 5-HT$_{2A}$ receptor trafficking, 119

## G

G-protein coupled receptor-interacting proteins, 62–64
  interaction of CaM with, 66–67
  putative CaM-binding domains, 68
G-protein coupled receptors (GPCRs)
  barriers to free protein diffusion, 55
  implication in diseases, 105
  implications of membrane organization for, 42
  lateral organization in membranes, 49
  membrane organization studies, 48
  modification by receptor-interacting proteins (RIPs), 62
  role of CaM in, 65–66
G-protein dependent cell surface dynamics, of serotonin$_{1A}$ receptor, 53–55
G-protein heterotrimer, 53
G-proteins
  analysis of mobility in receptor interaction studies, 55
  compartmentalized localization in cholesterol-rich membrane domains, 55
GAL4 activation domain (GAL4-AD), 93
GAL4-DBD hybrid system, for identifying transcription factor activity, 95–97
GAL4-DNA Binding Domain (DBD), 95
Gel retardation assay, for identifying DNA elements of transcriptional regulators, 90–92
Gel-shift analysis, 68, 71
Gene guns, 89
Gene knockout studies, of novel transcription factors, 98
Gene reporter assays
  interpreting, 89–90
  for promoter characterization, 87–88
Glutamate mGlu receptor, CaM binding to, 66

Gren fluorescent protein (GFP), tagging of membrane proteins with, 45
Guanine nucleotide exchange factor (Epac), 20

## H

Haloperidol, 142
HCN-channel, 22
  disadvantages as sensors, 27
  measuring intracellular cAMP levels using recombinant, 24
HEK293 cells, 5-$HT_{2A}$ receptor internalization in, 116, 117
HEKA software products, 21
HiPerfect, 97
His3/LacZ reporter system, 93
Homodimerization, GPCR, 114–115
Human calcitonin receptorlike receptor (hCRLR), 63
Hyperpolarization-activated cyclic nucleotide-gated channels (HCN), 22

## I

Igor-WaveMetrics software, 21
Illegal drugs
  and 5-HT receptor knock-out mice, 167–169
  5-HT receptor polymorphisms and addictions to, 161–162
  consumption after open-field stress challenge, 166
  and human 5-HT receptor gene polymorphisms, 160–162
Image-splitter, in cAMP measurement methodology, 32
Imaging depth, advantages with three-photon microscopy, 7–8
Imaging resolution, guidelines for multiphoton microscopy, 9
Immune system, role of 5-$HT_{1A}$ receptor in, 140
Immuno-electron microscopy, intracellular localization of 5-$HT_{2A}$ receptor by, 115
Immunostaining, in 5-$HT_{1A}$ receptor experimental procedures, 147–148
In situ hybridization, study of mRNA 5-$HT_{2A}$ receptor levels by, 122
In vivo assays, yeast one-hybrid analysis, 93
Inflammatory pain, 5-$HT_{2A}$ receptor in, 109
Intracellular localization
  of 5-$HT_{1A}$ receptor, 135
  of 5-$HT_{2A}$ receptor, 115–116
Intracellular signaling cascades
  5-$HT_{2A}$ receptor functional selectivity in, 111–115
  activated by 5-$HT_{2A}$ receptor, 109
  activation studies, 109–111
  $IP_3$ activation, 111
Intracellular trafficking, of 5-$HT_{2A}$ receptor, 116–120
Ion channels, monitoring receptor-mediated changes in cAMP levels with, 19–20
$IP_3$ activation, of 5-$HT_{2A}$ receptor, 111
$IP_3$/diacylglycerol signaling, with 5-$HT_{2A}$ receptor stimulation, 111
iXon camera, 33

## K

KS+ plasmids, 86

## L

Laser beam, spatial properties in multiphoton microscopy, 4–5
Learning, role of 5-$HT_{2A}$ receptor in, 106
Ligand-dependent pathways, 112, 135
  of 5-$HT_{2A}$ receptor function, 105–106
  for 5-$HT_{2A}$ receptor internalization, 116
Lipid-protein interaction, in cell membranes, 42
Lipid transfection protocols, 89
Lipofectamine 2000, 97
Lipofectamine Plus, 89
Liposome-mediated transfections, 88
Long-term regulation, of 5-$HT_{2A}$ receptor, 122–124
Lymphocyte proliferation, role of 5-$HT_{1A}$ receptor in, 140

## M

Maladaptive behavior
  in 5-HT receptor knock-out mice, 163–168
  complex interactions of genes and environmental challenges in, 158
Mammalian two-hybrid system, 97
MAP kinase
  activation by 5-$HT_{1A}$ receptor stimulation, 143
  activation of cyclin D1 by, 144
Matchmaker One-Hybrid System, 94
MDMA. *See* Ecstasy
Membrane dynamics
  bleach duration times for, 50–51
  cytoskeleton as barrier to free protein diffusion in, 54

experimental methodology, 50–53
fluid mosaic model, 42
G-protein dependent cell surface dynamics, 53–55
relevance to protein function, 42–43
of serotonin$_{1A}$ receptor, 42–43
study rationale, 49–50
Membrane organization
and detergent insolubility of serotonin$_{1A}$ receptor, 44–48
experimental methodology, 45–47
implications for signaling functions of G-protein coupled receptors, 42
lipid-cholesterol enrichment and detergent resistance, 44
relevance to protein function, 42–43
of serotonin$_{1A}$ receptor, 41
study rationale, 44–45
Memory, role of 5-HT$_{2A}$ receptor in, 106
Mental diseases
altered 5-HT neuromodulation in, 159
and limits of neuronal plasticity, 157–160
Meridian DASY Master Program, 47
Microarray analysis, of 5-HT$_{2A}$ receptor levels, 123
Mitochondria labeling, in live neurons, 14
Modulators, 134
Molecular diffusion, temperature dependence of, 50
MORA-OPO, 17
Mu-opioid OP$_3$ receptor, CaM binding to, 66
Multiphoton excitation (MPE), 2
energy scheme for, 7
localization near focus, 6
wavelength requirements, 5
Multiphoton microscopy (MPM)
advantages of three-photon, 7–8
basic concepts of multiphoton excitation, 2–3
coupling scanning optics with microscope for, 8–9
detection setup for, 9–11
epi collection for, 9–10
excitation setup for, 8–9
excitation wavelength selection for, 5–7
intensity dependence, 3
non-epi collection setup for, 10–11
optical setup for, 8–12
point scanning setup, 8
pre-imaging checklist, 12
pulsed laser requirements, 3–4
quantitative imaging of serotonin autofluorescence with, 1–2
specific optical setup for serotonin imaging, 11
tight focusing requirements, 4–5
Multiple fluorospheres, simultaneous excitation with three-photon microscopy, 8
Multistate models of receptor function, 113
*MUPP1,* interactions with GPCR RIPs, 64


## N


NADH, checking co-localization of bright structures with, 13–14
Nervous system
identifying novel transcriptional regulators in, 81–82global gene regulation in, 98
Neural development, role of serotonin$_{1A}$ receptor in, 43
Neuroblastoma glioma cells NIE-115, 34
measurement of cAMP levels in, 33
Neurogenesis, role of 5-HT$_{2A}$ receptor in, 106
Neuromodulator receptor families, GPCRs, 105–106
Neuronal excitation and inhibition, 134
Neuronal plasticity, 158
limits and mental diseases, 147–160
Neurotransmitters, intrinsic fluorescence of, 16
NGFIA transcription factor, role in reduced stress response, 82
NHERF, caveolin-1 binding to, 64
Nicotine withdrawal, 5-HT$_{1A}$ receptor therapeutic utility in, 138
Non-epi collection, 17
dopamine imaging using, 17
schematic diagram, 10
setup for multiphoton microscopy, 10–11
Nondescanned detection, 10
Novel transcriptional regulators, identifying in nervous system, 81–82
Novelty-induced locomotion, 159
in 5-HT receptor knock-out mice, 165, 167


## O


Obesity, in 5-HT receptor knock-out mice, 163
Obsessive-compulsive disorder (OCD)
5-HT$_{2A}$ receptor in, 108–109
SNRIs in treatment of, 139
Oligomerization, of 5-HT$_{2A}$ receptor, 114–115
One-photon excitation, 7
Ontogeny, of 5-HT$_{1A}$ receptor, 135–136
Open-field stress induction, 165
in 5-HT receptor knock-out mice, 164, 166
Optical parametric oscillators, 17


## P


Panic disorder, association of HTR2C polymorphisms with, 162

pcDNA3 plasmids, 86, 91
PDZ-binding domain, in dendritic targeting by 5-$HT_{2A}$ receptor, 115
PDZ proteins, interactions with GPCR RIPs, 64
pGEX, 91
Phosphodiesterases (PDE), cAMP level dependence on, 20
Phosphorylation
- antagonism with CaM binding, 69–70
- blunting by calmodulin, 67
- regulation by slow-acting receptors, 134

Photo multiplier tube (PMT), 10
Photoactivatable GFP (PAGFP), 119, 120
Photobleaching, in FRAP experiments, 50
Point bleaching, *vs.* area bleaching, 51
Point scanning, in setup for multiphoton microscopy, 8
Positron emission tomography
- of 5-$HT_{1A}$ receptor ligand, 135
- study of 5-$HT_{2A}$ receptor levels in vivo with, 122

Postsynaptic 5-$HT_{1A}$ receptor
- and long-term administration of SSRIs, 137
- signaling effects, 136–137

Prefrontal cortex, 5-$HT_{1A}$ receptor neuronal activity in, 148
Presynaptic 5-$HT_{1A}$ receptor, signaling effects, 136–137
Pretransformed libraries, 94
Primary neurons, transfection for promoter characterization, 89
Primer extension method, 85
- for TSS identification, 83

Progressive multifocal leukoencephalopathy, 5-$HT_{2A}$ receptor in, 109
Promoter characterization, 86–87
- bioinformatic analysis, 87
- gene reporter assays for, 87–88
- interpreting reporter assays for, 89–90
- transfection of cell lines for, 88–89
- transfection of primary neurons for, 89

Promoter location, 87
Promoter regions, 87
Protein kinase A (PKA)
- activation by cAMP, 27
- binding to GPCRs, 62

Protein kinase C (PKC), binding to GPCRs, 62
Proteins, associations with 5-$HT_{2A}$ receptor, 121
Prozac, 141, 142
PSD-95, interactions with GPCR RIPs, 63
Psychotropic drugs
- receptor functionality modified by, 106
- role of receptor down-regulation in, 124

pTriEX4, 91
Pulse-PulseFit 8.31 software, 21
Pulse shape, in time, 3
Pulse width, 3
- avoiding changes in, 12
- and excitation efficiency in two-/three-photon excitation, 4

Pulsed lasers, in multiphoton microscopy, 3–4


## Q


Quantitative imaging
- advantages with three-photon microscopy, 7
- with multiphoton microscopy, 1–2


## R


Raphe neurons, 12
Rapid amplification of cDNA 5-ends (5-RACE), 85. *See also* 5-RACE method
Ratiometric dyes, for quantitative 5-$HT_{2A}$ receptor measurement, 110
Rayleigh criterion, 8
Receptor associated modifying proteins (RAMPs), 62–63
Receptor down-regulation, 123, 124
Receptor-interacting proteins (RIPs)
- calmodulin as, 61
- G protein-coupled, 62–64
- GPCR signaling modification by, 62

Receptor internalization
- of 5-$HT_{2A}$ receptor, 112
- by HEK293 cells, 116, 117
- role in down-regulation, 124
- via serotonin and dopamine, 119
- visualizing with fluorescence detection, 120

Receptor multimerization, 114
Receptor recycling, 112, 113
Receptor up-regulation, 123
Recombinant protein, use in EMSA studies, 91
Repetition rate, for pulsed laser, 3
Reporter constructs
- limitations of, 89
- in yeast one-hybrid analysis method, 93

Repressor-operator systems, 95
Repressor regions, 87
Reverse transcriptase polymerase chain reaction (RT-PCR), estimation of 5-$HT_{2A}$ receptor levels with, 122, 123
Ribonuclease protection assay (RPA), estimation of 5-$HT_{2A}$ receptor levels with, 122, 123
RN-46A cells
- for serotonin imaging, 11
- three-photon image of, 13

RNA interference (RNAi), 97
RNase protection assays, 83–84

## S

Schizophrenia
 $5\text{-HT}_{2A}$ receptor in, 108
 clozapine as $5\text{-HT}_{1A}$ receptor agonist in, 134
 increases in $5\text{-HT}_{1A}$ receptor density with, 137–138
 and low dopaminergic neurotransmission, 142
Selective serotonin reuptake inhibitors (SSRIs)
 increased postsynaptic activity after long-term administration, 137
 mechanisms of $5\text{-HT}_{1A}$ receptor action, 141
Serotonergic cell differentiation, monitoring with autofluorescence imaging, 15–16
Serotonergic signaling, 20
Serotonin, spectral characteristics, 11
Serotonin and nor-adrenalin reuptake blockers (SNRIs), 139
 mechanisms of $5\text{-HT}_{1A}$ receptor action in, 141
 in treatment of OCD and depression, 139
Serotonin autofluorescence
 at 740 nm, 11, 13
 abbreviations, 2
 and amphetamine-induced serotonin release, 16
 applications of, 14–16
 catecholamine detection with, 16–17
 establishing signal source for, 13–14
 exocytosis with K+ depolarization, 15
 imaging in live neurons, 12–17
 imaging protocol, 12
 monitoring serotonergic cell differentiation with, 15–16
 optical setup for, 8–12
 pre-imaging checklist, 12
 quantitative imaging with multiphoton microscopy, 1–2
 specific optical setup for imaging, 11
Serotonin imaging
 of amphetamine-induced serotonin release, 16
 applications of, 14
 as assay for differentiation, 16
 exocytosis with K+ depolarization, 15
 in live neurons, 12–17
 monitoring serotonergic cell differentiation with, 15–16
Serotonin receptors. *See also* $5\text{-HT}_{1A}$ receptor; $\text{Serotonin}_{1A}$ receptor
 internalization and recycling of, 112
 ionotropic or coupled, 20
 paradoxical regulation of, 123
Serotonin system, transcriptional regulation in, 83
Serotonin vesicles, and mitochondria, 15
$\text{Serotonin}_{1A}$ receptor, 43–44. *See also* $5\text{-HT}_{1A}$ receptor
 cellular distribution with EYFP receptors, 46
 fusing with EYFP, 45
 G-protein dependent cell surface dynamics, 53–55
 membrane dynamics, 49–55
 membrane organization and dynamics monitored with fluorescence microscopy, 41
 membrane organization monitored using detergent insolubility, 44–48
 role of lateral mobility in receptor dynamics, 49
 study rationale, 44–45, 49–50
Setraline, 141
 counteraction of SNRI adverse effects by, 139
Shift Western blot, 92
Signal source
 checking co-localization of bright structures with NADH, 13–14
 establishing for serotonin autofluorescence, 13–14
 test for order of excitation, 13
Signal-to-noise ratio
 enhancing with excitation shielding, 10–11
 in FRAP studies, 51
 guidelines before imaging, 12
Signal transduction, 44, 133
 membrane organization and, 42
 quick and slow effects, 134
siRNA probes, for identifying transcription factor activity, 97–98
Sleep apnea, $5\text{-HT}_{2A}$ receptor in, 109
Slot blot, 68, 71
SNX27
 caveolin-1 binding to, 64
 interactions with GPCR RIPs, 63
Social defeat stress, and BDNF promoter sites, 82
Spectral characteristics, serotonin and optical elements, 11
Stress response
 deficits in adaptation, 159
 and NGFIA transcription factor, 82
Striatum, dopaminergic neurons in, 144
Supershift EMSA, for transcriptional regulator identification, 92
Surface plasmon resonance spectroscopy (SPRS), 68

## T

T- and B-cells, role of $5\text{-HT}_{1A}$ receptor in, 134

Targeted genetic inactivation, 182–183. *See also* 5-HT receptor knockout mice
  in study of anxiety, 182–183
TATA box promoters, 86
TATA-less promoters, 86
Temperature, and molecular diffusion, 50
Therapeutic agents, regulation of 5-HT$_{1A}$ receptor signaling by, 138–139
Three-photon microscopy
  advantages of, 7–8
  excitation efficiency and pulse width in, 4
  imaging depth with, 7–8
  minimizing cell damage with, 8
  multiple fluorosphere excitation with, 8
  proof of excitation, 14
  quantitative imaging advantages with, 7
TNT Coupled Reticulocyte Lysate, 91
TNT Coupled Wheat Germ, 91
Transcription Element Search System (TESS), 87
Transcription factor activity
  GAL4-DBD hybrid system for, 95–97
  siRNA or antisense probes of, 97–98
Transcription factors, cloning of, 82
Transcription start site identification, 82, 83
  5-RACE method, 85–86
  general strategies, 83–85
Transcriptional regulators
  and behavior, 82–83
  identifying DNA binding proteins, 92–95
  identifying DNA elements, 90–92
  identifying novel, 81–82
  promoter characterization, 86–90
  and serotonin system, 83
  transcription factor activity identification, 95–98
  transcription start site identification, 83–86
Transfection
  of cell lines for promoter characterization, 88–89
  of primary neurons for promoter characterization, 89
Transformed cell lines, 89
TransMessenger, 89
Triton X-100 detergent, 47
Tryptophan hydroxylase (TPH), interaction with 5-HT, 141
Two-photon microscopy, 3, 7
  excitation efficiency and pulse width in, 4


## U


Ultraviolet (UV), 2


## V


Ventral tegmental area (VTA), dopaminergic neurons in, 144


## W


WAY100635, in clozapine-evoked increase in PFC neuronal activity, 145
Western blotting, in 5-HT$_{1A}$ receptor experimental procedures, 147
Whole cell patch clamp recordings, 24
  cAMP measurement with, 21


## Y


Yeast one-hybrid analysis, 82, 92
  identifying transcriptional regulators via, 93–94
  screening protocol, 94


## Z


Zeiss LSM 510 Meta confocal microscope, 50